Grand Canyon Geology

Oxford University Press

Oxford New York Toronto
Delhi Bombay Calcutta Madras Karachi
Petaling Jaya Singapore Hong Kong Tokyo
Nairobi Dar es Salaam Cape Town
Melbourne Auckland

and associated companies in
Berlin Ibadan

Grand Canyon Geology

edited by
Stanley S. Beus and Michael Morales

New York Oxford
Oxford University Press
Museum of Northern Arizona Press
1990

DEDICATION

This book is dedicated to the memory of Dr. Edwin Dinwoodie McKee (1906-1985), who stands as one of the premier scientists of the Grand Canyon in this century. Eddie began his career in 1929 as a park naturalist at Grand Canyon where for eleven years he made the natural history of this marvelous place a labor of love. In that time, he made extensive collections and observations in all areas of natural history, increased the interpretive program of the park, and became an expert on the biology, archaeology, and geology of the Grand Canyon. He also initiated studies of Paleozoic rock units that would occupy him, at least part time, for the rest of his career and would result in the publication of seven books on Grand Canyon geological history and four other monographs, together with more than 200 shorter articles, on geology or other topics of natural history.

His hiking exploits in the Grand Canyon became legendary—especially his frequent weekend journeys on foot from the south rim to the north rim and back (42 miles round trip) to woo and win a young biology student, Barbara Hastings. Barbara and Eddie were married in 1929 to begin a union that would last more than 50 years.

Following his years with the Grand Canyon National Park, Eddie served for a time as Chairman of the Geology Department at the University of Arizona, Tucson, as well as assistant director of the Museum of Northern Arizona, Flagstaff. In 1955 he began a 30-year career with the U.S. Geological Survey in Denver. From there, he conducted worldwide and world-class scientific investigations in stratigraphy, sedimentation, and depositional processes. These investigations included frequent returns to the Grand Canyon, where he continued his work on the Paleozoic strata.

An entire generation of geologists who have studied the Grand Canyon, including the authors of this book, have benefited from the stimulation, encouragement, and inspiration of Edwin D. McKee. A number of us have spent long-remembered days with him in the field. His tireless efforts to develop new insights, stimulate discussion, ask the right questions, and share the excitement and curiosity of scientific investigation serve as a model for all who would pursue the hidden secrets of the Grand Canyon.

Published by Oxford University Press, Inc.,
200 Madison Avenue, New York, New York 10016

Oxford is a registered trademark of Oxford University Press

Library of Congress Cataloging-in-Publication Data
Grand Canyon Geology / edited by Stanley S. Beus
and Michael Morales.
p. cm.
Includes bibliographical references.
ISBN 0-19-505014-2 —ISBN 0-19-505015-0 (pbk.)
1. Geology—Arizona—Grand Canyon
I. Beus, Stanley S. II. Morales, Michael.
QE86.G73G73 1990
557.91'32—dc20 90-7163 CIP

2 4 6 8 9 7 5 3 1

Printed in the United States of American
on acid-free paper

TABLE OF CONTENTS

FOREWORD

It is strangely ironic that the Grand Canyon of the Colorado River, one of the most frequently visited natural wonders of the world, was for three centuries ignored by the Europeans who explored and exploited western North America. Of course, the Native Americans of the Southwest had known the canyon for thousands of years before some of them led a little band from the Coronado expedition to the south rim of the canyon in 1540. The Spaniards were impressed, and said so, but that was about the extent of their interest in this magnificent natural feature. During the long interval from that fateful day, four hundred and fifty years ago, when Don García López de Cárdenas and his companions gazed across the canyon until the third decade of the nineteenth century, very few people of European heritage visited or bothered to describe the canyon.

Then, in 1831 there appeared a description of the canyon by mountain man James Ohio Pattie, the first account to be written by an American. Yet his published portrayal of the canyon did not bring a rush of people to see it. It was hard to reach, and its breathtaking proportions made it a barrier to travel along the southwestern lines of latitude. Indeed, it was as a barrier that those Americans who knew anything about the canyon were prone to view it.

In the middle years of the nineteenth century, there were several United States Army surveying expeditions through the Southwest, in part to search for a feasible railroad route to the West Coast. One of the expeditions, led by Lieutenant Joseph Christmas Ives, reached the floor of the canyon near the mouth of Diamond Creek in 1858. The canyon seemed anything but inviting to Ives, who wrote that "the increasing magnitude of the colossal piles that blocked the end of the vista, and the corresponding depth and gloom of the gaping chasms into which we were plunging, imparted an earthly character to a way that might have resembled the portals of the infernal regions."

Fortunately, Ives had with him Dr. John Strong Newberry, one of the great pioneer American geologists, and it was Newberry who first looked at the canyon with a geologist's eye. To him the canyon did not look in the least like the gates of hell, for he wrote that it was "the most splendid exposure of stratified rocks that there is in the world."

Then, in 1869 and 1871 John Wesley Powell led his exploring parties down the Colorado River in their frail, wooden boats, penetrating the canyon from Green River to the Grand Wash Cliffs. Suddenly, people throughout the world were made aware of this great natural wonder. From then until the present day, the canyon has been scientifically explored and studied, so that there now exists a vast body of literature on all aspects of the canyon. All of this has taken place within two lifetimes.

This I know in a personal way, for one evening in the thirties I sat on a couch in the Explorers' Club, New York, with none other than Frederick Dellenbaugh, one of the members of Powell's second expedition. For an hour or more, I enjoyed a firsthand account of Powell and his party and of their adventures.

Grand Canyon Geology is dedicated to Edwin D. McKee— "Eddie" to all of us who knew him—no more appropriate dedication could be made. Eddie was the foremost modern authority on the geology of the Grand Canyon. Geologists today and in years to come will benefit from the profound studies of canyon geology that have come from Eddie's pen.

This book is not intended for the casual reader; he who turns its pages will need a certain amount of geological sophistication to appreciate the twenty chapters that describe and interpret Grand Canyon geology. Since each chapter is written by one of more authorities in their respective fields, there is here a storehouse of geological knowledge, much of it new. For geologists and those interested in geology, this is a book to be read, consulted, and treasured for years to come.

Edwin H. Colbert

CHAPTER 1

INTRODUCING THE GRAND CANYON

Stanley S. Beus
and
Michael Morales

The Grand Canyon contains a marvelous record of geologic and paleontologic events spanning most of the last two billion years, nearly one half of the life span of this planet. Although it is neither the deepest nor the longest canyon in the world, it is one of the few places where so many chapters of earth history are clearly displayed. The igneous and metamorphic rocks of the canyon's inner gorge are part of the basement of the North American continent. Two great packages of sedimentary and volcanic rocks—one of Middle Proterozoic age and one of Paleozoic age—make up most of the canyon's walls. Beyond the rim to the north and east are the giant stairsteps of Mesozoic and lower Cenozoic strata whose cliffs have been retreating from the canyon's edge for millions of years. In the central part of the canyon, upper Cenozoic volcanic rocks have poured intermittently into the gorge, temporarily blocking the Colorado River and altering its cutting action. During the past two billion years, many dramatic, earth-changing events have occurred in western

North America. We are incredibly fortunate that the rocks exposed in the Grand Canyon record this time period like no other place on the continent. This great chasm is truly a unique "window into the past."

The Grand Canyon (Fig. 1) lies entirely in the northwestern part of Arizona. It extends nearly 278 miles (448 kilometers) between Lake Powell on its eastern end and Lake Mead to the west. In 1919, the region became a national park, which today encompasses approximately 1900 square miles (4921 square kilometers) of land. This most famous of all canyons was formed by swiftly flowing waters of the Colorado River cutting into rock layers of the southwestern Colorado Plateau, a vast uplifted tableland that includes a large portion of the Four Corners states: Arizona, Colorado, New Mexico, and Utah. The land surrounding the Grand Canyon includes six local plateaus and one low lying platform, all of which are bounded by faults or monoclines (Fig. 2). In a west-east cross section to the north of the canyon

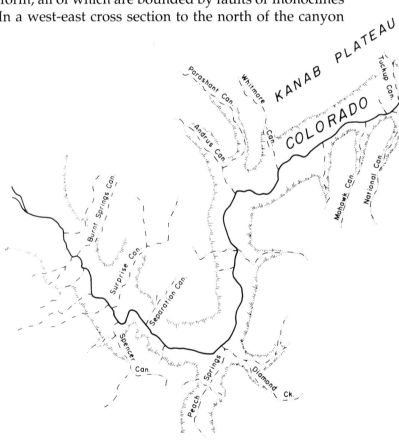

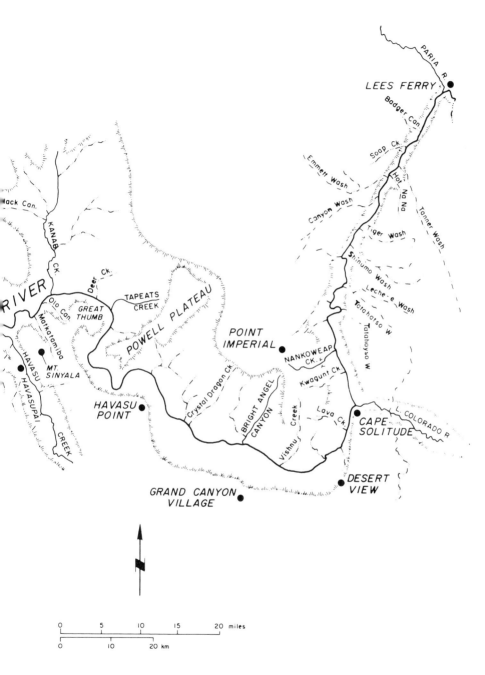

Figure 1. Map of the Grand Canyon

INTRODUCING THE GRAND CANYON   3

(Fig. 3), you can see how each of these blocks of land has been uplifted, downdropped, or tilted relative to its neighbor.

At its narrowest, the Grand Canyon is a little less than a mile (1.6 kilometers) across. Along the canyon's north and south rims, the relief is relatively gentle, except for incisions made by runoff waters flowing into the gorge. Within the canyon itself, however, the topography is quite varied and spectacular. The maximum depth of the canyon at any single place is about 6000 feet (1829 meters) from the rim to the floor. The maximum drop in elevation of the canyon as a whole, however, is approximately 6600 feet

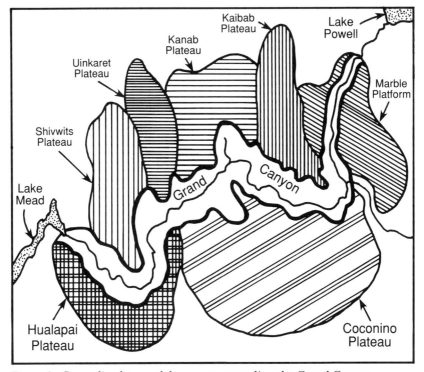

Figure 2. Generalized map of the area surrounding the Grand Canyon

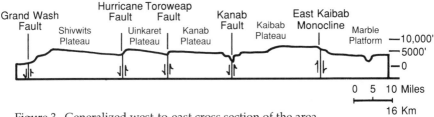

Figure 3. Generalized west-to-east cross section of the area just north of the Grand Canyon

GRAND CANYON GEOLOGY

(2012 meters) between Point Imperial on the north rim (8803 feet [2683 meters]) and the floor of the canyon near Lake Mead (1200 feet [366 meters]).

Many people are surprised to learn that there is a difference in elevation between the north and south rims. On the canyon's southern edge, the altitude ranges from 6000 to 7500 feet (1829 to 2286 meters) above sea level. The northern edge, however, is 1000 to 1200 feet (305 to 366 meters) higher even though both rims are capped by the same rock unit, the Kaibab Formation. The difference in height occurs because the various rock layers into which the canyon has been cut do not lie completely flat in this region. Instead, they arch or dome upward. The Colorado River carved the canyon through the southern flank of the dome, where the rock layers are tilted gently down to the south. Layers on the north rim, therefore, have been uplifted higher than the same layers on the south rim (Fig. 4).

Water that flows into and through the Grand Canyon comes primarily from four merging rivers—the Green, San Juan, Little Colorado, and Colorado—that drain hundreds of square miles of the Four Corners states (Fig. 5). As it courses through the canyon, the Colorado River drops about 2000 feet (610 meters) in elevation (Fig. 6). This steep gradient allows the river to continue its erosion of the chasm's floor.

Humans have been in the Grand Canyon for at least four thousand years, but the early records are sparse. Rare artifacts, remains of prehistoric dwellings, and the numerous mescal pits all attest to the early exploration and habitation by American Indians such as the Anasazi and the Cohonina. The first Euro-

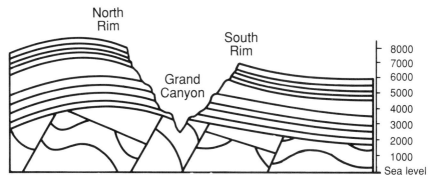

Figure 4. Generalized cross section through the Grand Canyon from the north rim to the south rim

Figure 5. Drainage area for runoff water that flows into and through the Grand Canyon

peans to see the canyon were a Spanish party of thirteen in search of the fabled lost cities of gold, under the command of Captain Don García López de Cárdenas. In 1540, Hopi Indian guides led them to the south rim in the eastern part of the Grand Canyon, but they were unable to reach the river below. In the next three

GRAND CANYON GEOLOGY

centuries, only two visits to the region have been reliably recorded. In 1776, Father Francisco Tomás Garcés, a Spanish missionary, explored Havasu Canyon in the south-central part of the Grand Canyon. In the same year, two Spanish priests, Father Silvestre Vélez de Escalante and Father Francisco Domíngues, led an expedition to the region and discovered a ford across the Colorado River (Crossing of the Fathers).

Among the earliest geological reports of the Grand Canyon country are those of Jules Marcou (1856) and John Strong Newberry (1861), who described the region's Paleozoic stratigraphy from their explorations of the canyon and the land to the south. Newberry was a geologist in the War Department-sponsored Ives expedition of 1857-58. Ives' rather discouraging and, as it turned out, unprophetic statement about the Grand Canyon was:

> Ours has been the first, and will doubtless be the last party of whites to visit this profitless locality. It seems intended by nature that the Colorado River, along the greater portion of its lonely and majestic way, shall be forever unvisited and undisturbed.

In 1869, John Wesley Powell led a party of ten men (reduced later to nine and finally only six) on an epic journey by boat down a thousand miles of the Colorado River from Green River, Wyoming, across Utah, and finally through the Grand Canyon to the mouth of the Virgin River at what is today the north end of Lake Mead. Powell's work was followed up by a small group of outstanding scientists through the turn of the nineteenth century. These included G.K. Gilbert, who was the first to apply formal rock unit names to Grand Canyon rocks; C.E. Dutton, who wrote

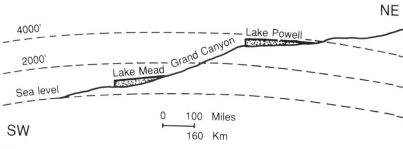

Figure 6. Diagrammatic profile of the Colorado River through the Grand Canyon, from Lake Powell to Lake Mead

the first monograph on the geology and geologic history of the Grand Canyon; A.R. Marvine, who participated in the U.S. Geographical Survey West of the 100th Meridian; and C.D. Walcott, who described both Paleozoic and Precambrian rocks in the canyon's central and eastern parts. These pioneer studies laid the groundwork for all subsequent research in the Grand Canyon.

Edwin D. McKee, to whom this book is dedicated, stands out as the premier research scientist of Grand Canyon geology in the twentieth century. Between 1933 and 1982, McKee either authored or coauthored five monographic publications on various Paleozoic rock units of the canyon—including the Redwall Limestone, Supai Group, Coconino Sandstone, the Toroweap and Kaibab formations, and Cambrian units. McKee also organized a special symposium in 1964 at the Museum of Northern Arizona to summarize the data and interpretations on the origin and evolution of the canyon. Much of our present understanding of the stratigraphy and age of Grand Canyon rocks (Fig. 7) is based on McKee's seminal work.

For more than a century, then, the Grand Canyon has attracted the extraordinary interest and effort of geologists to map its course, examine its rocks and fossils, and understand its record of earth history. In 1870, Powell wrote of the canyon country of the Colorado River:

> ... the thought grew in to my mind that the canyons of this region would be a Book of Revelations in the rock-leaved Bible of geology. The thought fructified and I determined to read the book.

Since then, several generations of earth scientists have searched the river and its canyon for answers to questions of when and how this part of our planet's crust developed. Although some questions have been answered, new ones are raised, and so the search goes on. This book is written for those who inquire about the revelations of the Grand Canyon. It is a compilation of the best efforts of a number of authors, all experts in their field, to summarize what is now known about the geology of the canyon. Much of the information presented here is an update and refinement of earlier works by pioneer scientists of the past. All of us who have written for this volume owe a great debt to those who preceded us in the search for the geologic secrets of the Grand Canyon.

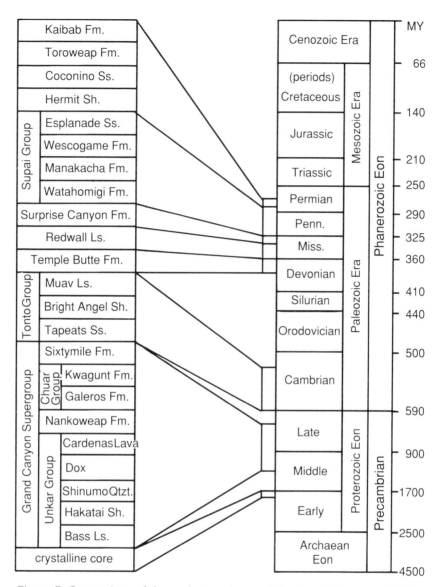

Figure 7. Comparison of the geologic column of the Grand Canyon with the Geologic Time Scale (After Haq and Van Eysinga, 1987)

CHAPTER 2

PRECAMBRIAN CRYSTALLINE CORE

R. S. Babcock

INTRODUCTION

The spectacular crystalline rocks that comprise the "grand gloomy depths" of the inner gorges of the Grand Canyon represent two of the most remarkable events in geologic history. The first was the accretion of a series of microplates or orogenic belts into Laurentia, the North American craton. This occurred with amazing rapidity between 2.0 and 1.6 billion years ago, as the area of the North American continent increased by more than fifty percent. After being assembled, the craton was perforated by an immense volume of plutonic rock, which apparently was generated by the heat associated with a mantle "superswell" or large zone of upwelling (Hoffman 1989).

Igneous and metamorphic rocks of Early and Middle Proterozoic age that are the product of these two events crop out along the Colorado River and its tributary canyons in the Upper Granite Gorge from mile 77 to mile 119, in the Middle Granite Gorge from mile 127 to 130.5, and in the Lower Granite Gorge from mile 216 to 261. Minor exposures also occur near miles 190-191, at Granite Park (mile 209), mile 212, Peach Springs Canyon, upper Spencer Canyon, and the Grand Wash Cliffs.

EARLY WORK

These spectacular rocks were first described by John Wesley Powell in 1874 following his legendary expeditions of the 1860s and 1870s. Powell's early work was continued by Charles Walcott, who described the unconformity between the Grand Canyon Series and the underlying metamorphics that he named Vishnu (Walcott 1894). The first petrologic study of these Precambrian rocks was done by Levi Noble and Fred Hunter (1916), who recognized that the Vishnu schists were sedimentary rocks that had been recrystallized by regional metamorphism. They observed that at least some of the gneissic rocks were granites that had intruded the Vishnu sediments prior to or during metamorphism.

Ian Campbell and John Maxson published a number of papers during the 1930s describing the petrology and structure of Precambrian rocks in the Grand Canyon. They subdivided the Vishnu section into the supracrustal Vishnu Schist and the metaplutonic Zoroaster Gneiss. In 1961, Maxon proposed the Brahma Formation as a name for amphibolites associated with the Vishnu Schist. He inferred that the amphibolites originally were mafic volcanics extruded over sedimentary strata that later were enclosed in a tightly folded syncline with an axis parallel to Bright Angel Canyon. However, Ragan and Sheridan (1970) demon-

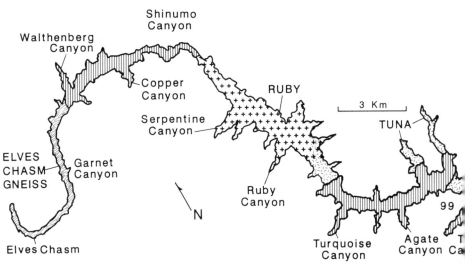

Figure 1 (a). Distribution of older Precambrian rock units along the Colorado River. Upper Granite Gorge

strated that this structural interpretation is incorrect and that metavolcanics generally are interbedded with metasediments.

During the late 1950s and early 1960s, attempts were made to date intrusive and metamorphic rocks in the vicinity of Phantom Ranch. Potassium-argon ages of 1240-1550 million years (m.y.) on pegmatites, granitic gneisses, and schists were reported by Aldrich and others (1957), Giletti and Damon (1961), and Damon (1968). These almost certainly represent recrystallization related to a late thermal event rather than the primary ages of intrusion of Vishnu metamorphism. Pasteels and Silver (1965) obtained more reliable uranium-lead ages of 1705± 15 m.y. for the Zoroaster Granite and 1675± 15 m.y. for migmatized Vishnu Schist.

The petrology, structure, geochemistry, and geochronology of Precambrian rocks are described in a series of papers by Babcock and others (1979), Brown and others (1979), and Clark (1979). Additional details can be found in MS theses by Boyce (1972), Lingley (1973), Walen (1973), and a Ph.D. thesis by Clark (1976). On the basis of these detailed studies, geologists have subdivided the rocks comprising the middle Precambrian section into four major units: the Vishnu Metamorphic Complex, the Zoroaster Plutonic Complex, the Trinity Gneiss, and the Elves Chasm Gneiss. The distribution of these units in the Upper and Lower Granite gorges is shown in Figure 1.

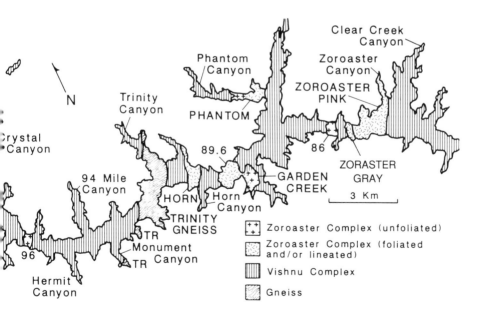

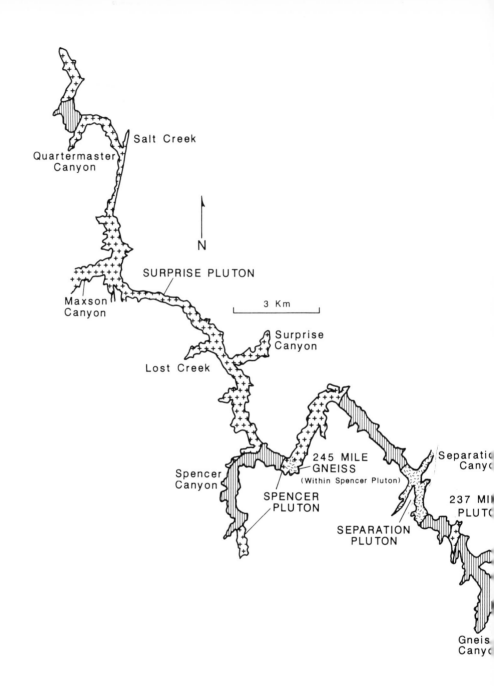

Salt Creek

Quartermaster
Canyon

N

SURPRISE PLUTON

3 Km

Maxson
Canyon

Surprise
Canyon

Lost Creek

Spencer
Canyon

245 MILE
GNEISS
(Within Spencer Pluton)

Separatic
Canyc

SPENCER
PLUTON

237 MI
PLUTC

SEPARATION
PLUTON

Gneis
Canyc

THE VISHNU METAMORPHIC COMPLEX

The Vishnu Metamorphic Complex includes metasedimentary and metavolcanic rocks exposed in various parts of the Upper Granite Gorge. Presumably, correlative rocks also are found in the Middle and Lower Granite gorges and in scattered outcrops elsewhere. Rocks of the Vishnu complex originated as sediments and volcanic rocks laid down on the sea floor, probably 2.0 to 1.8 billion years ago. The sediments consisted mostly of quartz-rich sand, silt, and clay, with lesser amounts of feldspar, mica, and iron oxide. Interspersed with the sediments were basaltic to andesitic lava flows and their feeder dikes, along with ash layers, which probably were associated with a chain of volcanic islands. It has been estimated that the total thickness of sediments and volcanics

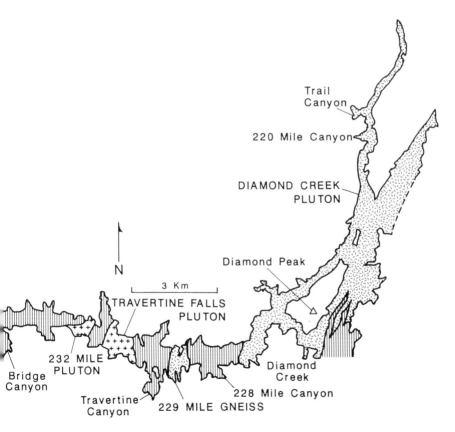

Figure 1 (b). Distribution of older Precambrian rock units along the Colorado River. Lower Granite Gorge

is greater than 40,000 feet (12,200 m) (Maxson 1961). However, the most interesting rocks in the Vishnu complex are discontinuous layers or lenses of carbonate that make up less than one percent of the total section (Fig. 2). This material possibly formed from mats of algae incorporated in the sediment and is the only evidence of life in the Proterozoic Vishnu sea.

After burial, the sediments and volcanics were subjected to at least two separate episodes of regional metamorphism. The first event converted the sediments into slates or phyllites and the volcanics into greenstones and greenschists. Because the metamorphism was low grade, many relict sedimentary and volcanic structures such as cross bedding or graded bedding can be found; good examples occur between mile 96 and 97 along the Colorado River and near the mouth of Boucher Canyon.

Figure 2. Calc-silicate lenses of possible algal origin interbedded with quartzofeldspathic schists in Clear Creek Canyon

In most places, evidence for this early metamorphism has been obliterated by a second, higher grade period of regional metamorphism. This main-stage event has produced the minerals and structures seen in exposures along the Bright Angel and Kaibab trails, as well as most other parts of the Grand Canyon. The metasediments formed during this metamorphism consist mostly of micaceous and quartzose schists or gneisses, while the metavolcanic rocks are amphibolites and hornblende schists. Some of the more common metamorphic mineral assemblages developed in these rocks are shown in Figure 3. The highest

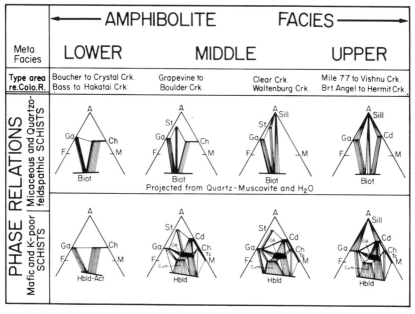

Figure 3. Phase relationships of quartzofeldspathic and mafic schists

temperature reached was about 700° C at a pressure of three to four kilobars. Such conditions apparently exceeded the melting point of quartz/feldspar/mica-rich rocks, which show evidence of partial melting in the form of migmatites (Fig. 4). Partially melted rocks can be seen near the mouth of Pipe Creek, in Phantom Canyon, and between miles 232.2 and 236.7 in the Lower Granite Gorge.

In addition to these regional metamorphic events, contact metamorphism also occurs in the vicinity of some plutons and

Figure 4. Outcrop of migmatitic gneiss along the margin of the Phantom pluton in Phantom Canyon

pegmatite dikes; this is best displayed at the contact of the Ruby pluton near Bass Rapids. Localized, retrogressive metamorphism also is found in the vicinity of faults and shear zones where high-grade schists and amphibolites have recrystallized to low-grade mineral assemblages characterized by the presence of chlorite. Notable occurrences of such retrogressive rocks are in Vishnu Creek, Bright Angel Creek, and along the Boucher fault at mile 95.5.

The most obvious metamorphic structure in the Vishnu complex is a foliation developed during the second period of regional metamorphism. In most places this foliation is nearly vertical and strikes towards the northwest. Three stages of deformation can be recognized clearly (and at least two more inferred from detailed structural analysis), but good fold structures are a bit difficult to find. Early folds cut by the dominant foliation can be seen near the mouth of Clear Creek and around Boucher Creek. Later folds that deform the foliation are displayed near the boat landing above Crystal Rapids. The largest scale structure mapped so far in the Grand Canyon is a dome extending from Cremation Creek to Zoroaster Canyon (Lingley 1973).

The age of Vishnu metamorphism has not been established definitely. Uranium-lead and rubidium-strontium dating suggests an age of 1680 to 1650 million years for the main-stage event and 1720 to 1710 million years for the earlier metamorphism.

ZOROASTER PLUTONIC COMPLEX

The Zoroaster Plutonic Complex is a term applied to the numerous igneous intrusive bodies that were emplaced over a time span of several hundred million years before, during, and after Vishnu metamorphism. The Zoroaster complex includes three types of intrusive rock: (1) granitic-to-dioritic plutons; (2) dikes and sills composed of fine-to-medium-grained granite or granodiorite; (3) dikes and sills composed of granitic pegmatite with aplite generally intermixed.

Plutons

Detailed mapping has revealed that twenty separate plutons are exposed along the Colorado River in the Upper and Lower Granite gorges. The location and some of the field characteristics of each of these plutons are summarized in Table 1. Based on age, composition, and textural/structural characteristics, the plutons can be assigned provisionally to one of three "plutonic superunits" (petrogenetic types), as indicated by Table 1.

The Surprise Canyon superunit consists entirely of medium-grained, anorogenic (unfoliated) granites with high initial strontium ratios (0.7100). This suggests an origin from the partial melting of sialic crustal rocks. These plutons probably are related for the most part to the 1500 to 1400 million-year-old transcontinental anorogenic plutonic belt of Silver and others (1977), though imprecise rubidium-strontium dating suggests that some of them are older, with emplacement ages of 1630 to 1580 million years.

The Phantom Canyon superunit consists of plutons that apparently were intruded during or just after the latest stages of deformation related to the main stage of Vishnu metamorphism. These plutons are mostly medium-to-coarse-grained granites and granodiorites that are characterized by migmatitic contacts or zones of interspersed migmatites. Foliation is highly variable. Initial-Sr ratios are about 0.7040, indicating a less radiogenic (more mantlelike) source material than the anorogenic plutons.

Table 1. Field Characteristics of Zoroaster Complex Plutons

PLUTON	LOCATION	COMPOSITION	TEXTURE	SUPERUNIT
Zoroaster Pink	84.6 to 85.5	Granite to Alkali Fsp Granite	Foliated	Ruby
Zoroaster Gray	Near 85.5	Tonalite	Foliated	Ruby
86 Mile	Near 86.2	Granite	Unfoliated	Surprise
Phantom	Phantom Canyon	Granodiorite Granite Granite	Foliated Folitated Unfoliated	Phantom
Garden Creek	Pipe and Garden Creeks	Granodiorite	Foliated	Phantom
89.6 Mile	89-90	Granite to Granodiorite	Foliated	Phantom
Horn	Near 90.6	Diorite	Foliated	Ruby
96 Mile	Near 96.0	Granodiorite to Tonalite	Unfoliated	Ruby
97.6 Mile	Near 97.6	Qz-Monzonite to Qz-Monzo-Diorite	Weakly Foliated	Ruby
99 Mile	Near 99	Granodiorite	Foliated	Ruby
Tuna	Near 99.2	Granodiorite	Foliated	Phantom
Ruby	102.7 to 107.9	Granodiorite to Tonalite	Foliated to Unfoliated	Ruby
Diamond Creek	216 to 227	Tonalite	Foliated	Ruby
Travertine	230 to 230.9	Granite	Unfoliated	Surprise
229 Mile	Near 228.7	Granite to Alkali Fsp Granite	Foliated	
232 Mile	231.7 to 232.2	Granite	Unfoliated	Surprise
237 Mile	236.7 to 237	Granite	Unfoliated	Phantom
Separation	238.5 to 239.9	Granite	Foliated	Phantom
245 Mile	Near 245	Tonalite	Foliated	Ruby
Spencer	242.3 to 245.3	Granite	Unfoliated	Surprise
Surprise	246.1 to 261	Granite	Unfoliated	Surprise

The Ruby Creek superunit is by far the most variable. Compositions range from granodiorite to diorite, and foliation is well developed to nonexistent. The unifying feature is that all of these plutons appear to have been intruded prior to or during the early part of the main stage of Vishnu metamorphism. They also have relatively low initial-Sr ratios, ranging from about 0.7012 to 0.7030. The pretectonic nature of these intrusives is most difficult to discern in the 96 and 97.6-mile plutons, which cut low-grade slates and phyllites but themselves appear to be almost entirely undeformed. However, careful field work has shown that only the first episode of metamorphism affected the wallrocks and that intrusion probably took place sometime between the first and second stages of metamorphism. Besides a pervasive foliation, the Diamond Creek, Ruby, and Zoroaster plutons also contain distinctive amphibolite dikes emplaced prior to the main stage of Vishnu metamorphism.

Dikes and Sills

Granitic dikes and sills of two types are included with the Zoroaster Plutonic Complex (Fig. 5). The first type consists of generally medium-grained granites and granodiorites, along with a few tonalites and alkali feldspar granites. There can be no doubt about the igneous origin of these rocks. In places such as the margins of the Travertine Falls and Spencer plutons, dikes can be traced directly into the magma chambers from which they originated. Concordant layers of granitic material also can be found in gradational contact with migmatitic gneisses, implying an origin by partial melting. Although the dikes and sills match the compositional range of the Zoroaster plutons, most are post-tectonic and show no evidence of foliation or metamorphic recrystallization. Foliated varieties are abundant only within exposures of the Trinity and Elves Chasm gneisses.

Pegmatite/aplite dikes and sills constitute one of the most abundant rock types in the middle Precambrian section. Percentages vary from zero in several places to almost 100 percent near mile 102—an area that originally was mapped as a small pluton but actually consists of a swarm of cross-cutting dikes. On the average, pegmatite/aplite material constitutes about twenty percent of the total rock volume. Most dikes and sills are undeformed and probably are related to post-tectonic Zoroaster plutons. In a few places (e.g., Monument Canyon), however, patches

Figure 5. Network of pegmatite/aplite dikes and sills cutting Vishnu schists in Tuna Creek Canyon

and layers of deformed pegmatite can be found sharing the foliation of the surrounding Vishnu schist. These must represent pretectonic dikes that were affected by the main stage of Vishnu metamorphism. Because of the coarse grain size and heterogeneity of the pegmatites, it is apparent that their origin involved a hydrothermal fluid in addition to a silicate melt phase. It is probable that most pegmatite/aplite material was generated during the late stages of magmatic crystallization, when molten silicates became saturated with water and exsolved (gave off) an aqueous fluid to coexist with the melt. The low nucleation rates and rapid diffusion in this aqueous phase account for the large grain size. The associated fine-grained aplites presumably represent crystallization of intermingled silicate melt. Some of the most spectacular mineral specimens in the canyon are found in pegmatites— including tourmaline, beryl, apatite, and feldspar crystals up to eight inches (20 cm) in length.

TRINITY AND ELVES CHASM GNEISSES

In the Upper Granite Gorge, two tracts of granitic gneiss form mappable units that can be distinguished clearly from the Zoroaster and Vishnu complexes (see Fig. 1). The Trinity Gneiss, which crops out in Trinity Canyon and between miles 91.5 and 92.9, typically is a well-foliated, medium-to-coarse-grained, quartz/plagioclase/orthoclase/biotite/garnet gneiss. Segregation layering is well developed, and cataclastic textures are common. Scattered throughout the gneiss are thin layers of metasedimentary quartzite and biotite schist, along with a few small, calc-silicate lenses. However, the composition of the gneiss suggests an igneous, rather than a sedimentary, parentage. It is possible that the protolith of the Trinity Gneiss consisted of dacitic to andesitic tuffs and flows with a few sedimentary layers deposited between eruptions. The age of the Trinity Gneiss is uncertain. The well-developed foliation and metamorphic grade, as well as the presence of amphibolite dikes, indicate deposition prior to the main stage of metamorphism. Noble and Hunter (1916) suggested that a "granitic gneiss" (presumably the Trinity Gneiss) formed the sea floor upon which the Vishnu sediments were deposited. This relationship, however, cannot be confirmed by the data presently available.

Further downstream, between miles 112.4 and 118.7, another granitic gneiss of somewhat different character is exposed. The Elves Chasm Gneiss consists predominantly of tonalitic gneiss (plagioclase/ hornblende/quartz) with minor portions of granodioritic to dioritic gneiss. Segregation is well developed only locally, and cataclasis is restricted to shear zones. Calc-silicate pods do occur, but metasedimentary interbeds are lacking. There also are zones of intrusive breccia and rotated xenoliths, which indicate an igneous parentage. The relationship between the Elves Chasm Gneiss and the Vishnu complex also is uncertain, but at the downstream contact near Walthenberg Canyon, the presence of unusual cordierite-anthophyllite gneisses suggests the possibility of a paleosol (soil horizon). If this is the case, then, the Elves Chasm rocks could be interpreted as the basement for Vishnu deposition.

Granite Park

The rocks at Granite Park are worthy of special mention because they apparently represent a layered mafic, intrusive complex that is quite different from any of the other Zoroaster plutons or the metabasalts of the Vishnu complex. Rock types include alternating layers of gabbro, anorthosite, and trondhjemite that have been metamorphosed to moderately foliated hornblende/ tremolite and plagioclase assemblages (Fig. 6). Layers and zones of metagabbroic pegmatite also can be recognized.

Tectonic Setting

The metasediments and metavolcanics of the Vishnu complex have chemical compositions most comparable to those found in modern volcanic-arc systems (Fig. 7). The Zoroaster complex plutons also have major and trace element compositions consistent with an origin in a tectonic setting similar to a volcanic arc (Fig. 8). With the exception of the anomalous Zoroaster and 229 mile plutons, there is a general trend over time toward an increased alkali and potassium content in the plutonic rocks. This may reflect the development of a thicker, more mature crust in the region.

Neodymium isotope evolution paths defined by Bennett and DePaolo (1987) indicate that magmatic rocks in the Grand Canyon were derived from materials that separated from the mantle about

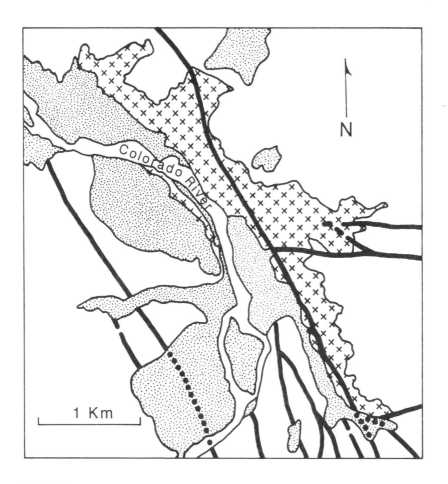

QUATERNARY DEPOSITS

PALEOZOIC ROCKS

PRECAMBRIAN GRANITE PARK
INTRUSIVE BODY

FAULT

Figure 6. Outcrop map of Granite Park Layered Mafix Complex

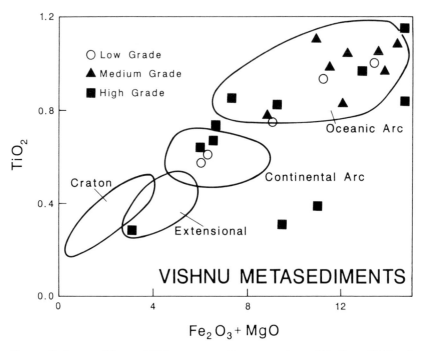

Figure 7. Compositions of Vishnu metasediments and amphibolite. Fields of tectonic environments from Copeland and Condie 1986. Note that the low-grade metasediments, which would give the most reliable indication of original composition, fall in the oceanic arc field.

2.0 to 1.8 billion years ago. This puts an upperlimit on the age of the initial orogenic activity in the region that is similar to the estimates of Copeland and Condie (1986) and Anderson (1986). Based on the data available, it can be inferred that the Early and Middle Proterozoic rocks of the Grand Canyon reflect a sequential development of the southwestern margin of the North American continent.

The Vishnu complex and premetamorphic, plutonic rocks represent a Proterozoic version of a progressively maturing island arc that was disrupted by two distinct orogenic events. The first, at 1720 to 1710 million years ago, involved intense compression but fairly low temperatures and pressures—at least in the Grand Canyon region. The second event (the main stage of Vishnu metamorphism) at 1680 to 1650 million years ago was caused primarily by high heat flow and occurred without high levels of compressional stress.

At least two periods of anorogenic intrusive activity followed deformation and metamorphism. The first probably was shortly after the culmination of Vishnu metamorphism; the second was part of the 1500 to 1400 million-year-old transcontinental anorogenic granite belt. It is tempting to surmise that the first metamorphism might have been related to collision of the Vishnu arc with the margin of change in the nature of subduction that resulted in extensive back-arc melting and metamorphism. The anorogenic intrusives could represent incipient, intracratonic rifting related to reorganization of plate motions. However, verification of such speculation must await further detailed geochronology in the Grand Canyon.

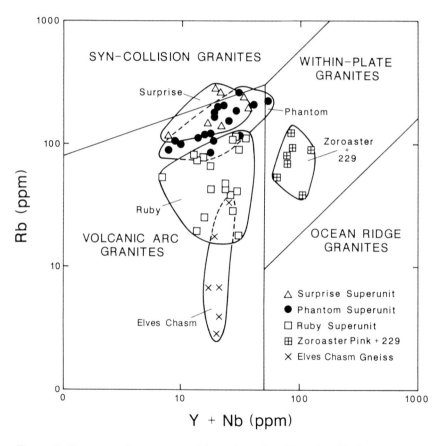

Figure 8. Zoroaster pluton compositions plotted on Rb vs Y + Nb Diagram of Pearce and Others (1985). Note the general progression towards increasing Rb content with decreasing age.

Thus, we see in the earliest part of the Grand Canyon section a replica of the geologic evolution of North America as a whole during the Early and Middle Proterozoic. The story begins with the deposition of sediments and volcanics 2.0 to 1.8 billion years ago and culminates with the emplacement of voluminous granites as part of an incipient transcontinental rift at 1.5 to 1.4 billion years. At least two major episodes of metamorphism and numerous magmatic intrusive events add to the complex history of this intriguing rock sequence that forms the crystalline basement upon which the younger stratified rocks of the canyon were deposited.

ACKNOWLEDGMENTS

This account is based on the joint studies of Ned Brown, Malcolm Clark, Don Livingston, and myself. Special thanks are due to numerous Western Washington University students who worked on the project—especially Bill Lingley and Mike Walen, who completed MS degrees in critical areas, and Mitch Bernardi, Leslie Lathrop, Chuck Lyon, and Wendy Walker, who were our primary field and laboratory assistants. Financial assistance was provided by the National Science Foundation (Grants GA31232 and GA41505), the Museum of Northern Arizona, Western Washington University, The Natural Environment Research Council of Britain, and the University of Leicester. Vital logistic support also was provided by the Inner Canyon rangers of Grand Canyon National Park and numerous commercial river touring companies.

CHAPTER 3

GRAND CANYON SUPERGROUP: UNKAR GROUP

J. D. Hendricks and G. M. Stevenson

INTRODUCTION

In Arizona's Grand Canyon, a series of gently tilted sedimentary and igneous rocks are exposed in isolated outcrops along the Colorado River and its tributary canyons. The rocks overlie the schists and granites of the inner gorge and occur below the flat-lying Paleozoic sedimentary units. These wedges of rock are quite noticeable both from the rim and from the river because of the angular difference of the beds (compared to those above and below) and the striking color and topographic variations within the sequence. Major outcrops of these rocks occur in seven separate locations within the Grand Canyon National Park (Fig. 1).

NOMENCLATURE

John Wesley Powell was the first person to note the geology of these rocks during his historic traverses of the Grand Canyon in 1869 and 1872. In his reports, he described their stratigraphic position and ascribed a tentative Silurian age to the sequence.

Charles D. Walcott conducted extensive field studies in the eastern Grand Canyon in 1882-83 and reported his findings some

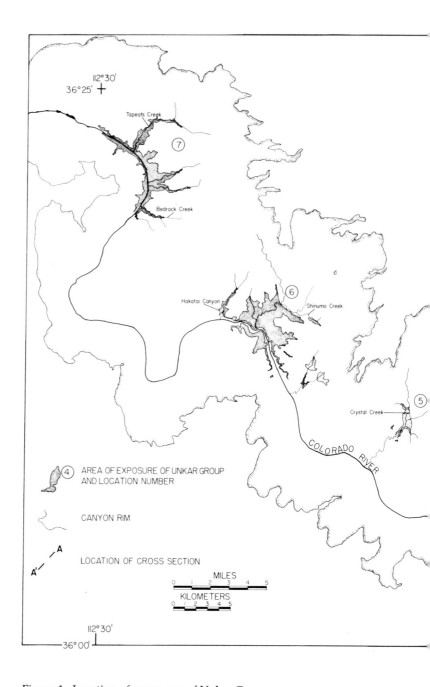

Figure 1. Location of exposures of Unkar Group
(1) "Big Bend" region of the eastern Grand Canyon; (2) Clear Creek;
(3) Bright Angel Creek; (4) Phantom Creek—Phantom Ranch; (5) Crystal
Creek; (6) Shinumo Creek; (7) Tapeats Creek

GRAND CANYON GEOLOGY

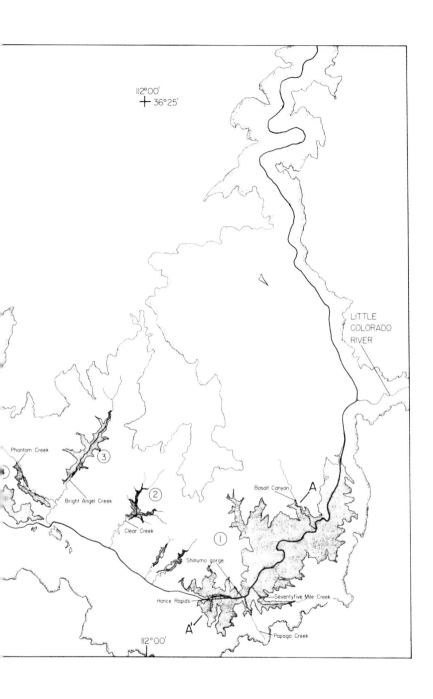

years later. In his 1894 report, Walcott divided this sequence of rocks into two terranes—the Chuar (upper) and Unkar (lower). The combined sequence was named the Grand Canyon Series. In the same report, Walcott provides the first geologic map of the eastern Grand Canyon and the measured stratigraphic thickness of the Grand Canyon Series. Walcott (1894) reports the series to be approximately 12,000 feet (3660 m) in thickness—with the Unkar terrane being 6800 feet (2073 m) and the Chuar 5200 feet (1587 m). A Precambrian age was assigned to the Grand Canyon Series by comparison with the "Keweenawan Series" of the north–central United States.

From studies conducted in the Shinumo area (Fig. 1), Noble (1914) subdivided the Unkar terrane into five formations and assigned group status to the Chuar and Unkar. The names applied to the formations of the Unkar Group, in ascending order, were 1) Hotauta Conglomerate, 2) Bass Limestone, 3) Hakatai Shale, 4) Shinumo Quartzite, and 5) Dox Sandstone. These names were all derived from local geographic features. Because of Precambrian erosional removal of the Grand Canyon Series above the middle of the Dox Formation in the Shinumo region, the upper part of the Unkar Group and all of the Chuar Group were not named by Noble. He did provide, however, a geologic map and structural description of the Shinumo area and detailed petrologic descriptions of the Unkar Group.

Van Gundy (1934, 1951) recognized two unconformities in the upper part of the Unkar Group from work done in the eastern Grand Canyon. These breaks were separated by approximately 330 feet (100 m) of sandstone and shale. He applied the name Nankoweap Group to this unit, thus removing it from Walcott's (1894) Unkar terrane. The Nankoweap overlies a series of basaltic flows and unconformably underlies sediments of the Chuar Group. Because the Nankoweap had not been subdivided into individual formations, Maxson (1967) classified this whole unit as the Nankoweap Formation.

Keyes (1938) used the name "Cardenas lava series" for the basaltic flow sequence at the top of the Unkar Group. Maxson (1961, 1967), however, in his geologic maps of the Grand Canyon, designated these flows as the Rama Formation and included it with intrusive rocks of similar composition found lower in the Unkar Group. Ford and others (1972) formally named the basaltic flows at the top of the Unkar Group the "Cardenas Lavas." This term has been modified to the Cardenas Lava by Lucchitta and Beus (1987). Because the term lava or lavas applies to a fluid and not a rock, this designation is not favored nomenclature, but the name Cardenas Lava is used commonly for these rocks in the current literature and is retained herein.

Dalton (1972) studied the Bass Limestone and Hotauta Conglomerate of Noble (1914) and suggested that the Hotauta be included as a member of the Bass. This designation is adopted in this discussion. Dalton (1972) also suggested that the Bass Limestone should be reclassified as the Bass Formation because of the variety of rock types within the unit and the fact that limestone is a minor lithology. Stevenson and Beus (1982) suggest the Dox Sandstone of Noble (1914) should be renamed the Dox Formation also because of the lithologic diversity. For the purpose of continuity of nomenclature, the original names of Noble, with the exception of the re-ranking of the Hotauta Conglomerate, naming

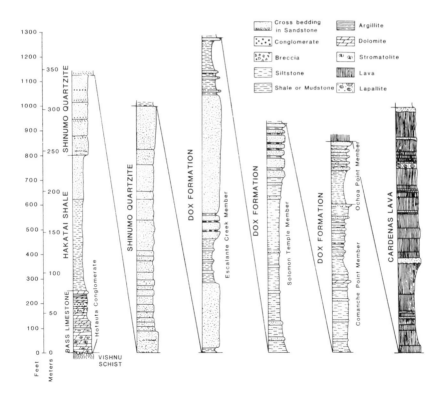

Figure 2. Columnar section of Unkar Group

of the Cardenas Lava by Ford and others (1972) and Lucchitta (1987), and the designation of the Dox Sandstone as the Dox Formation, will be used in the discussion of the individual formations of the Unkar.

As a result of these studies, the Unkar Group currently is subdivided into five formations: 1) Bass Limestone, 2) Hakatai Shale, 3) Shinumo Quartzite, 4) Dox Formation, and 5) Cardenas Lava. A stratigraphic section is presented as Figure 2, which depicts this nomenclature, the average thicknesses of the formations and members, and the general lithologies of the Unkar Group. Following the current code of nomenclature, the Unkar, Nankoweap, and Chuar comprise the Grand Canyon Supergroup.

All of the formations comprising the Unkar Group, except the Cardenas Lava, were named for localities in the Shinumo quadrangle. They occur in small, rotated, downfaulted blocks or slivers and commonly are partially exposed. Of the five recognized formations comprising the Unkar Group, only the Hakatai Shale is well represented in the Shinumo Creek vicinity without further consideration for an "alternate" area. However, in the case of the three other sedimentary units, sections should be selected where they are best preserved. In the case of the Bass Limestone, a much more complete marine section is present to the west at Tapeats Creek (Fig. 1). Both the Shinumo and Dox formations have good sections eastward of Shinumo quadrangle in the Vishnu quadrangle where complete stratigraphic intervals are preserved and well exposed. Perhaps the best locality for the Shinumo is in "Shinumo Gorge" near Hance Rapids, and the most revealing look at the Dox can be found in the "Big Bend" area (area 1 of Fig. 1).

AGE OF THE UNKAR

The Grand Canyon Supergroup overlies the metamorphic and granitic basement complex of Early Proterozoic age (1700 million years ago [m.y.a.] and underlies the middle Cambrian Tapeats Sandstone (550 m.y.a). Thus, the supergroup occupies a portion of the 1150 million-year interval between 550 and 1700 m.y.a. Two methods of age determination have been applied to establish the time of the Unkar Group's formation. The Cardenas Lava provides the only stratigraphically controlled lithology of the supergroup, discovered to date, suitable for radiometric age determinations. McKee and Noble (1974) obtained an age of 1100 m.y.a. for the Cardenas Lava using the Rb–Sr method. The K–Ar method has produced Cardenas ages that are considerably younger than 1100 m.y.a. Ford and others (1972) presented a single K–Ar age of 845± 20 m.y.a. for samples of the Cardenas Lava, and McKee and Noble (1974) obtained ages of 810 ± 20, 790 ± 20 mya, and 781 ± 20 m.y.a. (K–Ar) for samples of the Cardenas Lava. The lower ages obtained using the K–Ar method may reflect an episode of heating and resetting of the K–Ar clock about 800 million years ago (McKee and Noble 1974).

Extensive study of the paleomagnetic pole positions and polar wandering paths by Elston (summarized 1987) has led to the conclusion that the Unkar Group accumulated in the time interval 1250 to 1070 m.y.a. These results are in agreement with the Rb–Sr age of the Cardenas Lava (McKee and Noble 1974). It appears, therefore, that the unconformity ("greatest angular unconformity", Noble 1914) between the Early Proterozoic basement complex and the Unkar Group represents a time period of about 450 million years, whereas the unconformity between the Cardenas Lava and Nankoweap probably reflects a relatively short period of geologic time.

DESCRIPTIONS OF THE FORMATIONS OF THE UNKAR GROUP

The sedimentary sequence of the Unkar Group records a major west to east transgression of the sea. During the nearly 250 million-year time span postulated for Unkar deposition, the region apparently was at (or very near) sea level. Only one unconformity is documented within the Unkar—between the Hakatai Shale and Shinumo Quartzite. Minor fluctuations of sea level or sediment surface elevation is recorded by features suggesting both subaerial and marine deposition throughout the sequence. Apparently, the Unkar Group was deposited in a basin in which the rate of subsidence was approximately the same as the rate of deposition. The only suggestion of relatively deep-water deposition is noted by textural features in dolomites and mudstones in the middle part of the Bass Limestone in the western Grand Canyon (Dalton 1972).

The Bass Limestone and Hotauta Conglomerate Member

The Vishnu surface, over which the Unkar sea advanced from the west, was smooth with a local relief of probably no more than 150 feet (45 m). The Hotauta Conglomerate Member, lowermost unit of the Bass Limestone, was deposited in low areas of the Vishnu terrane. This conglomerate consists of rounded, gravel-sized clasts of chert, granite, quartz, plagioclase crystals, and micropegmatites in a quartz sand matrix. It is found in the eastern Grand Canyon. In the Unkar exposures of the western Grand Canyon, the lowermost part of the Bass contains intraformational breccias and small pebble conglomerates, suggesting that the source of these clasts was toward the east.

The lithology of the Bass Limestone is predominantly dolomite with subordinate amounts of arkose and sandy dolomite with intercalated shale and argillite. Intraformational breccias and conglomerates also are found throughout the sequence (Dalton 1972). One feature of note within the Bass is the presence of biscuit–form and biohermal stromatolite beds (Nitecki 1971). The thickness of the Bass Limestone shows a general increase to the northwest ranging from 330 feet (100 m) at Phantom Creek (Fig.1) to 187 feet (57 m) at Crystal Creek. The anomalously thin section at Crystal Creek probably reflects the presence of a Vishnu topographic high in this area during deposition. The Bass generally forms cliffs or stair-stepped cliffs—with the more resistant dolomites making the risers and the shale and argillite forming steep treads.

Sedimentary features common to all exposures of the Bass Limestone include symmetrical ripple marks, desiccation cracks, interformational breccias/conglomerates, and both normal and reversed small-scale, graded beds (associated with stromatolites). All of these features suggest a relatively low-energy intertidal to supratidal environment of deposition. Although no evaporites presently are recognized in the Bass, some of the interformational breccias may be the result of collapse of earlier formed gypsum. Dalton (1972) noted monoclinic crystal clasts in chert layers of the Bass Limestone. This is suggestive of a dolomitic replacement of gypsum.

The lithology and sedimentary structures observed in the Bass Limestone suggest deposition in an easterly transgressing sea. During the maximum incursion of the sea, carbonates and deep-water mudstones accumulated in the western Grand Canyon. In the east, stromatolites were forming, and shallow-water mudstones were being deposited. Following this period of transgression, the sea slowly regressed. Evidence for this includes ripple marks, mudcracks, and deposits of oxidized shales in the upper part of the Bass—all suggesting periods of subaerial exposure. Evaporite-forming conditions probably existed also during this regressive phase (Dalton 1972). Eventually, deltaic conditions predominated, which marked the beginning of Hakatai Shale deposition. The contact between the Bass Limestone and Hakatai Shale is gradational in the east and sharp, though conformable, in the west.

Hakatai Shale

The Hakatai Shale probably is the most colorful formation in the Grand Canyon, with colors that vary from purple to red to brilliant orange on outcrop. The colors result from the oxidation state of the iron-bearing minerals in the formation. The Hakatai is subdivided into three informal members based on lithology (Beus and others 1974, Reed 1974, Ford and Breed 1974). The lower two units are highly fractured argillaceous mudstones and shales that weather to gentle-to-moderate, granular slopes. The upper unit consists of cliff-forming beds of medium-grained quartz sandstone (Fig. 2). The Hakatai varies in thickness from about 445 feet (135 m) at Hance Rapids to nearly 985 feet (300 m) at the type section in Hakatai Canyon in the Shinumo Creek area.

Sedimentary structures, such as mudcracks, ripple marks, and tabular-planar cross bedding, suggest the Hakatai was deposited in a marginal marine environment. Mudstones and shales of the lower two members probably were deposited in a low-energy, mud-flat environment. The upper sandstones suggest a higher energy, shallow-marine environment (Reed 1974).

Although the contact beween the Hakatai Shale and Bass Limestone is gradational in the eastern Grand Canyon and sharp and easily located in the western exposures, the contact between the Hakatai and the overlying Shinumo Quartzite is evident in all exposures. It is marked by an unconformity that truncates cross beds and channel deposits of the Hakatai. From observations made in the canyon of Bright Angel Creek, Sears (1973) indicates that Hakatai deposition in the area ended in conjunction with tectonic activity along a series of northwest-trending, high-angle, reverse faults.

Shinumo Quartzite

In contrast to the slope-forming argillaceous beds of the Hakatai Shale below and the Dox Formation above, the Shinumo Quartzite is a series of massive, cliff-forming sandstones and quartzites. The color of the Shinumo ranges from muted red, brown, and purple to white.

Four or possibly five poorly defined members have been recognized within the Shinumo Quartzite (Elston 1987, Daneker 1974). The lower units, in ascending order, consist of conglomeratic subarkose and submature quartz sandstone; to mature quartz sandstone; to brown quartz sandstone with abundant cross beds, clay galls, and mudcracks.

Near Shinumo Creek, the uppermost unit is the thickest member. It consists of fine-grained, well-sorted, and rounded quartz grains in a siliceous cement. Beds in the upper part of the upper member are contorted by fluid evulsion, which suggests

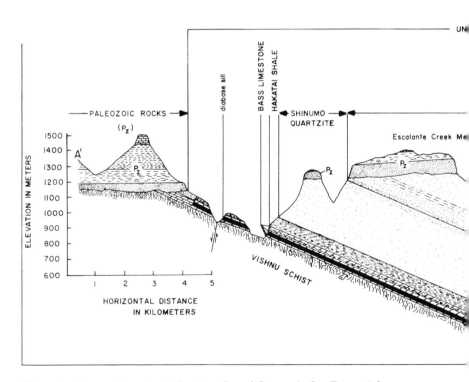

Figure 3. Cross section A - A' (eastern Grand Canyon). See Figure 1 for location of section.

that there might have been tectonic activity during this period.

The thickness of the Shimumo Quartzite shows a general increase to the west and ranges from 1132 feet (345 m) at Papago Creek in the east to 1328 feet (405 m) (Daneker 1974) or 1542 feet (470 m) (Noble 1914) in Shinumo Creek. Because the Shinumo is such a resistant unit, it formed hills where exposed during the pre-Tapeats erosional event.

This feature, the pinching out of the Tapeats Sandstone against Shinumo Quartzite highs, can be seen today in exposures near the bottom of the Grand Canyon. Analysis of the lithology and the sedimentary structures of the Shinumo suggests that the environment of deposition was near-shore, very shallow, marginal marine and part fluvial, part deltaic (Daneker 1974). The contact between the Shinumo and the overlying Dox Formation appears to be conformable in most locations and is marked by the lowermost shaley interval of Dox lithology.

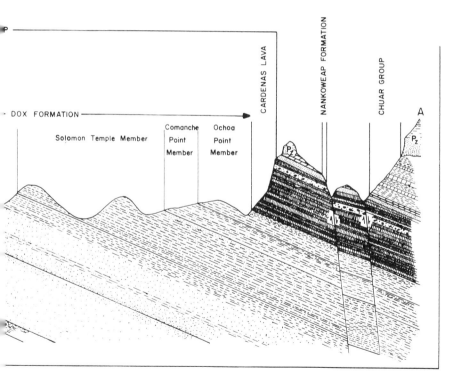

Dox Formation

The Dox Formation is the thickest unit of the Unkar Group. The only complete section presently exposed, however, is in the eastern Grand Canyon (area 1 of Fig. 3), with thicknesses variously reported to be 3020 feet (921 m), 3115 feet (950 m), and 3230 feet (985 m). The Dox consists of four members: (in ascending order) Escalante Creek, Solomon Temple, Comanche Point, and Ochoa Point. West of 75 mile Creek (western part of area l, Fig.1), only the Escalante Creek and Solomon Temple members are preserved; the Comanche Point and Ochoa Point have been removed by pre-Tapeats erosion.

The Escalante Creek Member is reported by Stevenson (1973) to be 1280 feet (390 m) thick where exposed in the eastern Grand Canyon. Here, it is a light-tan to greenish brown, siliceous quartz sandstone and calcareous lithic and arkosic sandstone that is 800+ feet (244+ m) thick, combined with an overlying 400 feet (122 m)-thick sequence of dark-brown-to-green shale and mudstone. The tan to brownish color of this lower member is in marked contrast to the characteristic red and red-brown color of the rest of the Dox Formation.

Sedimentary structures observed in the sandstones of the Escalante Creek included contorted bedding (within 100 feet [30 m] of the base), small-scale, tabular-planar cross beds, and graded beds (with shale interclasts at the base). The contacts beween members of the Dox Formation are gradational and are based mainly on topographic expression, depositional environments, and color changes. The Escalante Creek is tan to brown and forms a cliff-slope topograpy, as opposed to the more red-orange overlying Solomon Temple Member, which forms a rounded-hill topography that is more chaiacteristic of the remainder of the Dox.

The Solomon Temple Member is a cyclical sequence of red mudstone, siltstone, and quartz sandstone. It is 920 feet (280 m) thick in the eastern Grand Canyon. The lower 700 feet (213 m) is a slope-forming series of predominantly red-to-maroon shaley siltstone and mudstone with subordinate quartz sandstone. The upper 220 feet (67 m) of the member is primarily maroon quartz sandstone with numerous channel features. These channels, along with common low-angle, tabular cross beds, suggest a floodplain environment.

The Comanche Point Member occupies more than half of the Dox outcrop area and is distinguished from enclosing members by its slope-forming and color-variegated character. This member varies in thickness from 425 to 617 feet (130 to 188 m) in the eastern Grand Canyon. The lithology is primarily shaley siltstone and mudstone with minor amounts of sandstone. Five pale green-to-white, leached red beds, some up to 40 feet (12 m) in thickness, provide the variegated appearance of this unit. Stromatolitic dolomite layers are found within or directly adjacent to the leached beds. Sedimentary features found in this member include ripple marks; mudcracks and curls; salt casts; and wavy, irregular bedding.

The Ochoa Point Member is 175 to 300 feet (53 to 92 m) thick and forms steep slopes and cliffs below the Cardenas Lava. It consists of micaceous mudstone that grades upward into a predominantly red quartzose, silty sandstone. Sedimentary structures of this member also include salt crystal casts in the mudstone and asymmetical ripple marks and small-scale cross beds in the sandstones.

The lithology and sedimentary structures found within the Dox Formation suggest, in ascending order, a subaqueous delta, floodplain, and tidal flat environment during deposition. According to Stevenson and Beus (1982), features of the lowermost member, the Escalante Creek, record a rapid transgression of the Dox sea followed by gradual basin filling.

This basin was filled by the close of Escalante Creek time, and the region was at or very near sea level for the remainder of Dox time. Stevenson and Beus (1982) also suggest a possible westerly source for some of the sediments of the Dox, which is opposite to the inferred source direction for other formations of the Unkar Group.

The Cardenas Lava

Cardenas Lava is the name given to a series of basalt and basaltic andesite flows and sandstone interbeds that are stratigraphically above the Dox Formation and below the Nankoweap Formation (Ford and others 1972). This sequence of rocks is exposed only in the eastern Grand Canyon, where the thickness of the formation ranges from about 785 feet (240 m) to nearly 985 feet (300 m). The contact between the Cardenas Lava and the overlying Nankoweap Formation is unconformable, with an unknown amount of the Cardenas being removed prior to Nankoweap deposition.

The contact between the Dox Formation and the Cardenas is conformable and interfingering. This is highlighted by the presence of a thin, discontinuous basaltic flow in the upper Dox a few meters below the top of the formation. The Dox near the contact is mildly baked and locally displays small folds and convolutions that are suggestive of soft sediment deformation. At one location, the basalt of the lowermost flow sequence contains rounded masses of the igneous rock, up to 3.3 feet (1 m) in diameter, completely surrounded by a thin layer of siltstone of Dox lithology. The uppermost Dox is fine-grained sandstone and siltstone deposited in a tidal flat environment. The region at the time of initial Cardenas eruption was at or very near sea level (Stevenson and Beus 1982). One interpretation is that the lava flowed over unconsolidated sandy and silty Dox sediments that were wet at the time. Whether these sediments were slightly above or slightly below water level is unknown.

Strictly on the basis of topography, the Cardenas Lava can be divided into two units (Hendricks and Lucchitta 1974). The lower unit forms granular slopes and is from 245 to 295 feet (75 to 90 m) thick. Although this unit is highly altered and weathered, many of the primary features are preserved. This "bottle-green member" (Lucchitta and Hendricks 1983) is a composite of many, thin, discontinuous flows and sandstone interbeds. The basalt of this unit is broken and weathers into nodules typically 3 to 10 inches (10 to 30 cm) in diameter. Near the top of the unit, some 230 feet (70 m) above the base, the basalt is more massive and less altered. Petrographically, the lower unit is an olivine-rich basalt with a subophitic texture. Lower in the bottle-green member, the rock is highly altered but has a texture that suggests it may have been quite glassy originally.

Chemically, rocks from the lower unit are high in sodium and magnesium and depleted in potassium, suggesting spilitic alteration. The nodular character, glassy texture, and anomalous chemistry of rocks of the lower unit suggest rapid quenching in sea or brackish water. The thin, discontinuous sandstone interbeds also indicate that water flowed over or was ponded on the lavas during periods of volcanic quiescence.

Approximately 328 feet (100 m) above the base of the Cardenas is a continuous sandstone bed some 16 feet (5 m) in thickness that overlies the bottle-green member. This sandstone is laminated and forms vertical cliffs. In a number of locations, the sandstone occupies channels in the upper surface of the bottle-green flow member. The petrology of this sandstone suggests quiet-water deposition; the channels, therefore, probably are lava channels that were left as extrusion temporarily ceased and basin subsidence continued—lowering the lava surface below sea level.

The upper unit of the Cardenas is a series of cliff-forming basalt and basaltic andesite flows along with additional sandstone interbeds. Features preserved in the individual flow units suggest that the volcanic pile accumulated at a slightly greater rate than basin subsidence. In ascending order, these are: an autoclastic breccia directly above the 328 foot (100 m) sandstone, a fan-jointed unit, ropey lava, and, finally, a lapillite unit at the 754 foot (230 m) level. Following eruption of the lapillite, volcanic activity ceased for a short period of time. The pyroclastic surface was smoothed as subsidence continued until the surface again was lowered below sea level. This is noted by a planar upper surface on the lapillite unit with a continuous sandstone layer directly above. At least two more eruptive events followed deposition of this sandstone layer.

Following cessation of volcanic activity, the sediments and igneous rocks of the Unkar were tilted gently toward the northeast. An unknown amount of Cardenas Lava was eroded prior to Nankoweap deposition. Elston and Scott (1976) suggest that major tectonic movement occurred along the Butte fault in the eastern Grand Canyon during the post-Cardenas, pre-Nankoweap interval.

Intrusive Rocks of the Unkar Group

Diabase sills and dikes intrude all formations of the Unkar Group below the Cardenas Lava. Sills are restricted to the Bass Limestone and Hakatai Shale. Dikes intrude the Hakatai Shale, Shinumo Quartzite, and Dox Formation above the sills. Feeder dikes or vents for the sills are not exposed. Above the sills, dikes can be traced, discontinuously, to within a few meters of the base of the Cardenas Lava.

Unkar sills range in thickness from 75 feet (23 m) at Hance Rapids in the eastern Grand Canyon to 985 feet (300 m) in Hakatai Canyon (Shinumo Creek area). Fine-grained, chilled margins suggest that the magma was highly fluid at the time of intrusion. Only the sills of the Shinumo Creek area show extensive differentiation and segregation products. Here, a picritic layer of unknown thickness is present near the base of the sill, while the top of the intrusion is marked by a 20-foot (6-m)-thick granophyre layer.

Contact metamorphism caused by intrusion of the diabase resulted in the formation of chrysotile asbestos above the sills where the magma intruded the Bass Limestone. Adjacent to the sills, the Hakatai Shale is a knotted hornfels containing porphyroblasts of andalusite and cordierite that have been replaced by muscovite and green chlorite, respectively.

The relation between the sills, dikes, and Cardenas flows is not self-evident. The mineralogy of the sills is uniform throughout the region and is identical to that of the unaltered parts of the bottle-green member of the Cardenas. Chemical variation diagrams (Hendricks and Lucchitta 1974) indicate that the flows are more felsic than the sills, but the evidence does not preclude the possibility of a common parentage. Paleomagnetic evidence (Elston 1987) suggests that the majority of the sills were intruded at the same time that the Comanche Point Member of the Dox Formation was being deposited. The Hance sill, the easternmost of the Unkar sills, produces an anomalous paleomagnetic pole position that Elston (1987) interprets as an indication of a slightly greater age for this intrusion. If the majority of the sills were emplaced during Comanche Point time, approximately 330 feet (100 m) of sediment of the Comanche Point and Ochoa Point members of the Dox was deposited during the time interval between intrusion of the sills and initial eruption of the Cardenas Lava. Some of the dikes in the eastern Grand Canyon have paleomagnetic pole positions similar to the Cardenas and may represent feeders. These dikes cannot be seen connecting to the nearby Hance sill.

The sills, dikes, and flows of the Unkar Group may represent a single volcanic episode. If so, the earliest phases were intrusions of diabase sills followed by a period of igneous quiescence long enough for accumulation of approximately 330 feet (100 m) of Dox sediment. Later phases were eruptions of basalt and basaltic andesite flows via a network of thin dikes. Elston and McKee (1982) indicate that initial Strontium-87, Strontium-86 ratios are slightly different for a sill in the Shinumo Creek area and the Cardenas Lava and conclude that either the sill and flows are not comagmatic or the magma acquired ^{87}Sr from passing solutions. These ratios, 0.7042 $\pm$ 0.0007 (sill) and 0.7065 $\pm$ 0.0015 (flows), do not preclude a common parentage for the intrusive and extrusive rocks. This might occur if the magma ponded in the crust long enough for derivation of the basaltic andesite of the Cardenas Lava from the basaltic magma of the sills, provided the intruded crust was assimilated and that it had $^{87}Sr/^{86}Sr$ greater than 0.7065.

OVERVIEW OF GEOLOGIC HISTORY

Prior to the beginning of deposition of the Unkar sediments, the region of the Grand Canyon was along the southwestern margin of the North American craton. The surface was a smooth terrain composed of Vishnu granitic and metamorphic rocks. Pre-Unkar relief on this surface was undulating with a relief on the order of 30 to 500+ feet (a few tens to a few hundreds of meters). Relatively high terrain may have existed to the east and northeast, with probable open oceans to the west. As the land surface subsided, the ocean advanced eastward, marking the beginning of Unkar deposition about 1250 million years ago.

The Hotauta Conglomerate Member of the Bass Limestone, lowermost unit of the Unkar Group, was deposited in relatively low areas of the Vishnu terrain. The clast size suggests an easterly source area. During this initial transgression, the sea advanced at least as far as Hance Rapids in the eastern Grand Canyon—and probably much farther. Sediments in the middle of the Bass Limestone in the western Grand Canyon suggest that this area was below wave base and away from any strong currents. At the same time, indicators in the east suggest an intertidal to supratidal environment.

The initial transgression of the sea was followed by a gradual regression, which resulted from sediment accumulation. Sedimentary features found in the upper part of the Bass Limestone and the Hakatai Shale indicate that conditions varied from subaerial to subaqueous throughout the region during this period of time. Tectonic activity along northeast-trending, high-angle reverse faults marked the end of deposition of the Hakatai Shale and resulted in a period of erosion prior to the accumulation of sands of the Shinumo Quartzite.

Near-shore fluvial and deltaic conditions predominated in the region during Shinumo and early Dox time, marking a further subsidence of the region and the second transgression of the sea;

marine conditions returned by the close of this period. Following deposition of about 165 feet (50 m) of Dox sediments, rapid subsidence caused the sea to advance further eastward followed by a sustained period of basin filling. Subaerial conditions returned by the close of Solomon Temple time. The contact between the Solomon Temple and Comanche Point members of the Dox Formation marks another transition to marine conditions with the remainder of the Dox environment fluctuating between marine and nonmarine conditions.

Features found in the lowermost Cardenas Lava suggest the outpouring of the basalt onto wet, probably shallow-water Dox sediments. Sporadic accumulation of the lava pile and continued basin subsidence resulted in conditions that varied between marine and nonmarine, but, generally, the flows accumulated at a greater rate than that of regional subsidence. Following extrusion of more than 985 feet (300 m) of Cardenas Lava, tectonic uplift raised and tilted gently the region of the eastern Grand Canyon toward the northeast. Subsequent erosion removed an unknown amount of the lavas prior to deposition of sediments of the Nankoweap Formation.

INTER-REGIONAL CORRELATIONS

The only other unmetamorphosed sequence of rocks that is younger than the 1700 million-year basement complex and older than Cambrian in the region is found in central Arizona. These rocks, the Apache Group and Troy Quartzite, are similar to those of the Unkar Group.

Shride (1967) has suggested a possible correlation of the Unkar Group with the Apache Group. This is based on similarities in the age and lithology of the two units. Elston (1987) reviews the correlation of various Middle and Late Proterozoic sequences on a paleomagnetic basis. He concludes that the Mescal Limestone of the Apache Group correlates with the middle part of the Dox Formation and that sills in the Unkar Group are similar paleomagnetically to sills that intrude both the Apache Group and Troy Quartzite. Elston (1987) also provides possible correlations of the Unkar Group with other Middle Proterozoic sequences of North America. These include the Uinta Mountain Group of Utah and Colorado, the Belt Supergroup of Montana and Idaho, the Sibley Series of Ontario, and the Keweenawan Supergroup of the Lake Superior Region.

SUMMARY

Sediments and lavas of the Unkar Group of the Grand Canyon Supergroup were deposited in a basin along the western edge of the North American craton during the period 1250 to 1070 million years ago. Features preserved in the Unkar record a west to east transgression of the sea with minor sea-level variations resulting from basin filling and subsidence. About 5800 feet (1770 m) of sediment was deposited in the Unkar basin before the onset of volcanic activity in the area. Unkar sedimentary rocks presently are exposed in isolated outcrops in the Grand Canyon and have been subdivided into formations; in ascending order they are, the Bass Limestone, Hakatai Shale, Shinumo Quartzite, and Dox Formation. A volcanic episode marked the end of sedimentation in the Unkar basin. During this volcanic period, igneous material formed sills in the lower parts of the Unkar and dikes in the upper parts. Lava was erupted onto sediments of the Dox Formation and accumulated to a thickness of nearly 1000 feet (305 m). These extrusive rocks have been named the Cardenas Lava. Following cessation of volcanic activity, the Unkar Group rocks were tilted slightly and the top of the Cardenas Lava eroded before the onset of further sedimentation in the region.

CHAPTER 4

GRAND CANYON SUPERGROUP: NANKOWEAP FORMATION, CHUAR GROUP, AND SIXTYMILE FORMATION

Trevor David Ford

INTRODUCTION

The upper half of the younger Precambrian strata of the Grand Canyon presents a sequence of some 6800 feet (2100 m) of rocks not seen anywhere else in the southwestern United States. They have not been metamorphosed, and the sedimentary rocks include an unparalleled assemblage of late Precambrian fossils. For these reasons, the Nankoweap and Chuar deserve a chapter to themselves. Furthermore, in the Sixtymile Formation they provide unique evidence of sedimentation during tectonic activity, with examples of massive block falls from a fault scarp.

The younger Precambrian rocks of the Grand Canyon first were recognized as "Algonkian" by Powell (1876), who named them the Grand Canyon Series. Now redesignated as the Grand Canyon Supergroup (Elston and Scott 1976), the upper half comprises the Nankoweap Formation, overlain by a thick Chuar Group and a thin Sixtymile Formation. Walcott (1894) gave us the first detailed descriptions but divided the Supergroup into the Unkar and Chuar "terranes." Since the Unkar Group is treated elsewhere in this volume, only the younger part of the younger Precambrian is covered in this chapter.

The Nankoweap and Chuar are exposed in the wide eastern part of the Grand Canyon, clearly visible from the Desert View Tower overlook (Fig. 1). The former division is present in cliffs

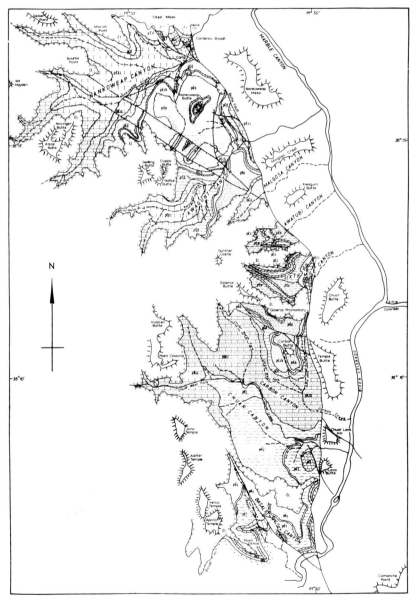

Figure 1. Geological map of the Chuar Group

overlooking Basalt, Tanner, and Comanche canyons. It is visible from the river, but the Chuar Group has been eroded away from the immediate course of the Colorado River and crops out only in the upper parts of several right-bank tributaries—notably Nankoweap, Kwagunt, Carbon, Chuar, and Basalt canyons, where it has been folded into a prominent north-south syncline, west of and parallel to the Butte fault.

The north rim overlooks at Point Imperial and Cape Royal provide fine views of the Chuar outcrops in these tributary canyons, and observers at Cape Royal can see clearly the synclinal structure (Fig. 1). The outcrops of the Chuar Group are bounded on the east by the Butte fault. Displacement on this fault was of the order of 10,000 feet (3000 m) down to the west in late Precambrian time, but this has been canceled out partly by an easterly downthrow in post-Paleozoic times of some 2700 feet (880 m). Both the Nankoweap and Chuar strata are unconformably overlain by the Cambrian Tapeats Sandstone at Powell's Great Unconformity.

THE NANKOWEAP FORMATION

In the middle portion of the younger Precambrian sequence of the eastern Grand Canyon, a distinctive group of red-brown sandstones lies unconformably on top of the Cardenas Lava. The rocks crop out in the cliffs from west of Basalt Canyon eastwards to near Carbon Canyon on the north bank and appear intermittently around Tanner Canyon and Comanche Creek on the south bank. Here, outcrops are broken up by branches of the Butte fault. These sandstones were included partly in the top of the Unkar Group and partly in the basal part of the Chuar Group by Walcott (1894) but were separated as a new unit, the Nankoweap Group, by Van Gundy (1937). Later, in 1951, Van Gundy gave a fuller description and noted the presence of unconformities at both upper and lower contacts. Maxson (1961, 1967) introduced the term "Nankoweap Formation" on his geological maps of the Bright Angel Quadrangle and eastern Grand Canyon but made no comment on the unconformities. The name is taken from the fault-bounded block of Nankoweap surrounded by younger rocks in Nankoweap Canyon, but more extensive outcrops occur above the Cardenas Lava in Basalt Canyon (Fig. 3), Comanche Creek, and Tanner Canyon.

The Nankoweap Formation is divided into two members. An upper member, 330 feet (100 m) thick, is composed mainly of fine-grained quartzitic sandstones that are shaly and silty towards the top. These sandstones are thin-to-medium bedded, with cross beds, ripple marks, mudcracks, and occasional salt pseudomorphs. A lower (ferruginous) member is 40 feet (13 m) thick and composed of red, fine-grained sandstones and siltstones with hematite laminae and lenses of volcanic detritus derived from the Cardenas Lava. A disconformity separates the two members so that only the upper member is present in Basalt Canyon. Details are given in Elston and Scott (1976).

While exposures in Basalt Canyon and in the cliffs immediately north of Tanner Rapids show well-bedded, red-brown sandstone dipping evenly at about five degrees to the northeast, the upper parts of Tanner Canyon reveal a more complex story. If the effects of several faults are removed, it seems that there was a considerable interval of erosion after the eruption of the Cardenas lava. As a result of this erosion, the lava was beveled by a northeast-facing scarp that was perhaps as much as 650 feet (200 m) high. The lava itself was deeply weathered and affected by ferruginous alteration before the lower part of the scarp was buried by the lower member of the Nankoweap. After a hiatus of unknown duration, the sediments of the upper member completely buried this scarp. After further warping in post-Chuar times, the Cambrian Tapeats Sandstone transgressed over both upper and lower members.

No direct dating is possible on the Nankoweap Formation, so its age can be bracketed only between the 1090 ±70 million-year age of the Cardenas Lava below (Elston and Scott 1976) and that of the Chuar Group above, which Elston (1979) has argued was terminated by the Grand Canyon orogeny approximately 823 million years ago.

The limited paleomagnetic evidence available (Elston and Grommé 1974; Elston and McKee 1982) supports these arguments for a provisional date of 1050 million years on the beginning of deposition of the upper member of the Nankoweap Formation. However, a recent re-estimation of the age of the Cardenas Lava by Larson and others (1986) suggests that they are a little younger than previously determined at 950 million years. If this is accepted, then the age of the Nankoweap Formation may be somewhat less than 1000 million years.

Paleontology

A structure found in a sandstone of the Nankoweap Formation in Basalt Canyon (eastern Grand Canyon) was identified as a trace fossil impression of a stranded jellyfish (Van Gundy 1937, 1951). It comprises a series of lobes rounded at the extremities. Some lobes have a median groove radiating from a small, irregular hollow, and the whole structure is approximately five inches (12 cm) in diameter. This also was considered to be a jellyfish impression by Hinds (1938) and Bassler (1941). It was named *Brooksella canyonensis* by Bassler as a new species of a genus well known in the Paleozoic, though the interpretation of the genus as a jellyfish is still in considerable doubt.

Cloud (1960, 1968) subsequently obtained a partial second specimen but claimed that the structures were of inorganic origin formed by "compaction of fine sands deposited over a compressible but otherwise unidentifiable structure, possibly a small gas blister." Glaessner (1969) was unconvinced by Cloud's explanation and drew comparisons with a Mesozoic stellate trace fossil, *Asterosoma*, deducing that it was of organic origin and that its "possible originator was a sediment feeder able to burrow into the sediment... worm-like in shape... probably an annelid."

Accordingly, Glaessner renamed the structure *Asterosoma? canyonensis* (Bassler 1941) and apparently still accepts the trace-fossil interpretation (Glaessner 1984) in spite of its age. In a 1981 study, Kauffman and Steidtmann supported Glaessner's interpretation as a burrow made by a sediment-feeding, wormlike organism.

If the age of the Nankoweap Formation is approximately 1000 million years, this fossil would be one of the earliest records of a burrowing sediment-feeder on Earth. In view of the profusion of trails and burrows in Cambrian rocks, their absence in the Nankoweap Formation arouses suspicion.

An examination of both specimens revealed a similarity to small "sand-volcanoes" formed by the expulsion of gas or fluid from sands deposited in shallow-water, turbidite conditions. In view of this, it is difficult to support Glaessner's interpretation of the phenomenon as trace fossils.

A more intensive search of Nankoweap sandstone outcrops for comparable structures obviously is needed. Sinuous markings on the sole of some beds, though very much like worm-trails, probably are the truncated bases of shallow mudcrack structures.

THE CHUAR GROUP

The Chuar Group has been subdivided by Ford and Breed (1972a, 1973a) into three formations and seven members. The present nomenclature is summarized in Table 1 and shown graphically in Figure 2.

Table 1. Stratigraphic Subdivisions of the Chuar Group

GROUP	FORMATION	MEMBER	THICKNESS FEET (METERS)
	Sixtymile 200 feet (60 m)		
Chuar 6610 feet (2013 m)	Kwagunt 2218 feet (675 m)	Walcott	838 (255)
		Awatubi	1128 (344)
		Carbon Butte	252 (76)
	Galeros 4272 feet (1302 m)	Duppa	570 (174)
		Carbon Canyon	1546(471)
		Jupiter	1516 (462)
		Tanner	640 (195)

Reynolds and Elston (1986) have revised the thicknesses since they found the Galeros Formation to be only 3000 feet (915 m) thick and the Kwagunt 2083 feet (635 m). Full details have not yet been published, but their abstract suggests that the environments of the four subdivisions of the Galeros are from highest to lowest:

4. (Duppa?) alluvial plain
3. (Carbon Canyon) mixed coastal or paludal swamp
2. (Jupiter?) coastal or alluvial plain
1. (Tanner?) sediment-starved basin, rich in organic material.

Thin sequences of stromatolitic and cryptalgal carbonates occur at the bases of the lower two. In the Kwagunt Formation, Reynolds and Elston (1986) noted a shoreline sandstone at the base followed by organic-rich mudstone and siltstone with a variety of stromatolitic, oolitic, and pisolitic intercalations as described below. Fossil microorganisms are present in the darker beds of both the Galeros and Kwagunt formations.

GALEROS FORMATION

The Tanner Member, which overlooks the Tanner Rapids on the Colorado River, consists of forty to eighty feet (12 to 14 m)

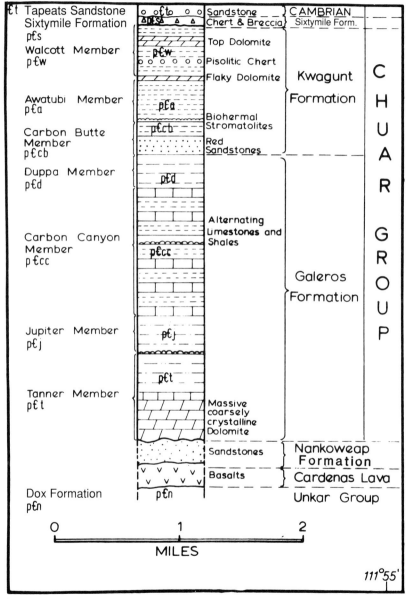

Figure 2. Stratigraphic column of the Nankoweap Formation and Chuar Group

of thick-bedded, coarse, crystalline dolomite at the base and 580 feet (177 m) of other material (almost entirely shales) above. No primary sedimentary structures have been reported from the dolomite. It forms a massive ledge capping the cliffs around Basalt Canyon (Fig. 3) and an inlier (older rocks surrounded by younger) on the southern flank of Chuar Canyon, which also is known as Lava Canyon. A faulted patch also occurs in Nankoweap Canyon.

Figure 3. The upper part of Basalt Canyon showing cliffs of Cardenas Lava in the bottom left corner, overlain by the flat-topped terrace of Nankoweap sandstones and Tanner Member Dolomite, with shales up to the basal stromatolitic layers of the Jupiter Member, extending upwards toward the unconformable base of the Cambrian. Photograph by Parker Hamilton

GRAND CANYON GEOLOGY

The overlying shales crop out throughout much of Basalt and Chuar canyons. The Tanner Dolomite was included in the Unkar by Walcott, but it was transferred to the Chuar by Van Gundy (1951).

The Jupiter Member also consists of carbonates below and shales above. The basal division is about 40 feet (12 m) of stromatolitic limestones—including undulating and broad-domed forms of algal laminate within a mass of dolomitized tufa-like rock, with flat-pebble conglomerates at the base. The upper part of the carbonate member has layers with abundant casts of gypsum crystals and a few, poorly defined and solitary stromatolite columns. These are similar to the form *Inzeria*, as well as undulating stromatolites of the form *Stratifera*. The remainder of the 1516-foot (462-m)-thick member is dominantly argillaceous, but includes frequent beds of thin sandstones and siltstones. Rarely more than a few inches thick, these beds have abundant ripple marks and mudcracks as well as occasional raindrop prints and salt pseudomorphs. Commonly, hematite cement occurs in patches and sometimes weathers to goethite box-stones. The shales are variable in color—from red-purple through ocherous yellow to pale green—and have scattered, blue-black, micaceous bands.

The Carbon Canyon Member consists of a rapid alteration of carbonate, shale, and sandstone (each a few feet thick) attaining a total thickness of 1546 feet (471 m) (Fig. 4). The carbonates commonly are three to six feet (1-2 m) thick, and almost all are fine-grained, dolomitic micrite with scattered chert nodules and lenses of quartz silt. In places, the carbonates grade into calcareous siltstones. Most of these show an irregular lamination, possibly of algal origin. For the most part, the tops of the carbonate beds are ripple-marked and mudcracked.

The intervening shales vary from blue-black, micaceous shale to red mudstones with scattered green bands. Sandstones are fewer in number and rarely more than two feet thick. They generally are green-grey in color and have subangular quartz grains set in a yellow-green, carbonate matrix. Scattered lenses of large, round quartz grains could suggest brief episodes of aeolian sedimentation. Mudcracks are common in the sandstones, and some laminae show truncated, incipient cracks that look misleadingly like worm tracks on weathered joint faces. The stromatolites take the form of rapidly widening, irregularly branching columns

Figure 4. The Chuar syncline seen looking north along its axis. Chuar Canyon in lower left; Carbon Canyon, too, in the right center. The ridge in the middle foreground is of alternating limestones and shales of the Carbon Canyon Member, while the synclinal cuesta in the middle distance is of the Carbon Butte sandstone. The Butte fault trends north through the shadows on the right. Photograph by John Shelton

that show strongly convex laminae. Dolomitization has destroyed internal detail, but they appear to fall within the form *Baicalia* (probably *B. aff. rara*) Semikhatov.

The Duppa Member is over 570 feet (174 m) thick and marks a return to the argillaceous character with only a few thin, scattered limestones. Siltstone beds up to three feet (1 m) thick are scattered throughout with well-rounded grains. The rest is shale (generally micaceous) that grades into red mudstones towards the top.

KWAGUNT FORMATION

Since the Carbon Butte Member has at its base the only thick sandstone in the Chuar Group, it provides a distinctive marker for the base of the Kwagunt Formation. The member is 252 feet (75 m) thick. The basal 80 feet (24 m) are a thick, red sandstone, which makes a scarp surrounding Carbon Butte. A micaceous purple shale parting in the middle covers mudcracks (Fig. 5), and there is cross bedding in the highest beds. Red and purple mudstones make up the upper part of the member, with a bed of mottled, carious-weathering sandstone in the middle.

Figure 5. Mudcracks in the Carbon Butte sandstone

The Awatubi Member, again, is dominantly argillaceous with shales and mudstones of varying color diversified only by thin, ferruginous siltstones. At the bottom of the 1128 feet (344 m) member, however, is a massive stromatolite layer 12 feet (3-5 m) thick. This consists of biohermal domes, each 8-10 feet (2.5-3 m) in height and width and made up of a complex of columns 2 to 3 inches (5-7.5 cm) in diameter, alternating with confluent domes (Fig. 11). The columns show nearly flat laminae and generally are parallel sided with rare bifurcation. Dolomitization has destroyed most of the internal detail, but a clearly defined wall is present, indicative of the form *Boxonia* Koroljuk. The matrix between columns generally is crystalline dolomite whereas that between bioherms is coarsely granular dolomite of a highly porous nature. Flat-pebble conglomerates occur at the bases of some bioherms.

Some 30 feet (9 m) from the top of the member, black, finely fissile shales yield abundant *Chuaria circularis* on both the eastern and western slopes of Nankoweap Butte. The former is thought to be Walcott's type locality.

The Walcott Member forms the topmost subdivision of the Kwagunt Formation and is much more diverse in character than those below it. It is 838 feet (255 m) thick and forms the upper part of Nankoweap Butte (Fig. 6). At the base is a remarkable flaky dolomite bed some eight feet (2.5 m) thick (Fig. 7). Throughout its outcrop, this bed consists of randomly oriented silicified flakes set in a fine-grained dolomite matrix. The flakes appear to be disrupted algal-laminate, and occasional masses with a domed

Figure 6. Nankoweap Butte, showing the top of the Awatubi Member, the Walcott Member with its ledges of pisolitic chert and flaky dolomite, capped by an outlier of Sixtymile Formation. Seen looking northeast towards the lower part of Nankoweap Canyon. Photograph by John Shelton

Figure 7. The "Flaky dolomite" bed in the Walcott Member, Nankoweap Butte

shape occur. Scattered layers of dolomitized oolite also are present. Possible causes for this disrupted bed could be earthquake shock or storms followed by mass flows.

Above the flaky dolomite are black shales, with *Chuaria* and several silicified, cherty pisolite beds. The lowest is about two feet (60 cm) thick, whereas the higher ones are only about six inches (15 cm) in depth. These show pisoliths completely replaced by chert and with a matrix of iron-rich carbonate. The pisoliths' outer surfaces contain a mat of algal filaments of at least two types, as well as spheroidal bodies (Schopf and others 1973). Occasional lenses of nonsilicified oolite are entirely dolomitized. Towards the top of the member are two thick, dolomitized, oolitic limestones with occasional algal laminae.

SIXTYMILE FORMATION

In complete contrast to the underlying Chuar Group, the Sixtymile Formation is composed largely of breccias and coarse sandstones and has only subordinate siltstones and occasional mudstones. Slump rolls are common at some levels.

Approximately 200 feet (60 m) thick, the Sixtymile Formation caps Nankoweap Butte, where it has been misidentified in the past as an outlier of Cambrian Tapeats Sandstone (Van Gundy 1951; Maxson 1967). However, magnificent exposures of this unusual formation are present in the little-visited Awatubi and Sixtymile canyons. The formation has taken its name from the latter canyon, and its northwestern cliffs are well worthy of detailed study (Fig. 8).

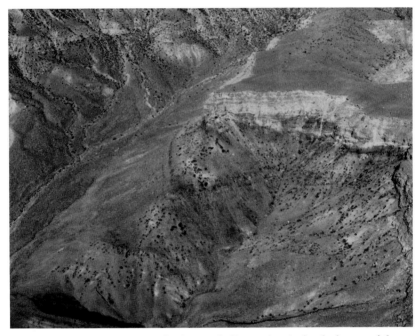

Figure 8. Sixtymile Canyon: the Sixtymile Formation is the dark part of the cliffs on the right.

Elston (1979) divided the Sixtymile Formation into three unnamed members. He also provided a comprehensive and detailed description of the rock types, together with an analysis of the events resulting in this unusual formation. The lower member is a landslide deposit that slid into the eastern side of the contemporarily growing Chuar syncline. Here, it cut into a series of thin-bedded sandstones resting disconformably on Chuar shales. Parts of the member are coarse breccias full of many different types of clasts—including chert and dolomite derived from the underlying Chuar Group (Fig. 9). Among these are large derived blocks, one being a dolomitic limestone block 26 x 130 feet (8 x 40 m) that matches the dolomite horizon in the Walcott Member of the Kwagunt—some 230 feet (70 m) stratigraphically lower.

Figure 9. Breccia in the Sixtymile Formation, Nankoweap Butte

The lower member is up to 90 feet (27 m) thick and has slid off the uplifting anticlinal block to the east. The middle member is a laminated, fine-grained quartzitic sandstone 80 feet (25 m) thick, with common chert lenses. Slump folds are oriented into the axial part of the Chuar syncline. The upper member is comprised of both coarse sandstone and conglomerates. Preserved only in the core of the syncline, the lithofacies change from mainly conglomeratic in the east to arenaceous in the west.

It is the clasts in the breccia and conglomerate units that reveal the unusual history of the Sixtymile Formation. They are derived from the underlying Chuar Group sediments in such a way that they demonstrate, first, that the Chuar syncline was developing during sedimentation and, second, that the Butte fault was active at the time and that the block to the east of it was being progressively uplifted. Elston (1979) has argued that the total relative uplift was of the order of five miles (3 kilometers). Elston and McKee (1982) called this phase of folding and block-faulting the Grand Canyon disturbance. They dated it at about 823 million years ago on the basis that K-Ar "clocks" were reset by alteration of the Cardenas Lava at this time. Larson and others (1986) have suggested that the diagenesis of the lavas was later—occurring about 725 million years ago.

The Sixtymile Formation is covered unconformably by the Cambrian Tapeats Sandstone. Renewed movement on the Butte fault in post-Paleozoic times has reversed the sense of displacement, canceling out the upthrow of the Chuar Group and, in fact, throwing down the Paleozoic column by more than half a mile (1 kilometer) to the east.

Correlation of the Nankoweap and Chuar

Younger Precambrian rocks crop out widely in central and southern Arizona where the nonmetamorphic rocks are the Apache Group. Shride (1967) suggested a correlation of the Unkar Group with the Apache Group, and Wilson (1962) put forward a correlation of the Troy Quartzite of the Apache Group with the Chuar Group of the Grand Canyon. However, the dominantly arenaceous character of the former contrasts markedly with the argillaceous Chuar though they may have been deposited in separate basins. Otherwise, it seems that there is no equivalent to either the Nankoweap Formation or the Chuar Group in Arizona, and we must look further afield for strata of the same age.

In the Death Valley region of California, a thick Infra-Cambrian sequence lies beneath the equivalent of the Tapeats Sandstone and rests on the Pahrump Group. This latter group may be equivalent to the Unkar. Thus, the Chuar could be equivalent to some part of the Infra-Cambrian sequence since both are dominantly argillaceous and contain stromatolitic limestones, but no direct comparison of units has yet been made. Also, toward the top, the Infra-Cambrian contains a tillite, which is unknown in the Grand Canyon sequence, making correlation less certain.

Further north, on the border of Utah and Colorado in the Uinta Mountains, the Red Pine Shale of the Uinta Mountain Group has yielded microfossils almost identical with the Chuar. Several geologists have proposed a correlation between these two groups (Hofmann 1977; Vidal and Ford 1985).

Still further north in the Cordillera, Beltian strata are possibly coeval with the Apache and Unkar groups. The overlying Windermere Group appears to have been deposited during the period from 850 million years ago to the start of the Cambrian (Harrison and Peterman 1971). The presence of *Chuaria* in the Hector Formation of the Windermere Group provides support for a correlation with the Chuar (Ford and Breed 1973a and b). The Little Dal Group of the Mackenzie Mountains in northwest Canada, which has yielded similar, if fewer, microplankton fossils, appears to be of roughly the same age (Hofmann and Aitken 1979).

Although the ages of these formations cannot be ascertained by radiometric dating, the microfossil evidence provided by acritarch assemblages strongly suggests intercontinental correlation with the Visingsö Group of Sweden and, thus, with the latest Riphean of Russia. Radiometric age determinations in Sweden and Svalbard indicate the age of these beds to be around 800-700 million years (Vidal and Ford 1985).

Support for the micropaleontological correlation has come from paleomagnetic studies of the Cardenas Lava and Unkar Group (Elson and Scott 1973; Elston and Grommé 1974) and of the Chuar Group (Elston 1986). The results indicate pole positions in what now is the central Pacific Ocean, showing broad agreement with determinations on other North American rocks of this age.

As noted earlier, the age of the Grand Canyon disturbance has been dated at about 823 $\pm$ 26 million years, and even if revised downward to 725 million years, it correlates reasonably well with the structural disturbance that terminated the Little Dal Group in northwestern Canada at 770 million years. Similar pole positions have been obtained from associated sills (see discussion in Elston and McKee 1982, p. 695).

Taking correlation by long distance lithological comparison, micro-paleontological assemblages, paleomagnetism, and very limited age dating, it seems that the Nankoweap Formation and Chuar Group were deposited within the period 1000-700 million years ago.

Environment of Deposition

Ripple-marked and mudcracked surfaces present throughout the Chuar Group indicate shallow-water sedimentation with intermittent desiccation. Limestones with stromatolites suggest subtidal to intertidal conditions. Pisolite horizons near the top of the Chuar again suggest subtidal shallow waters. The presence of megaplankton (*Chuaria*) and abundant microplankton, stromatolites, and filamentous algae are indications of at least partial marine influences.

Sandstones in the Carbon Butte Member occurred during a relatively short-lived invasion of the shallow-water basin by fluvial currents, resulting in coarse, clastic sedimentation close to the shore. The sediments of the Chuar Group suggest that in general they were deposited in a quiet, nonturbulent embayment on the marine platform bordering a continental mass. The outcrops are too limited, however, to indicate the direction of the coastline around this embayment. An alternate explanation by Reynolds and Elston (1986) proposed coastal swamp conditions for the Carbon Butte Member. A drastic change in environment took place in the Sixtymile Formation, with folding and block-faulting occurring at the same time as sedimentation—yielding fanglomerates, breccias, and sandstone sheets, probably under largely terrestrial conditions.

Paleontology of the Chuar Group

During the past century, a number of fossils or possible fossils have been described from the Precambrian rocks of the Grand Canyon. In recent years, most of these either have been reas-signed to biological groups other than those in which they first were placed, or their biologic affinities have been questioned. With the increasing interest in the early stages of life on our planet, several reviews of Precambrian life have been published, and these have presented conflicting interpretations of some of the fossils from the Grand Canyon. A review is presented here of the state of the knowledge concerning fossils from the Chuar Group.

Chuaria circularis

This small, disc-like, carbonaceous fossil first was found by Walcott during his pioneer work on the Grand Canyon Series (Precambrian) in 1882-1883. It was formally named in 1899 and described as a primitive brachiopod allied with *Orbiculoidea* or *Discina*. As such, it was the first Precambrian brachiopod to be described, and it is somewhat surprising that it was largely overlooked by subsequent writers. The exceptions include Wenz (1938), who assigned *Chuaria* to the gastropods without giving very clear reason for so doing; Häntzschel (1962), who regarded *Chuaria* as inorganic; and both Glaessner (1966, 1984) and Cloud (1968), who tentatively regarded it as algal in nature, without going into details. A full description was given by Ford and Breed (1972b and 1973b).

Others, including Vidal and Ford (1985), re-examined the problem and showed that this smooth, organic-walled, spheroidal microfossil was best classified with the Acritarcha (algal cysts of unknown affinity). It has been found at many horizons in the Chuar Group, where it is associated with assem-blages of other late Riphean to early Vendian acritarchs.

The size range of *Chuaria circularis* revealed by palynological techniques was 70-712 μm and, taken together with the megascopic specimens, this gives a total range of 70 μm to about 5 mm. The large forms are most common at what is presumed to be Walcott's type locality, high in the Awatubi Member about 30 feet (9 m) below a prominent ledge of pisolitic chert on the eastern flank of Nankoweap Butte.

The large forms of *C. circularis* are crushed, carbonized, spheroidal objects commonly 2-3 mm in diameter, lying either alone or in small clusters (they are never seen to lie on top of another), indicating their globular shape at the time of deposition (Fig. 10). Specimens extracted from the shale with acids are hollow with a narrow marginal thickening. The central area is wrinkled like a crushed prune. No apertures have been seen and no ornamentation recognized. It clearly is not a brachiopod or gastropod. Other tentative suggestions included the possibilities that *Chuaria* might be a trilobite egg or a non-calcareous foraminiferan; however, no evidence to support such suggestions can be found. The size range down to 70 μm, together with the composition and features noted above, make it clear that *Chuaria* is an acritarch.

Figure 10. *Chuaria circularis* - a mega-acritarch from the Walcott Member. Each carbonized disc is about 2 mm in diameter.

Microfossils

Samples of the darker shale beds throughout the Chuar Group have yielded organic detritus, and most horizons have yielded acritarchs. These are spheromorphic, cystlike objects presumed to be the resting stage of some form of algae. Among the genera recovered are *Stictosphaeridium, Trachysphaeridium, Leiosphaeridia, Kildinosphaera, Cymatiosphaeroides, Tasmanites,* and small *Chuaria.* A new form, *Vandalosphaeridium walcotti,* also was discovered.

Assemblages with either megascopic *Chuaria* or the above microplankton have been recovered from many late Precambrian strata around the world, and they can be used for stratigraphic correlation. On these micropaleontological grounds, the Chuar Group can be correlated with sequences in Utah, northwest Canada, Greenland, Scandinavia, Svalbard, Russia, Iran, India, China, and Australia. In both Scandinavia and Russia, the sequences are earlier than the Vendian glacial deposits and are placed in the latest Riphean and earliest Vendian divisions of the Proterozoic, i.e., around 800-700 million years ago. This is slightly younger than the date of c. 823 million years deduced for the Grand Canyon disturbance by Elston and McKee (1982), but dating of both deposits and events around this period still has many uncertainties.

Vase-shaped microfossils

A shale horizon in the Walcott Member, about three feet (one m) below the prominent pisolitic chert, has yielded an assemblage of vase-shaped, organic-walled microfossils (from 32-170 μm in length) that Bloeser (1985) has named *Melanocyrillium.* Up to 10,000 specimens per cubic centimeter of shale have been found. The teardrop-shaped, vaselike, noncolonial microfossils have an ornamented aperture at the narrow end, and three species have been erected on the basis of ornamental detail. They are of uncertain biological affinity and are not chitinozoa, as first thought (Bloeser and others 1977). Comparable forms have been found in rocks of this age in Sweden, Greenland, and Brazil.

An assemblage of well-preserved filamentous and spheroidal plant microfossils have been found in a cherty pisolite bed of the Walcott Member on the slopes of the Nankoweap Butte in eastern Grand Canyon (Schopf and others 1973). The spheroidal forms probably are related to coccoid blue-green algae; the filaments bear a striking resemblance to *Eomycetopsis,* a thallophyte from bedded cherts in the Bitter Springs Formation of central Australia. Preliminary studies have revealed similar filamentous forms at other horizons in the Chuar Group.

Stromatolites

Stromatolites are common in the Precambrian of the Grand Canyon, where they first were noted as "concretionary lime-stones" by Walcott (1894). They were given the now obsolete name *Cryptozoon occidentale* by Dawson (1897), though their nature was not then understood. Ford and Breed (1969, 1972a, 1973a) have recorded three conspicuous stromatolite horizons in the Chuar Group, as well as a number of less important occurrences. They generally have shapes such as gentle undulations or low domes with confluent laminae and only occasionally show the more distinctive columnar forms that Russian workers regard as diagnostic fossils. Forms regarded as *Inzeria, Baicalia,* aff. *B. rara,* and *Boxonia* have been recognized by comparing the external form with those illustrated by Cloud and Semikhatov (1969), though diagenetic alteration has obscured internal detail and made more specific identification impossible. Recognition of these genera allows correlation of the Chuar Group with the Russian Upper Riphean, i.e., late Precambrian.

The three beds of stromatolites in the Chuar Group are well exposed. The highest forms a bed of stromatolite "reefs", or bioherms, cropping out around the foot of Nankoweap Butte in Nankoweap and Kwagunt canyons (Fig. 11) and just above the prominent sandstone bench on Carbon Butte. The middle horizon is a single bed of limestone within the upper part of the Carbon Canyon Member—outcropping low on the flanks of Carbon Butte and also about one hundred feet above the creek in Nankoweap Canyon at the foot of Nankoweap Butte. The lowest horizon forms a prominent ledge and waterfall about a mile up Chuar Canyon from its mouth. It also is seen as a prominent ledge in the middle reaches of Basalt Canyon.

The paleoenvironment of stromatolites in the Chuar Group, judged by the associated sediments, is one of quiet, shallow waters. Ripple marks and mudcracks are common, suggesting intermittent desiccation. Thin layers of flake-breccia indicate turbulence of brief duration, but no evidence of an intertidal environment close to a shoreline has been demonstrated. Horodyski (1986) has proposed a possible hypersaline environment.

Figure 11. Stromatolite bioherm from the Awatubi Member, Kwagunt Canyon

SUMMARY

The late Precambrian paleontological record from the Chuar Group consists of a variety of microfossils from filamentous algal sheaths, coccoid algae, acritarchs (including the megascopic *Chuaria*), vase-shaped microfossils of unknown affinity, and stromatolites. The lobate markings from the Nankoweap Formation can be regarded at best as a dubio-fossil and may not be of organic origin at all.

CHAPTER 5

GEOLOGIC STRUCTURE OF THE GRAND CANYON SUPERGROUP

James W. Sears

INTRODUCTION

The Grand Canyon Supergroup is a thick sequence of tilted Precambrian strata sandwiched between overlying, nearly horizontal Paleozoic rocks and underlying crystalline rocks of the Granite Gorge. John Wesley Powell first observed these rocks in 1869 on his historic journey down the Colorado River, while he and his crew were subsisting on rations of spoiled bacon, resifted flour, dried apples, and coffee (Powell 1895).

Powell recognized that these tilted rocks record an important chapter in the tectonic history of the region. He concluded that the beds were folded, uplifted, and deeply eroded before the overlying Paleozoic rocks were "spread over their upturned edges" (Powell 1875, p. 213). Charles Walcott outlined the geologic structure and stratigraphy of the sequence in a series of reports (Walcott 1883, 1890, 1894). He showed that it is about 12,000 feet (3660 m) thick by measuring stratigraphic sections and correlating units from canyon to canyon. Walcott recognized that the Grand Canyon Supergroup is Precambrian in age and suggested a correlation with the Belt strata of Glacier National Park in Montana. Radiometric dating has since shown that the Grand

Canyon Supergroup accumulated during Middle and Late Proterozoic time upon the deeply eroded Lower Proterozoic Vishnu crystalline complex. It was folded, faulted, and beveled by erosion before deposition of the Middle Cambrian Tapeats Sandstone.

Geologists now divide the Grand Canyon Supergroup, from the base to the top, into the Unkar Group, Nankoweap Formation, and Chuar Group. The groups are subdivided further into formations and, in some cases, members. Chapters three and four detail the stratigraphy and depositional history of the Grand Canyon Supergroup. This chapter summarizes its geologic structure and tectonic history.

Figure 1 identifies Precambrian rock exposures in the eastern Grand Canyon. The exposures are limited to the area below the

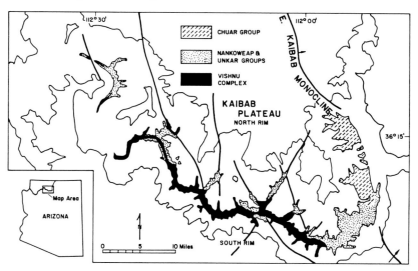

Figure 1. Precambrian rocks in eastern Grand Canyon. Major structures with Precambrian ancestry also are shown.

Kaibab Plateau, Powell's "Kaibab division" of the Grand Canyon, where the canyon is deepest and the Paleozoic section rises to its greatest elevation. Precambrian rocks emerge from beneath the Paleozoic cover in deep canyons along the East Kaibab monocline, the great, steplike fold on the east edge of the Kaibab Plateau, where Paleozoic rocks rise almost 3000 feet (900 m) from east to west. Exposures of Precambrian rocks end fifty miles (80 km) to the west, where regional tilt carries the Paleozoic rocks below the level of the Colorado River. Farther west, Precambrian metamorphic rocks emerge again in the lower Granite Gorge, but the Grand Canyon Supergroup is absent.

The Grand Canyon Supergroup occurs in isolated, wedge-shaped remnants as shown in the cross section of Figure 2. The underlying Vishnu metamorphic rocks rise to the southwest to intersect the base of the Paleozoic section at the left edge of the section. The unconformity at the base of the Paleozoic cuts out the Precambrian section from northeast to southwest. The Grand Canyon Supergroup, therefore, is found only in fragments where down-dropped geologic structures protected it from pre-Paleozoic erosion. We can piece these fragments together to decipher the late Precambrian tectonic history of the Grand Canyon region.

Figure 3 provides a hypothetical view of the geology of the Grand Canyon region as it might have looked at the close of Precambrian time—before Paleozoic sediments buried the region. This figure covers the same area as Figure 1 and was constructed by matching Precambrian structural trends from one exposure to the next. The Precambrian rocks occupy several great fault blocks, each tilted down to the northeast. The Vishnu Metamorphic Complex emerges on the southwestern edges of the blocks, and Unkar, Nankoweap, and Chuar beds appear successively toward the northeast. The youngest part of the Grand Canyon Supergroup forms a synclinal downwarp against the Butte fault, where it accumulated during a period of faulting (Elston 1979).

Three main types of structures disrupted the Grand Canyon Supergroup during late Precambrian time. These occurred as successive episodes of igneous intrusion, crustal contraction, and crustal extension affected the Grand Canyon region.

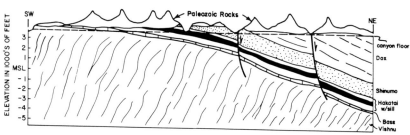

Figure 2. Cross section along the course of Bright Angel Creek, illustrating the wedgelike character of remnants of the Grand Canyon Supergroup. Great thicknesses of the tilted strata are exposed between the canyon floor, shown by the dashed line, and the base of the Paleozoic section.

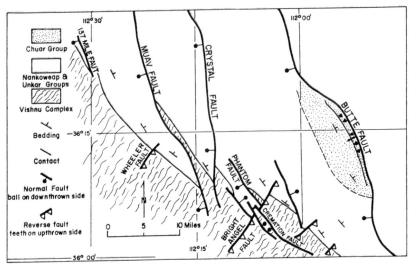

Figure 3. Schematic map of the Precambrian structure as it might have looked before burial by Paleozoic sediments. Triangles near Butte fault represent deposits shed from fault scarp as the blocks shifted. This figure shows the same area as Figure 1.

STRUCTURES RELATED TO IGNEOUS INTRUSION

The Unkar Group contains abundant diabase—a dark, intrusive igneous rock much like basalt in composition. The diabase magma spread out along the beds to form thick sills. In some places, it followed fractures across beds to form dikes. The sills are most common in the lower two units of the Unkar Group: the Bass Formation and Hakatai Shale. Diabase is most abundant in the western exposures, where sills are nearly 1000 feet (300 m) thick. In Galloway Canyon, two thick sills inflate the lower Unkar Group. In eastern exposures, however, sills are thin or absent. None exist in the Unkar beds near Phantom Ranch. Near Hance Rapids, a single 80-foot (24-m)-thick sill lies within the Bass Formation.

When the hot magma invaded the cooler sedimentary rocks, it produced zones of contact metamorphism—baking the ordinarily soft, red Hakati shales into hard, bluish hornfels and generating asbestos seams in the Bass dolomites.

The invasion of the diabase caused structural adjustments within the Unkar strata (Fig. 4). At some localities (Fig. 4a), sills end abruptly against steep contacts. The strata above the sills moved upwards relative to neighboring beds, forming faults in the overlying rocks. Without exposures of the underlying sills, such faults could be mistaken for tectonic features. In some cases, as Figure 4b shows, sills invading favorable bedding planes encountered older faults that prevented further lateral spreading. The faults moved again as the sills inflated. The fault shown in Figure 4b illustrates opposite senses of displacement above and below the sill; the rock on the right originally was downthrown. Figure 4c is an example of an expansion fault that served as a conduit for magma rising through the section. In Figure 4d, a sill steps from one bedding plane to another. A fault crosses the step, but beds above and below it are unbroken.

The diabase sills and dikes were the "plumbing system" for the Cardenas Lava, which erupted into the Unkar basin about the same time the diabase was crystallizing underground. McKee and Noble (1974) dated the diabase at 1150 million years. Radiometric dates reported by Ford, Breed, and Mitchell in 1972 and

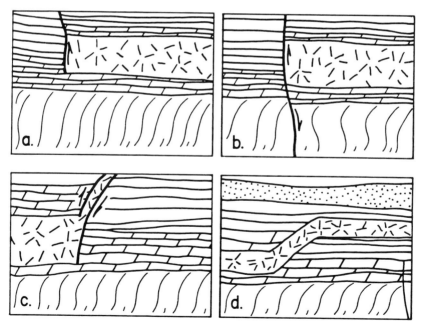

Figure 4. Types of structures formed by igneous intrusion

paleomagnetic pole positions of the Cardenas flows documented by Elston and Scott (1976) show that the lavas are about 1100 to 1200 million years old.

CONTRACTION FAULTS

Contraction faults horizontally shorten (and vertically thicken) a section of rocks. A number of steeply dipping contraction faults of Precambrian age cut the lower parts of the Grand Canyon Supergroup and the underlying Vishnu complex. They shifted the southeastern blocks toward the northwest—approximately straight up the fault planes. The faults have associated gouge and breccia because they formed at shallow crustal levels. They die out into steplike, monoclinal folds upward in the Unkar strata, where slip was taken up along bedding planes.

Figure 5 shows the Bright Angel fault and monocline. The monocline is very tight in the lower beds, which are vertical and even overturned adjacent to the fault, but it opens steadily upwards in the Unkar section. In the Dox Formation, it is a broad, subtle deflection. The contraction faults parallel the metamorphic grain of the Vishnu crystalline rocks. It seems apparent that the faults follow the older weaknesses of the earth's crust.

The Bright Angel fault was active during deposition of the Shinumo Quartzite, which thickens by about 200 feet (60 m) from southeast to northwest across the structure. When diabase was injected later into the Unkar Group, it encountered a barrier along the Bright Angel structure. It then formed an expansion fault and crossed to a different level on the southeastern side of the structure. The Bright Angel fault moved again after the diabase crystallized. That later movement folded, fractured, and sheared the diabase. Throughout this part of its history (Precambrian time), rocks southeast of the fault rose about 800 feet (240 m). The Bright Angel fault moved again after deposition of the Paleozoic section. This time, however, the southeastern side moved down (Maxson 1968).

Northwest-trending extension faults offset the Bright Angel fault later in Precambrian time and tilted the entire system toward the northeast in a great fault block. This tilt causes deep levels of the structure to crop out in the southwest. Although Precambrian rocks make up only hundreds of feet of the lower walls of Bright

GRAND CANYON GEOLOGY

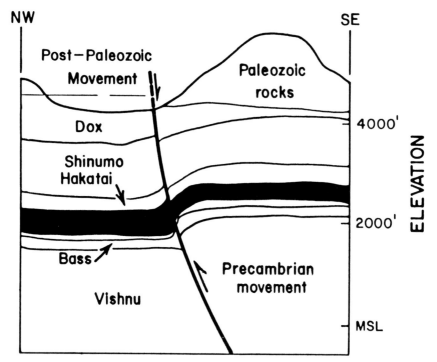

NW   SE

Post–Paleozoic Movement

Paleozoic rocks

Dox

Shinumo
Hakatai

Bass

Vishnu

Precambrian movement

4000'

2000'

MSL

ELEVATION

Figure 5. Cross section of Bright Angel structure, northeastern Bright Angel Creek. Because the rocks are tilted into the plane of the paper, the deeper units emerge successively to the southwest. Post-Paleozoic movement was opposite in sense to Precambrian.

Angel Canyon, we can trace the Bright Angel structure through thousands of feet in the Vishnu complex and the Unkar Group along the course of the canyon.

Other Precambrian reverse faults and monoclines are known from Bass, Vishnu, and Red canyons. These structures are known only in the Unkar Group. Outcrops of the Nankoweap and Chuar beds exist only in the eastern Grand Canyon, away from the known contraction faults. For this reason, we do not know if those units also were deformed.

The crustal compression may have been an isolated and relatively minor disturbance in western North America. On the other hand, it may have been associated with Middle Proterozoic tectonic activity reported in other widely separated parts of the western continent as far north as Canada's Yukon Territory (Douglas and others 1970).

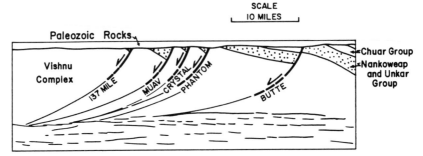

Figure 6. Schematic cross section showing major extensional faults of the Grand Canyon Supergroup. Rotation of the beds suggests the faults flatten with depth.

EXTENSION FAULTS

Extension, or "normal," faults form during horizontal stretching of the earth's crust. Five major Precambrian extension faults form an array about thirty miles (50 km) wide in the depths of the Grand Canyon. Their combined vertical displacements of about 20,000 feet (6000 m) make them much larger than the compression faults. They trend generally northwesterly along curving traces. Figure 6 shows that the major faults dip to the southwest. The blocks slipped down the fault planes toward the southwest to form large, asymmetric troughs, or half-grabens. The blocks rotated as they slid down. The beds on the downthrown sides typically tilt toward the faults, and in many cases, the bedding is steepest near the fault planes. The rotation implies that the faults, which are steep where exposed, curve and flatten downward—a pattern documented in other areas of crustal extension.

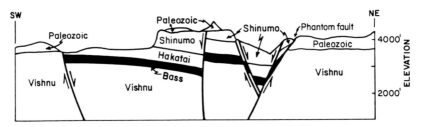

Figure 7. Cross section showing details of geologic structure near the Phantom fault

GRAND CANYON GEOLOGY

Smaller structures formed near the faults as the blocks shifted. Figure 7 shows small blocks and slivers that moved up and down near the Phantom fault. The Butte fault dragged the Chuar beds into a major syncline.

The extension structures of the Grand Canyon Supergroup formed during deposition of the uppermost units of the Chuar Group. These units occupy small areas near the Butte fault and contain slump structures that formed as material was eroded off the rising block and dumped onto the adjacent subsiding block (Elston 1979). Walcott (1899) and Schopf and others (1973) discovered primitive fossils in the upper Chuar Group, which suggests it is 800 ± 200 million years old, a Late Proterozoic age. Elston (1979) suggested that the faulting coincided with an event that reset the radiometric clocks of the Cardenas Lava about 845 to 810 million years ago.

After faulting and tilting, erosion reduced the Grand Canyon Supergroup to a nearly smooth plain. Ridges of Shinumo Quartzite rose up to 600 feet (180 m) above the generally level plain along the trends of the tilted blocks. The sea advanced over this plain from west to east and drowned the ridges in sands and muds by Middle Cambrian time, about 550 million years ago.

What might have formed the normal fault system? Its structural style is much like that of modern areas experiencing crustal extension. In the Great Basin of Utah and Nevada, tilted half-graben blocks form individual mountain ranges and neighboring basins. That region apparently has doubled its width since extension faulting and crustal thinning began.

The Atlantic coastal plain contains structures very similar to those of the Grand Canyon Supergroup. There, large half-grabens collapsed in early Mesozoic time when the present Atlantic Ocean basin began to rift open. Tilted fault blocks wore down to a flat surface, and lower Mesozoic beds were preserved in wedges faulted against the older, crystalline complex of the Appalachian Piedmont. The sea advanced from the new Atlantic Ocean basin and drowned the remnants of the lower Mesozoic beds in sand and mud.

The tectonic history of the Atlantic coastal plain follows the general pattern for rifted continental margins. The earth's crust heated and bulged up—and then began to slide laterally off the bulge. This formed extension features, such as the tilted half-grabens, and was accompanied by deep erosion because the crust was standing high. Continued extension eventually opened a new ocean basin. As the newly rifted continental margin slowly cooled, it contracted and sank below sea level.

A similar model can explain the extension structures of the Grand Canyon Supergroup. Evidence from many localities in western North America shows that an ancient form of the Pacific Ocean opened during Late Proterozoic time as the continent rifted from California to the Yukon. A large continental fragment drifted away as the ocean basin widened. This fragment may be part of Asia today (Sears and Price 1978). The rifting formed extension structures that later were drowned as the sea advanced over the new continental shelf from the new ocean basin. Elston (1979) correlated the extension faulting of the Grand Canyon Supergroup with rifting elsewhere in western North America.

In Figure 8a, the crust is greatly uplifted, extended, and deeply eroded near the site of the continental rift. The Grand Canyon Supergroup drops into half-grabens, where it is preserved. Farther west, more than 12,000 feet (3660 m) of material is removed by erosion. In Figure 8b, the rift cools, contracts, and sinks, and the Cambrian marine deposits gradually transgress across the shelf, partly in response to rising worldwide sea levels, to reach the Grand Canyon region by Middle Cambrian time. Therefore, west of Fishtail Rapids, on the site of the old uplift, the Cambrian beds rest directly on the older crystalline rocks.

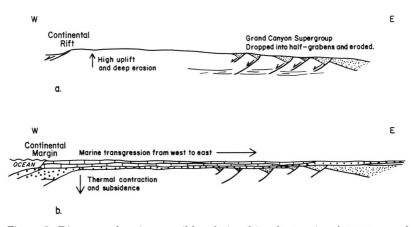

Figure 8. Diagrams showing possible relationship of extensional structures of the Grand Canyon Supergroup and continental rifting

Northern Arizona was part of the continental shelf during Paleozoic time. The rifted continental margin was farther west, in Nevada and California. That margin was destroyed during the growth of the western mountains in Mesozoic and Cenozoic time.

REACTIVATION OF PRECAMBRIAN STRUCTURES

After Paleozoic and Mesozoic strata buried the Grand Canyon region, many of the Precambrian structures shifted again and generated faults and folds in the younger beds. Nearly all of the major structures of the Paleozoic rocks that cross the Grand Canyon have a Precambrian ancestry. During the post-Paleozoic

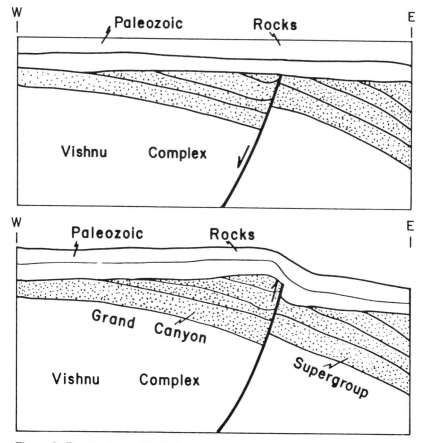

Figure 9. Reactivation of Butte fault after deposition of Paleozoic section over truncated Precambrian rocks

period, most of the fault blocks shifted in the *opposite* direction from their Precambrian movements. Walcott observed this pattern along the Butte fault in 1890. Figure 9 sketches the history of the Butte fault. The Grand Canyon Supergroup first dropped about 5000 feet (1500 m) on its west side, as shown in the upper diagram. Paleozoic rocks buried the truncated beds and Butte fault. The western block later rose 2700 feet (800 m), forming the East Kaibab monocline shown in the lower diagram. This indicates that Butte fault was first a normal fault and then a reverse fault. Net displacement for the Precambrian beds is about 2300 feet (700 m), with the west side down.

In many areas of the continent, fundamental Precambrian structures shifted many times over long geologic periods. Less energy is required to reactivate these flaws in the earth's crust if they have favorable orientations than is needed to create entirely new faults. In the Grand Canyon region, the Precambrian deformations produced sets of criss-crossing faults. This means that whatever later forces should be applied, some older faults are suitably oriented to allow the crust to adjust. The ancient crust could be thought of as a mosaic of blocks (bounded by faults) that shift to accommodate changing stress patterns.

The post-Paleozoic fault movements produced zones of crushed and broken rock. These zones strongly influenced the erosion of side canyons in the Grand Canyon—helping to form the vistas we enjoy today.

SUMMARY

The Grand Canyon Supergroup experienced three distinct phases of Precambian deformation, but its dominant geologic structure reflects Late Proterozoic crustal stretching—perhaps related to formation of the ancestral Pacific Ocean.

CHAPTER 6

TONTO GROUP

Larry T. Middleton and David K. Elliott

INTRODUCTION

The Cambrian of the Grand Canyon is, without question, one of the classic sequences of sedimentary rocks exposed in North America. These strata crop out along a prominent, essentially horizontal surface known as the Tonto Platform in the central part of the canyon and near the banks of the Colorado River in western areas of the canyon. The surface of the Tonto Platform roughly coincides with the top of the oldest Cambrian formation, the Tapeats Sandstone. Above the Tapeats, a series of small cliffs are separated by thicker intervals of slopes composed of finer-grained deposits of the Bright Angel Shale. These, in turn, are overlain by cliffs of resistant carbonate of the Muav Limestone, the youngest Cambrian formation of the Tonto Group.

This three-part system of sandstone, mudstone, and limestone is well known to most geologists and to a great number of Grand Canyon hikers and tourists. Despite this, research into the origin of these rocks has not kept pace with the developments in the last twenty years concerning the dynamics of nearshore and shelf depositional systems. The classic work of McKee and Resser (1945) endures as the most comprehensive study of the Cambrian system in the Grand Canyon.

To date, only a few studies have attempted to document carefully the lateral and vertical facies associations recorded in these strata (McKee and Resser 1945; Wanless 1973; Hereford 1977; Martin 1985; and Middleton 1988). A major objective of this chapter is to present new data that will help us recreate the depositional systems that existed during the Cambrian in northern Arizona. Both sedimentologic and ichnologic data will be used to examine the depositional history of the Tonto Group.

Cambrian deposits in the Grand Canyon and throughout the Rocky Mountains long have been cited as representing a classic transgressive sequence of sandstone, mudstone, and limestone that accumulated on the slowly subsiding Cordilleran miogeocline and adjacent craton (McKee and Resser 1945; Lochman-Balk 1970, 1971; Stewart 1972; Stewart and Suczek 1977). During Early and Middle Cambrian time, a north-south trending strandline migrated progressively eastward across the craton. This shoreline was characterized by numerous embayments and offshore islands that affected sedimentation in nearshore areas. Shoreline migration for the most part was eastward, resulting in deposition of coarse clastics in shallow water areas to the east and finer clastics and carbonates in more offshore areas to the west.

Cambrian strata in Arizona and along the entire western margin of North America thicken to the west, presumably reflecting more subsidence in the miogeoclinal or offshore shelf areas. Continued subsidence and/or sea-level rise, interrupted by a number of sea-level retreats or regressions, resulted in the complete submergence of the western cratonic margin by the Late Cambrian.

REGIONAL STRATIGRAPHIC RELATIONSHIPS

The Tonto Group comprises three formations (Fig. 1) that are, in ascending order: Tapeats Sandstone, Bright Angel Shale, and Muav Limestone. The term "Tonto Group" was used first by G.K. Gilbert (1874) to describe this sandstone-shale-limestone sequence though he considered these rocks to be of Silurian age. Subsequent stratigraphic and paleontologic work by Walcott (1890) established a Cambrian age for the Tonto Group, and Noble introduced the now accepted formational names in 1914 during his mapping of the Shinumo quadrangle in the Grand Canyon.

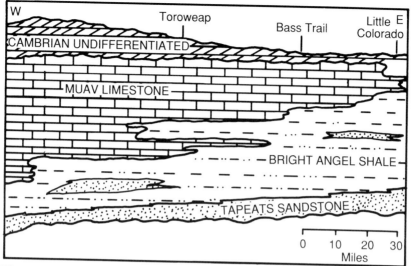

Figure 1. West to east cross section of Tonto Group in Grand Canyon illustrating stratigraphic relationships and eastward younging of the group (from McKee and Resser 1945)

Strata of the Tonto Group also crop out along the Grand Wash Cliffs in western Arizona and to the east in the Juniper Mountains and the Black Hills in west-central Arizona. In these areas, the Tapeats Sandstone is overlain disconformably by the Devonian Martin Formation or the Chino Valley Formation, the age of which is uncertain (Hereford 1975). Presumably, the Bright Angel Shale and the Muav Limestone were removed during extensive pre-Devonian erosion. In central Arizona, scattered outcrops of the Tapeats Sandstone occur along the East Verde River and in the Sierra Ancha Range north of Young, Arizona. Tonto Group equivalents in southeastern Arizona include the Bolsa Quartzite and part of the overlying Abrigo Formation (Hayes and Cone 1975; Middleton 1988).

Cambrian strata overlie a variety of Precambrian lithologies throughout the Grand Canyon. In the eastern part, the Tonto Group rests on tilted beds of the 1.4 - 1.1 billion-year-old Grand Canyon Supergroup; whereas, in western areas, the Tonto Group nonconformably overlies older Precambrian (circa 1.7-1.6 billion-year-old) rocks of the Vishnu Group and other metamorphic units. This major unconformity between Precambrian and Tonto Group rocks, which has been recognized for a long time, obviously represents a considerable period of time during which the

TONTO GROUP

region was subjected to episodes of mountain building and extensive erosion. Walcott (1910) applied the name "Lipalian interval" to the period of time represented by this unconformity. Although dating of the younger Precambrian strata establishes a minimum age, it is impossible to measure precisely the time from cessation of uplift to production of the nearly flat erosion surface onto which sediments of the Tapeats Sandstone were deposited. Clearly, we are dealing with a long period of time.

The surface upon which the Tonto Group accumulated was quite irregular. It was characterized by a rolling topography of resistant bedrock "hills" and lowlands. The Precambrian bedrock was weathered extensively in places and eroded during prolonged periods of subaerial exposure. Walcott (1881) and Noble (1914) were among the first to recognize that the Precambrian surface represented paleotopography and that sedimentation patterns were influenced by the relief and lithologies of these "hills." Other workers likewise have documented the influence of Precambrian topography on Cambrian sedimentation in other areas of the Rocky Mountains and in the midcontinent. There are numerous places in the canyon where the Tapeats Sandstone thins across or pinches out against these Precambrian highs. Where the Tapeats pinches out, the Bright Angel Shale overlies the Precambrian surface. The influence of these Precambrian highs will be discussed as the depositional environments of the Tonto Group are reconstructed.

A highly weathered horizon occurs on top of the Precambrian surface in several places in the canyon. The only effort to understand the genesis of this potentially very significant unit is that of Sharp (1940). His study suggested that extensive chemical weathering of Precambrian rocks occurred prior to deposition of Cambrian sediments. In places, this highly weathered surface, or regolith, is up to 50 feet (15.3 m) thick but generally is less than 10 feet (3.1 m) thick. Sharp speculated that where the Tapeats rests on unaltered basement, the regolith probably was removed by wave erosion associated with the initial Cambrian transgression. Sharp, McKee, and Resser and McKee have suggested in separate studies that the presence of such a thick, weathered horizon indicates that dominantly humid conditions existed during the early Paleozoic prior to deposition of the Tapeats Sandstone. Unfortunately, there have been no petrologic and geochemical studies that could substantiate this hypothesis. Considering that the time represented by the unconformity could have been several hundred million years, that the climate could have changed numerous times during this period, that this horizon was buried and exhumed numerous times prior to deposition of the Tapeats Sandstone, and that in the absence of terrestrial vegetation, weathering processes in soils would have been different (Basu 1981), a humid climate interpretation is quite tenuous. Obviously, considerable research needs to be done in this area.

STRATIGRAPHY OF THE TONTO GROUP

Tapeats Sandstone

The Tapeats Sandstone was named for exposures along Tapeats Creek in the western part of the Grand Canyon National Park. For the most part, the formation is a medium- to coarse-grained feldspar and quartz-rich sandstone with granule and pebble-size conglomerate present locally near the base. The percentage of feldspar is highest at the base and decreases upwards through the formation. The composition of the basal Tapeats reflects to varying degrees the mineralogy of the underlying Precambrian rocks. To date, however, there have been no petrologic studies of the Tapeats aimed at documenting the changes in mineralogy with respect to facies changes or evaluating the influence of basement lithology and paleotopography on the composition of the Tapeats.

The formation can be divided into two generalized packages. The majority of the Tapeats crops out as a cliff consisting of beds typically less than three feet (1 m) thick. Sedimentary structures include planar and trough cross stratification and crudely developed horizontal stratification. Both the scale of the bedding and the cross stratification decrease upwards. Overlying the main cliff is a thinner zone of interbedded fine- to medium-grained sandstone and mudstone. Stratification is of a smaller scale in these beds and is largely trough and ripple cross stratification and horizontal stratification.

The significance of the upper unit is that it marks a major facies transition into the upper Bright Angel Shale (Fig. 2.). An increase in fine-grained material indicates a reduction in the bedload to suspension load ratio. The concomitant changes in bedding thickness and scale of sedimentary structures are consistent with the

Figure 2. Thin bedded, horizontally and ripple laminated upper portion of the Tapeats Sandstone. Overlying slope marks gradational contact with the Bright Angel Shale.

above interpretation. The contact between the two formations, therefore, is arbitrary and probably should be placed at the top of the thickest sandstone bed within the transitional interval.

The Tapeats varies considerably in thickness throughout the Grand Canyon and also in areas to the south and west. A thickness of 393 feet (120 m) was reported by Noble (1922) along the Bass Trail. It is possible that this represents a maximum thickness in the canyon. Typically, the formation is between 100 to 325 feet (30 and 100 m) thick. The thickness of the Tapeats clearly is controlled by relief of the underlying Precambrian surface. As previously stated, there are areas where the Tapeats thins across and/or pinches out against these Precambrian highs.

Except for trace fossils, which in places are quite common, body fossils are rare and only occur within the transition zone (McKee and Resser 1945). However, these fossils establish a late Early Cambrian age for the upper parts of the Tapeats Sandstone in the Grand Wash Cliffs in the western part of the canyon and an early Middle Cambrian age for the formation in the eastern canyon. These ages are based on trilobite assemblages (*Olenellus-Antagmus*) in the overlying Bright Angel Shale. This diachroneity is a reflection of the west-to-east sense of strandline migration.

Bright Angel Shale

The Bright Angel Shale is perhaps the least studied formation in the Grand Canyon. It was named by Noble (1914) for exposures of slope-forming, interbedded, fine-grained sandstone, siltstone, and shale just above the Tonto Platform along Bright Angel Creek. Conglomerates and coarse-grained sandstones of the Bright Angel Shale contain quartz, minor feldspar and sedimentary rock fragments, and glauconite. The latter is responsible for imparting the green color to many of the siltstones and sandstones. A number of the sandstones and siltstones contain a high percentage of hematitic ooids and iron oxide cements. These appear reddish brown on outcrop. The dominant lithology is greenish shale composed largely of illitic clay with varying amount of chlorite and kaolinite. Inarticulate brachiopods, trilobites, and *Hyolithes* are locally abundant. Trace fossils are extremely abundant and varied.

McKee and Resser (1945) recognized one member in the Bright Angel Shale, which they termed the Flour Sack Member. This unit consists of shale, siltstone, and limestone and forms the uppermost part of the formation in the western canyon. Limestone decreases in abundance toward the east until the entire member is shale at its easternmost outcrop near Quartermaster Canyon. A number of rusty-brown dolomite tongues that occur in the Bright Angel represent carbonate extensions of the Muav Limestone (McKee and Resser 1945).

Sedimentary structures are numerous in the coarser-grained lithologies in the Bright Angel Shale throughout the Grand Canyon. These include horizontal laminations, small- to large-scale planar tabular and trough cross stratification, and wavy and lenticular bedding (Fig. 3). Locally, structureless and crudely stratified conglomeratic sandstones typically overlie a scoured surface. Martin (1985) documented a number of coarsening-upward and fining-upward sequences in the Bright Angel in the central canyon.

The Bright Angel is over 450 feet (137 m) thick in the western Grand Canyon, only 270 feet (82 m) at Toroweap in the central canyon, and 325 feet (99 m) along Bright Angel Creek (McKee and

Figure 3. Wavy bedded sandstone in the Bright Angel Shale along Pipe Creek. Internal structures suggest deposition by storm-enhanced currents.

Resser 1945). Thickness is quite variable because of complex intertonguing relationships with the Muav Limestone. The Bright Angel Shale thins toward the south and is only a few feet thick in the Juniper Mountains north of Prescott, Arizona. South and east of the Black Hills, the formation is absent—presumably the result of extensive erosion.

Like the rest of the Cambrian of the Grand Canyon, the Bright Angel crosses time lines, becoming younger toward the east. In the western part of Grand Canyon, the base of the formation lies below the *Olenellus-Antagmus* assemblage zone. It is, therefore, late Early Cambrian, whereas in the eastern part of the canyon, the lower third of the Bright Angel lies below the Middle Cambrian *Alokistocare-Glossopleura* assemblage zone.

Muav Limestone

The Muav Limestone is the youngest formation of the Tonto Group. It forms resistant cliffs above the Bright Angel Shale throughout the Grand Canyon. Noble (1914) named the formation for exposures in Muav Canyon. Contact with the Bright Angel Shale is gradational and characterized by complex intertonguing of the two formations. McKee and Resser (1945) defined seven members within the Muav (Fig. 1). The members are differentiated on the basis of key marker horizons defined by fauna, intraformational conglomerate, or persistent beds of shale and/or thin-bedded limestone. The upper three members can be correlated throughout the entire canyon, whereas the lower four members are confined to areas west of Fossil Rapids.

The Muav consists of thin- to thick-bedded, commonly mottled, dolomitic, and calcareous mudstone and packstone, as well as beds of intraformational and flat-pebble conglomerate. Thin beds of micaceous shale and siltstone, minor amounts of fine-grained sandstone, and silty limestone occur at numerous horizons in the Muav, where they form small recesses and/or benches in the cliff-forming carbonate. The amount of siliciclastics increases toward the east, concomitant with a decrease in carbonates. Bedding thickness, in general, increases toward the west.

Most of the Muav comprises beds of structureless or horizontally laminated carbonate. Small-scale (less than 5 cm thick) trough, planar tabular, and low-angle cross stratification occurs at many localities. Fenestral fabrics and desiccation cracks also are reported (Wanless 1975). Trace fossils, though not as abundant as in the Bright Angel Shale, are present throughout the canyon.

As a result of the intertonguing relationships between the Muav Limestone and Bright Angel Shale, thickness trends within the Muav are somewhat variable. The unit as a whole thickens towards the west. McKee and Resser reported that the Muav is 827 feet (252 m) thick in the Grand Wash Cliffs near Lake Mead, 439 feet (134 m) thick at Toroweap in the central canyon, and only 136 feet (42 m) thick at the confluence of the Little Colorado and Colorado rivers at the eastern end of the canyon.

In the western part of the canyon, the Muav lies above the *Alokistocare-Glossopleura* assemblage zone and is Middle Cambrian in age. In eastern Grand Canyon, the upper part of the Muav contains the *Bathyuriscus-Elrathina* zone and is late Middle Cambrian. The decrease in age of the Muav towards the east parallels the age trends of the Tapeats Sandstone and the Bright Angel Shale and reflects the west- to-east nature of the Cambrian transgression.

TONTO GROUP

Cambrian Undifferentiated

In the western part of Grand Canyon, a thick (up to 426 feet [130 m]) sequence of dolostone overlies the Muav Limestone. McKee and Resser referred to this unit as the undifferentiated dolomites and considered it to be Upper Cambrian, though there is no paleontological evidence. Wood (1956) proposed the term "supra-Muav" for this unit, and Brathovde (1986) has suggested that these dolostones be named the Grand Wash Dolomite since the best exposures occur along the Grand Wash Cliffs in western Arizona.

McKee and Resser recognized three lithofacies: white-to-buff massive dolomite; white-to-yellow, very fine-grained, thick-bedded dolomite; and gray, fine-grained, thick-bedded dolomite. Brathovde (1986) reported thick beds of oolitic grainstones and stromatolites that are interbedded with the fine-grained dolostones. Sedimentary structures include wavy and asymmetric ripple laminations and small-scale cross stratification. Biogenic structures include a variety of horizontal traces.

PALEONTOLOGY

Since the early work of Resser, there has been virtually no work done on the taxonomy and biostratigraphy of Cambrian strata in the Grand Canyon. Therefore, the systematics of the invertebrate fauna remain the same and will be reviewed only briefly here. Fossils have been described from the transition interval of the Tapeats Sandstone and from the Bright Angel Shale and Muav Limestone. There have been a few trace fossil studies of these strata, however, that have added to our understanding of the paleoecologic and environmental conditions during periods of deposition. These will be used in conjunction with the depositional environmental reconstructions to characterize the depositional systems.

Invertebrate Fossils

Despite the paucity of well-preserved invertebrate fossils, analysis of the fauna has provided some information concerning the paleoecology and certainly has facilitated the biostratigraphic zonation of the Tonto Group. Brachiopods and trilobites are the most common invertebrates reported from the Tonto Group though preservation typically is poor. Fragments of sponges, primitive mollusks, echinoderms, and algae occur in the Bright Angel Shale and the Muav Limestone; however, these fossils are not very abundant. In addition to establishing the time-trangressive nature of these deposits, the biostratigraphic reconstructions have aided in documenting the numerous transgressive and regressive cycles that characterize the Cambrian of the Colorado Plateau and the Rocky Mountain regions (Lockman-Balk 1971; Aitken 1978).

Trilobites are the most abundant fossils in the Tonto Group, and common genera include *Olenellus, Antagmus, Zacanthoides, Albertella, Kootenia, Glossopleura,* and *Bolaspis.* Most specimens are poorly preserved and occur in the coarser-grained sandstones of the Bright Angel Shale and also in the mudstones of the Bright Angel Shale and the Muav Limestone. Resser reported 47 species of trilobites from the Tonto Group and suggested that these arthropods were, to some degree, facies-specific. More research needs to be done to evaluate this interesting relationship between depositional environment and trilobite distribution.

Brachiopods are locally abundant in the coarse-grained sandstones of the Bright Angel Shale. They also occur in some of the mixed siliciclastic/carbonate facies of the Muav Limestone. The most common genera in the Tonto Group are *Lingulella, Paterina,* and *Nisusia.* In general, the brachiopods tend to occur in beds containing few other invertebrate taxa.

Paleontologists have reported a number of species of primitive mollusks (Conchostraca) from the coarser-grained, hematitic sandstone of the Bright Angel Shale. The association of these fossils in coarse-grained sandstones led Resser to speculate that these mollusks occupied shallow-water habitats. Documentations of this environmental zonation, however, remains to be substantiated.

Resser reported sponge spicules from the Muav Limestone in the western Grand Canyon. These consist of thick, six-rayed spicules that Resser suggested were similar to purported sponge spicules of *Tholiastrella? hindei* Walcott reported from the Cambrian of British Columbia. Elliott and Martin (1987) described six-rayed sclerites, which they assigned to the genus *Chancelloria,* from the Bright Angel Shale along Horn Creek in the Grand Canyon. Although Walcott (1920) considered *Chancelloria* to be a sponge (Phylum Porifera), Rigby (1976) and Elliott and Martin

(1987) have questioned the assignment of this genus to this phylum and have suggested that *Chancelloria* represents a separate, yet unknown, phylum.

Algae, echinoderms, and gastropods have been described from the Tonto Group though they certainly are the rarest taxa reported. Algae in the Muav Limestone consist of convex-upward laminae of calcite and/or dolomite and also small nodules composed of concentric laminations that have been termed *Girvanella* (McKee and Resser 1945). The environmental significance of the algae has yet to be established. Two well-preserved specimens of *Eocrinus* have been reported from the Bright Angel Shale. The excellent preservation of these echinoderms suggests relatively quiet water environments. Gastropods are represented by one species of *Scenella* from the Muav Limestone and several well-preserved species of *Hyolithes* from the Bright Angel Shale.

ICHNOLOGY

Trace fossils are common in all formations of the Tonto Group, particularly the Bright Angel Shale, and include a diverse array of tracks, trails, and burrows (Fig. 4 and Table 1). Despite this, the ichnofauna has been described in only a few studies (McKee 1932; McKee and Resser 1945; Seilacher 1970; Hereford 1977; Martin and Elliott 1987). Consequently, much remains to be done to establish the taxonomic affinities and the relationships between certain physical processes such as current strengths and substrate stability and the mode of infaunal and epifaunal behavior. Recent studies of other ancient shelf sequences, e.g., Crimes (1970), have demonstrated the benefits of integrating ichnologic and sedimentologic data.

Trace fossils are more abundant in the upper half of the Tapeats Sandstone, particularly in the transition interval into the Bright Angel Shale. These consist of single and paired vertical tubes and several types of horizontal traces.

Unbranched, straight vertical burrows assigned to the ichnogenus *Skolithos* are common at many localities. These sand-filled burrows occur near the top of beds. Burrows of this type probably functioned as dwelling and/or temporary resting structures of suspension-feeding organisms. Their occurrence in fine- to coarse-grained sandstones suggests an environment characterized by currents capable of active bedload transport. This is

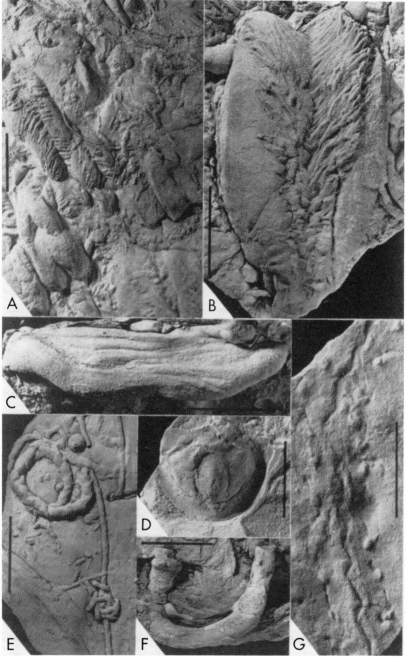

Figure 4. Trace fossils from the Bright Angel Shale in the central Grand Canyon. a) *Cruziana* and *Rusophycus;* b) *Rusophycus;* c)*Teichichnus;* d) *Glossopleura;* e) *Phycodes pedum;* f) *Diplocraterion;* g) *Angulichnus alternipes*

TABLE 1. Common Invertebrate Trace Fossils

Ichnogenus	Formation
Angulichnus	Bright Angel Shale
Arenicoloides	Tapeats Sandstone
Corophioides	Tapeats Sandstone
Cruziana	Bright Angel Shale
Diplichnites	Bright Angel Shale
Diplocraterion	Bright Angel Shale
Palaeophycus	Bright Angel Shale
Phycodes	Bright Angel Shale
Rusophycus	Bright Angel Shale
Scalarituba	Bright Angel Shale
Scolicia	Bright Angel Shale
Skolithos	Tapeats Sandstone, Bright Angel Shale
Teichichnus	Bright Angel Shale

further substantiated by their occurrence in cross-bedded sandstones. Similar structures are common in many modern nearshore settings.

U-shaped burrows perpendicular to bedding also occur in the fine- to coarse-grained sandstones of the Tapeats Sandstone and the Bright Angel Shale. These tubes appear as paired holes on bedding planes or as concave-upward scours, where they have been eroded to the base of the burrow. These abundant traces, assigned to the ichnogenus *Corophioides*, occur in shallow-water deposits (Hereford 1977). The traces probably represent dwelling structures of suspension-feeding organisms, such as certain groups of annelids, and are common in many modern nearshore deposits.

Horizontal traces first were reported by Walcott (1918) from green shales in the Tonto Group and by McKee (1932) from the Tapeats Sandstone. These so-called "fucoides" are smooth-sided, curving traces several inches in length that typically occur in large numbers covering entire bedding surfaces. Presumably, they were formed by detritus-ingesting annelids moving through the sediment.

Trilobite crawling (*Cruziana*) and resting (*Rusophycus*) traces occur in the transition interval and throughout the Bright Angel Shale (Figs. 4a, b). Seilacher (1970) provided the first detailed

GRAND CANYON GEOLOGY

description of *Cruziana arizonensis* from the Tapeats Sandstone, and Martin (1985) reported trilobite trace fossils from the Bright Angel Shale. Martin (1985) noted the common occurrence of these traces in interbedded sandstones and mudstones in the Bright Angel Shale. Elliott and Martin (1987) suggested that the *Cruziana* traces were formed during fair-weather periods as these arthropods moved across the muddy shelf sediments. They also suggested that *Rusophycus* marks formed during storms.

Trace fossils are relatively uncommon in the coarser-grained, cross-stratified sandstones of the Bright Angel Shale (Martin 1985). Only the U-shaped trace *Diplocraterion* (Fig. 4f) is common, attesting to a relatively mobile substrate where the infauna frequently had to relocate their burrows (Martin 1985). *Diplocraterion* also occurs in upward-fining sequences where preservation of the complete burrow is common. In these sequences, the animals evidently recolonized the substrate and then evacuated the sediments as silt and clay were deposited from suspension following the passage of storms (Elliott and Martin 1987).

Interbedded sandstones and mudstones constitute the most abundant facies sequence in the Bright Angel Shale and also contain the most diverse trace fossil assemblage. Horizontal traces, which dominate these beds, include *Cruziana, Rusophycus, Palaeophycus, Diplichnites, Scalarituba, Scolicia, Angulichnus, Teichichnus,* and *Phycodes* (Elliott and Martin 1987). These traces were produced by organisms burrowing in mud, crawling and feeding on the sediment surface, or moving across the top of a sand bed covered with a thin layer of mud (Elliott and Martin 1987).

Trace fossils are not common in the Muav Limestone. Wanless (1975) described several horizons of burrowed, fine-grained carbonate, yet there has been no attempt to provide systematic descriptions of these trace fossils. Nor have we examined their potential enviromental significance. Horizontal burrows, which appear to be the most abundant variety in the Muav, consist of relatively thin, sinuous traces.

DESPOSITIONAL SETTINGS OF THE TONTO GROUP

Cambrian strata in the Grand Canyon accumulated in marine settings though fluvial deposits occur locally near the base of the group. Environments range widely to include beach and intertidal flats; shallow, subtidal sand wave complexes (Tapeats Sandstone); offshore sand sheet deposits; open-shelf, fine-grained sandstones and mudstones (Bright Angel Shale); and subtidal and possibly intertidal carbonate buildups (Muav Limestone and Grand Wash Dolomite). The purpose of this section is to provide new detail on the specifics of these depositional systems and to provide a brief review of previous sedimentologic and stratigraphic studies.

Tapeats Sandstone

Deposition of the Tapeats Sandstone was influenced by a variety of geomorphic factors, as well as processes inherent in fluvial and marine depositional environments. The basal sediments of the Tapeats were deposited on a Precambrian surface that had been exposed for long periods of time. Since this surface had weathered extensively, it showed considerable relief. This is particularly true in the eastern canyon, where the bedrock was more resistant than the crystalline rocks of the western canyon (McKee and Resser 1945). In the eastern areas, these "hills" have relief as great as 800 feet (244 m) (McKee and Resser 1945). Generally, however, the relief is considerably less. Large blocks of the younger Precambrian Shinumo Quartzite occur in the basal deposits of the Tapeats Sandstone near Bright Angel Canyon. Walcott (1883) reported large basement blocks mantling the sides of the areas of high relief. Dott (1974) has shown that similar deposits in Cambrian strata in Wisconsin almost certainly were eroded by storm waves.

McKee and Resser (1945) attempted to reconstruct the depositional environments of the Tapeats Sandstone based on texture of the sediments, paleocurrent trends, and, to some degree, the types of sedimentary structures. These authors concluded that most sedimentation occurred below the beach zone and seaward for tens of miles from the coast in water depths up to 100 feet (33 m). The prevalent west-to-southwest dip of the cross bedding indicates net offshore transport of sediment. McKee and Resser believed that wide channels filled with cross stratification and other large-scale scour-and-fill structures represented rip channels oriented perpendicular to the coast.

McKee and Resser also concluded that the monadnocks, or islands, had relatively little impact on sedimentation—other than serving as a local source of coarse, clastic sediment. They based their conclusions on the relatively consistent dip directions of the cross stratification. The major influence of these islands, according to the authors, was in modifying sedimentation patterns in the inter-island embayments. They reported southeast-directed current trends in the major embayment that developed between the Shinumo Quartzite islands in Bright Angel Canyon.

A far more detailed facies analysis of the Tapeats Sandstone is that of Hereford (1977). This study was concentrated in Chino Valley and the Black Hills of north-central Arizona. Hereford recognized six environmentally specific lithofacies that he related to physical and biological processes operative on modern tidal flats, beaches, and in braided river systems.

This study documented a continuum of tidal flat deposits ranging from lower tidal flats to upper tidal flats dissected by channels. Lower tidal flat sandstones are characterized by complex cross stratification, reactivation surfaces, and herringbone cross stratification. The complex cross bedding reflects the passage of smaller bedforms across the surface of larger dunes or sand waves. Reactivation surfaces are common in tidal systems where reversals of flow and/or erosion during periods of bedform immobility result in the scouring of lee-side avalanche deposits. The herringbone cross stratification reflects bimodal-bipolar flow of the tidal currents. Despite the polymodality of current directions indicated in this study, there is a dominantly southwestern component of sediment transport that is in agreement with the data of McKee and Resser. Tidal systems typically are characterized by asymmetry in flow velocities and durations. In the case of the Tapeats, it is apparent that the ebb phase was the strongest and that it resulted in a preservational bias toward the structure produced during offshore flow.

Deposits of the high intertidal flats are characterized by inter-bedded sandstones and mudstones exhibiting a variety of features that attest to exposure and late-stage emergent runoff. Additionally, Hereford (1977) was able to document the presence of tidal channels that drained the flats.

Large channels occur near the top of the Tapeats at several localities in the central and western canyon (Fig. 5). The channels are up to 13 feet (4 m) deep and 60 feet (18 m) wide. At several localities, up to three laterally contiguous channels form a complex of southwest-oriented channel systems. Channel fill is variable and consists of thick sets of planar tabular cross stratification, co-sets of planar tabular and trough cross stratification, and/or simple vertical fills that conform to the shape of the channel. Flow tends to parallel the southwestern strike of the channel axis, and in some instances, there is a well-developed bimodal-bipolar orientation to the cross-bed dip directions. Vertical trace fossils occur in the upper parts of the channel fills.

Although the geometry of these channels is similar to that of both fluvial and tidal flat channels, the internal stratification differs from that found in fluvial sequences in the Tapeats Sandstone (Middleton and Hereford 1981). The presence of trace fossils, bimodal-bipolar foreset dips, and the absence of exposure features suggest a subtidal channel complex dominated by offshore flow, with minor preservation of flood-oriented structures. Although subtidal channels occur on modern tide-dominated coasts, comparatively little is known of their sedimentologic characteristics. In the lower intertidal and shallow subtidal zones, bedload transport and erosion can be intense, particularly along meso- and macrotidal coasts, because of the concentration of flow in these areas.

Figure 5. Large subtidal channel complex near top of the Tapeats Sandstone. Transition zone is indicated by covered slope below first Bright Angel Cliff.

GRAND CANYON GEOLOGY

Johnson (1977) documented similar deposits in late Precambrian shallow-marine, quartz arenites in Norway. He demonstrated that these subtidal channels were oriented perpendicular to the coast and separated (dissected) subtidal sand bodies. To date, we have not gathered enough data to identify definitively the sandstone bodies that surround these channels as subtidal ridges or sandwaves. Nor have we established the fair-weather or storm-generated origin of these channels.

Two other facies reported by Hereford (1977) were not documented in the study of McKee and Resser. One comprises low-angle, cross-laminated sandstone that likely formed on beaches. These tend to occur most frequently around Precambrian highs, where beach and upper foreshore sediments should have been common.

The second facies association represents braided stream deposits that grade into the marine units. Fluvial deposits in the Tapeats occur in the basal portions of the formation. Typically, they are less mature texturally and mineralogically than the associated marine deposits, which reflects a lack of extensive reworking that is common in the high-energy nearshore. These deposits are characterized by broad, shallow channels filled by horizontally stratified, coarse-grained sandstone and conglomerate that alternate with thick sets of planar tabular and trough cross-stratified sandstone (Middleton and Hereford 1981). This sequence of structures, which is consistent with processes operative in coarse-grained, braided fluvial systems, has been reported from other pre-vegetation fluvial systems (Cotter 1978; Middleton and others 1980; Cudzil and Driese 1987). Deposition occurred in wide, shallow streams where in-channel transport of sediment was accomplished by the movement of sheets of coarse-grained sediment along the bed and by migration of dunes and slightly sinuous transverse bars.

Bright Angel Shale

The Bright Angel Shale comprises a variety of lithologies and sedimentary and biogenic structures that indicate deposition in open shelf environments. McKee and Resser concluded that the Bright Angel Shale accumulated in waters below wave base at depths intermediate between the shallow water represented by the Tapeats Sandstone and the deeper waters of the Muav Limestone. More recent work by Wanless (1973), Martin and others

(1986), and Elliott and Martin (1987a) have provided new data that permit more precise environmental reconstructions. Although generally supporting the conclusions of McKee and Resser, these workers have documented shallow-water deposits in the Bright Angel Shale—as well as providing important information concerning the roles of fair-weather and storm-related processes in controlling depositional patterns in the Bright Angel Shale.

Martin (1985) recognized eight facies in the Bright Angel Shale and grouped these into three genetically significant facies sequences. These include cross-bedded, upward-coarsening, and upward-fining sequences and a heterolithic sequence consisting of interbedded sandstone and mudstone. These facies sequences reflect deposition in subtidal areas influenced by tidal and meterologic processes. They also show transgressive and regressive movements of the strandline.

Upward-coarsening sequences are up to 25 feet (8 m) thick and typically can be traced for several tens of kilometers. The lower parts of these sequences are characterized by laminated, bioturbated mudstones that were deposited during fair-weather suspension settling of silt and clay. The coarser grained portions accumulated as sand waves, dunes, and ripples that migrated over sand sheets. These portions are characterized by thick sets of planar tabular cross stratification (Middleton 1989). The presence of reactivation surfaces and abrupt changes in the dip of many foresets indicates periodic movement of the large bedforms. It also may indicate lee-side erosion during tidal reversals and/or storms. Deposition was entirely subtidal though the upper portions of many of these sequences were deposited in relatively shallow waters, as evidenced by eroded burrows of *Diplocraterion* (Elliott and Martin 1987a).

Sequences that fine upwards are common and consist of a lower, normally graded small-pebble conglomerate or sandstone overlying an erosive base. This, in turn, is overlain by interbedded and fine-grained sandstone and mudstone. The basal coarse-grained facies represent deposition from high-energy, storm-induced currents that transported coarse material from nearshore areas. The tops of these sequences contain symmetrical ripples, as well as appreciable amounts of laminated mudstone. They were deposited from waning flows following the passage of storms (Fig. 3). These beds are very similar to those reported from both modern and ancient storm deposits. Complete vertical and horizontal traces occur at the top of many beds, indicating that the substrate was recolonized soon after deposition (Elliott and Martin 1987a).

Lenticular beds of interbedded sandstone and mudstone constitute the majority of the Bright Angel Shale. This sequence consists of very fine-grained sandstone lenses and micaceous shale. Most beds are graded normally and contain an abundant and diverse trace fossil assemblage (Elliott and Martin 1987a). These deposits represent post-storm suspension settling of muds and sands and, possibly, remobilization during fair-weather periods. *Cruziana* and *Rusophycus* indicate a substrate inhabited by trilobites. Other trace fossils also indicate a relatively stable substrate colonized by a variety of infaunal and epifaunal organisms.

Muav Limestone

McKee and Resser considered the Muav Limestone to have been deposited in subtidal environments. The subtidal origin of much of the Muav is based on faunal and textural characteristics. These include an open-marine fauna, the very fine-grained nature of the mottled limestone and dolostone facies, and the fact that the Muav grades eastward into a shallow-water facies of the Bright Angel Shale. These authors also indicated that many of the flat-pebble conglomerates occurring throughout the formation were deposited in relatively deep water.

Intraformational or flat-pebble conglomerates are an extremely important facies in the Muav Limestone. These deposits, which are abundant from the Bass Trail eastward, consist of disc-like clasts of micrite and, occasionally, silt-size quartz and glauconite grains. The orientation of these clasts is variable. Some are oriented parallel with the bedding; some clasts are imbricated, and in some instances, clasts are vertical.

McKee and Resser described two associations of these conglomeratic beds. One variety consists of intraformational conglomerates that occur as scattered, discontinuous lenses within thinly bedded limestones. The other variety consists of one to several thin, conglomeratic beds that extend up to 45 miles. The great lateral persistence of these beds makes them ideal stratigraphic markers, and McKee and Resser used them to correlate over great distances in the canyon. These workers considered the widespread conglomerates to represent subtidal deposits formed during regressions.

The origin of the clasts obviously requires early lithification by cementation and/or compaction since they are derived from sediments within the basin. Where this induration takes place is controversial. Opinions range from ripups of carbonate muds exposed on tidal flats by storms and/or tidal channels to submarine lithification and subsequent erosion during storms.

Dew (1985) documented the occurrence of intraformational conglomerates similar to those reported from the Muav Limestone in the Upper Cambrian DuNoir Limestone in Wyoming. Based on facies associations, her study showed that both intertidal and subtidal limestone conglomerates can occur over a short stratigraphic interval. Sepkoski (1982) demonstrated that storm-induced currents were mostly responsible for the widespread distribution of flat-pebble conglomerates in Montana's Cambrian strata. In this case, the conglomerates are interbedded with shales that lack any evidence of subaerial exposure. Considering the stratigraphic importance that has been made of these conglomerates, it is clear that they require restudy in light of recent work.

Regardless of their mode of origin, it is particularly interesting that they are abundant only in Cambrian and Ordovician rocks. Sepkoski (1982) speculated that with the proliferation of organisms living within the sediment during the Ordovician, the potential for early submarine cementation of carbonate shelf deposits was reduced substantially. This hypothesis has yet to be tested and, of course, assumes a subtidal origin for these deposits.

Although many of the limestone and dolomite beds in the Muav are subtidal, Wanless (1973, 1975) reported intertidal and supratidal facies from outcrops in the western Grand Canyon. Many of the textures and structures reported by Wanless are similar to those found in modern tidal flats on Andros Island in the Bahamas. In particular, the laminated dolostones in the Muav have many characteristics in common with laminated dolomites that occur on supratidal levees adjacent to tidal channels on Andros Island. In these areas, fine-grained carbonate sediment is deposited during periods of overbank flooding following storms. Algae that inhabit the levees trap the sediment, resulting in the generation of continuous laminae of carbonate mud and pellets. Aitken (1967) referred to these laminated horizons as cryptalgal laminations since the evidence of algal binding had to be inferred.

Discontinuous laminae also occur and are produced by traction transport of pellets and other grains over the algally bound sediment. Wanless (1975) reported that these units are up to 66 feet (20 m) thick in the Muav Limestone. This suggests that there were prolonged periods of supratidal sedimentation far offshore from the Cambrian strandline.

It is clear that the Muav Limestone records episodes of both subtidal and peritidal deposition. A reasonable depositional model, therefore, might be one of offshore shoals surrounded by deeper water areas. Pratt and James (1986) proposed a tidal flat island model for Lower Ordovician shelf carbonates of New-foundland (Fig. 6). In this model, small, localized carbonate islands occurred far offshore and were separated by subtidal areas. Middleton and others (1980) documented similar facies distributions from Cambrian strata in Wyoming.

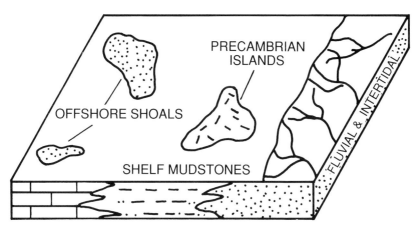

Figure 6. Block diagram illustrating the distribution of depositional environments represented by the Tapeats Sandstone, Bright Angel Shale, and Muav Limestone

Undifferentiated Dolomites

The undifferentiated dolomites that overlie the Muav are not well understood in terms of their temporal and environmental significance. Brathovde (1986), however, has documented thick beds of ooolitic grainstones and stromatolites interbedded with fine-grained carbonates. This association clearly indicates shallow subtidal and, possibly, intertidal environments.

SUMMARY

Facies analyses of the Tonto Group indicate deposition in a variety of fluvial, nearshore, and shallow shelf environments. Braided stream and intertidal-to shallow-subtidal deposits of the Tapeats Sandstone grade seaward into a complex array of shallow shelf sands and muds of the Bright Angel Shale. Shelf sedimentation was influenced by both tidal and storm currents. Sand ridges, sand waves, and broad areas where fine-grained siliciclastics were deposited from suspension settling following storms and during fair-weather periods characterized the shelf. Farther offshore, carbonate islands dotted the shelf. Here, the carbonate buildups were characterized by intertidal and possible supratidal zones separated by deeper water areas where tidal currents were active and where finer-grained carbonate sediments were deposited.

A number of transgressions and regressions resulted not only in the intertonguing of the formations of the Tonto Group but also in the vertical juxtaposition of facies belts that probably were not laterally adjacent. As Curry pointed out in his 1964 study, rapid migration of the strandline can result in the overstepping of offshore facies over nearshore deposits with no record of intervening environments. Numerous examples of such transitions occur in the Tonto Group, but most are poorly documented.

More detailed facies mapping and, in particular, documentation of lateral facies changes are needed. Only through such studies can the nature of the transgressive and regressive stratigraphies preserved in the strata of the Tonto Group be reevaluated and related to regional paleogeography.

CHAPTER 7

TEMPLE BUTTE FORMATION

Stanley S. Beus

INTRODUCTION

Strata of the Temple Butte Formation of early Late Devonian and possibly late Middle Devonian age are exposed through most of the Grand Canyon but are relatively inconspicuous. Outcrops in the east are thin, discontinuous lenses, and in central and western Grand Canyon the exposures, though continuous, tend to merge within the much thicker overlying Redwall Limestone cliffs. Temple Butte lithology is predominantly dolomite or sandy dolomite with minor sandstone and limestone beds.

NOMENCLATURE

Rocks of Devonian age in the Grand Canyon region first were reported by Walcott (1880, 1883), who recognized Devonian strata beneath the Redwall Limestone in Kanab Canyon and Nanko-weap Canyon. He applied the name Temple Butte Limestone (Walcott 1889) to a thin band of Devonian dolomite along Temple Butte (on the west side of the Colorado River about three miles southwest of the mouth of the Little Colorado River in eastern

Figure 1. Approximate location of the type section of the Temple Butte Formation, on west side near south end of Temple Butte. Cu, Cambrian Muav Limestone; Dtb, Temple Butte Formation; Mr, Mississippian Redwall Limestone

Grand Canyon). Although geologists have not designated a specific type section, the Temple Butte site (Fig. 1) has gained universal acceptance. McKee (1939) and others have extended the name, as Temple Butte Formation, to the much thicker and more extensive outcrops in central and western Grand Canyon. West and north of the Grand Canyon, Devonian strata equivalent to the Temple Butte are designated the Muddy Peak Limestone (Longwell and others 1965), a name derived from the Muddy Mountains of southern Nevada (Longwell 1921). South of the Grand Canyon, along the Mogollon Rim of central Arizona, equivalent Devonian rocks are recognized as the Martin Formation (Teichert 1965), a name originally applied to Devonian strata in the Bisbee area of southeastern Arizona (Ransome 1904).

DISTRIBUTION

In eastern Grand Canyon and upstream in Marble Canyon, the Temple Butte crops out as scattered, lens-shaped exposures that fill channels eroded into the upper surface of the Muav Limestone or into the Cambrian undifferentiated dolomite. These channel-fill lenses commonly are less than 100 feet (30 m) thick, but they may be up to 400 feet (120 m) wide. Numerous exposures of these lenses are visible near river level in Marble Canyon beginning just below mile 37 (see Fig. 2), and some occur in the lower Little Colorado Gorge as well. From Hermit Creek westward throughout central and western Grand Canyon, the Temple Butte forms a continuous band of dolomite above local channel-fill deposits at the base. It gradually thickens to more than 450 feet (140 m) at Iceberg Ridge, five miles west of the mouth of Grand Canyon (Fig. 3). Earlier descriptions of an Iceberg Ridge Devonian section more than 1200 feet (365 m) thick mistakenly included several hundred feet (220 m) of unnamed Cambrian dolomite beds (the Cambrian[?] undifferentiated dolomites of McKee and Resser) as part of the Devonian section.

Figure 2. Temple Butte channel-fill outcrop on right bank of Marble Canyon at approximately mile 39.4

CAMBRIAN-DEVONIAN UNIFORMITY

In a broad, regional sense, Devonian strata truncate successively older rocks from west to east across northern Arizona (Fig. 4). The unconformity at the base of the Temple Butte is one of the major stratigraphic breaks in the Paleozoic sequence of the Grand Canyon. It probably represents latest Cambrian, all of Ordovician

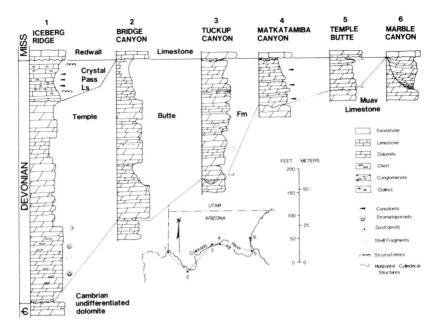

Figure 3. Selected stratigraphic sections of the Temple Butte Formation

and Silurian, and most of Early and Middle Devonian time. The time gap involved probably is somewhat less in western Grand Canyon because the uppermost Cambrian strata were deposited by a sea regressing westward, and the Temple Butte was formed in a sea trangressing eastward.

In western Grand Canyon, a sequence of light-gray dolomite beds up to 500 feet (150 m) thick overlies, and appears conformable with, the Muav Limestone. These strata were referred to as undifferentiated Cambrian(?) dolomites by McKee and Resser (1945). The beds are truncated gradually eastward by the Temple Butte but are present near Lava Canyon in easternmost Grand Canyon, where McKee and Resser (p. 141) reported 163 feet ((50 m) of dolomite and siltstone. Although no diagnostic fossils are known from these unnamed dolomite beds, Brathovde (1986) has presented compelling evidence that they are conformable with the Muav Limestone below—and probably are equivalent to Upper Cambrian strata nearby in southern Nevada and southeastern Utah. They form a mappable unit and have been named informally the Grand Wash dolomite (Brathovde 1986, p. 1).

The unconformable surface at the base of the Temple Butte is marked locally by considerable relief in the form of channels and depressions cut into the underlying Cambrian strata. These channels, which are up to 100 feet (30 m) deep in Marble Canyon, occur throughout most of the Grand Canyon. Where present, they clearly mark the Devonian base. They first were observed in Kanab Canyon by Walcott (1880) and were studied and described in detail between Garnet and Cottonwood Creek in eastern Grand Canyon by Noble (1922, pp. 49-51). In parts of the Grand Canyon, including the type section on Temple Butte (where the channels are absent), the Cambrian-Devonian strata appear in local exposures to be without angular discordance, and the contact is planar with gray dolomite beds below and above. Here, the unconformity, even though representing more than 100 million years, may be difficult to locate.

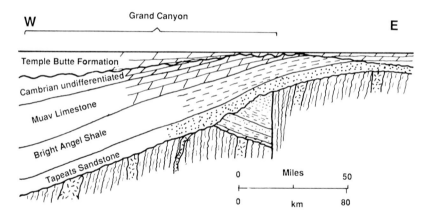

Figure 4. Regional pattern of Devonian-preDevonian unconformity and angular discordance of beds across northern Arizona. Vertical scale greatly exaggerated, relative thickness of units only approximate

STRATIGRAPHY

The strata forming the basal channel-fill part of the Temple Butte commonly are a distinct pale, reddish purple dolomite or sandy dolomite. The bedding generally is irregular, and it is gnarly in places. In some localities, the beds are horizontal, but elsewhere they may conform to the walls of the channel they fill and be truncated with angular discordance by the overlying beds (Fig. 5). Rarely, there are basal conglomerate beds composed of subrounded dolomite pebbles. The upper beds in the channel-fill illustrated in Figure 2 contain about twenty percent insoluble residue consisting of detrital quartz grains, clay, and hematite. They also exhibit a peculiar columnar pattern of pale gray and purple dolomite—perhaps as the result of leaching or weathering.

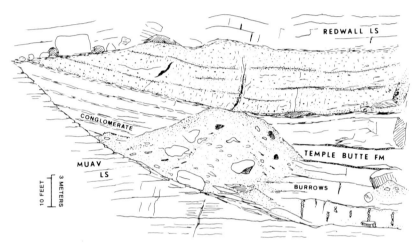

Figure 5. Field sketch by James Obey of portion of Temple Butte channel-fill lens in Marble Canyon on right bank at approximately mile 38.4

The more continuous strata above the basal channel-fill beds of the Temple Butte are exposed as uniformly medium-to-thick, blocky ledges outlined by parting planes or thin recesses (Figs. 6, 7). The predominant lithology is a dark-to-light olive gray, fine-to-medium crystalline dolomite that commonly weathers to a sugary texture. A subordinate, though extensive, second carbonate lithology is that of thin-bedded, very fine-grained or aphanitic dolomite that is grayish orange/pink, which weathers to a very light gray or yellow, exhibits a conchoidal fracture, and commonly appears porcelaneous. Rounded, frosted quartz sand grains occur either scattered or as thin lenses or laminae in some of the aphanitic dolomite beds. Locally, they form marker beds of quartz sandstone up to seven feet (2 m) thick, as at Havasu and Matkatamiba Canyon (Fig. 3). The dolomite beds commonly crop out as a uniform series of steep, receding ledges, but in central

Figure 6. Basal part of Temple Butte section at Tuckup Canyon showing channel cut into underlying Cambrian undifferentiated dolomite beds. Arrow indicates 6-ft. figure.

Figure 7. Paleozoic strata in western Grand Canyon near the mouth of Quartermaster Canyon, about mile 265. Mr - Redwall Limestone; Dtb - Temple Butte Formation; Cu - Cambrian undifferentiated dolomite

TEMPLE BUTTE FORMATION

Grand Canyon, they may form part of a vertical 1500-foot (460 m) carbonate wall that appears unbroken from the Muav Limestone up through the Redwall Limestone.

At Iceberg Ridge, five miles west of the mouth of the Grand Canyon, the upper 85 feet (26 m) of the Devonian carbonate section consist of light-gray lime mudstone to oolitic wackestone with a basal conglomerate or breccia. These beds previously were treated as part of the basal Whitemore Wash Member of the Mississippian Redwall Limestone. Ritter (1983, p. 6) has recovered diagnostic latest Devonian (Famennian) conodonts from these strata and assigned them to the late Devonian Pinyon Peak Limestone. This unit apparently pinches out eastward and thus far has not been recognized in the Grand Canyon.

PALEONTOLOGY AND AGE

Although marine fossils are abundant in the Upper Devonian Jerome Member of the Martin Formation in central Arizona (Teichert 1965; Beus 1978), the Temple Butte Formation has yielded surprisingly few identifiable organic remains. Walcott (1883, p. 221) reported indeterminate brachiopods, gastropods, corals, and "placoganoid" fish from the walls of lower Kanab Canyon. Noble (1922) reported fish plates identified as *Bothreolepis*, a fresh or brackish water form, from Sapphire Canyon. Denison (1951) confirmed the identification and the Late Devonian age assignment for the fish plates. Rare silicified corals, gastropods, crinoid plates, and massive stromatoporoids occur in the Temple Butte section at Iceberg Ridge, but none so far are identifiable to the generic level.

Peculiar cylindrical forms, somewhat resembling the trace fossil *Palaeophycus*, occur in dolomite beds near the base of the Temple Butte at the type section and at Tuckup Canyon. These forms are subhorizontal, straight to gently arcuate, and have a micritized central core within a cylindrial rim about three mm. in diameter (Fig. 8). They may be trace fossils, or possibly some sort of algal or stromatoporoid structure, but to date are indeterminate.

Conodont microfossils reported from the Temple Butte Formation at Matkatamiba Canyon (about mile 148 on the Colorado River in Grand Canyon) by Elston and Bressler (1977) are the most

Figure 8. Subhorizontal cylindrical structures near the base of the Temple Butte type section. Scale is in inches.

diagnostic and significant fossils yielded by the Temple Butte. These conodonts, identified by D. Schumacher (1978, written communication), include the following: 1) *Polygnathus pennatus, P. xylus,* and *Icriodus* cf. *I. subterminus*—possibly a late Givetian lowermost *Polygnathus assymetricus* Zone assemblage—occur near the base of the section; 2) *Pandorinella insita* and *"Spathagnotus"* cf. *S. gradatus*—possibly an early Frasnian assemblage—

TEMPLE BUTTE FORMATION 115

occur between 20 and 40 feet (6 to 12 m) above the base of the Temple Butte; and 3) *Polygnathus* cf. *P. angustidiscus*—probably an early Late Frasnian form—occurs about twenty feet (6 m) below the top of the formation. The upper twenty feet (6 m) of the Temple Butte in the Matkatamiba section are devoid of fossils. The conodonts in the Temple Butte Formation suggest a latest Givetian to late Frasnian age (latest Middle Devonian through early Late Devonian) for most of the formation in central Grand Canyon. The Temple Butte Formation thus is the approximate age equivalent of the Jerome Member of the Martin Formation in central Arizona, the Muddy Peak Formation of southern Nevada, and the Elbert Formation of the subsurface in northeastern Arizona (Knight and Cooper 1955).

DESPOSITIONAL ENVIRONMENT

Devonian strata in the Grand Canyon are perhaps the least understood of the Paleozoic rock units studied there. The dolomitization of original limestone and a general lack of recognizable fossils have made refined environmental interpretations difficult. The carbonate facies, and the few known fossils—crinoids, corals, stromatoporoids, and conodonts typical of nearshore biofacies (D. Schumacher 1984, written communication)—indicate accumulation in shallow, subtidal, open-marine conditions for most of the Temple Butte in central and western Grand Canyon. The aphanitic dolomite beds in the Temple Butte may record local supratidal conditions. Similar modern, supratidal dolomites have been formed through evaporative pumping of magnesium-enriched seawater moving through porous, supratidal sediments, as described in the Persian Gulf (Illing and others 1965; Shinn and others 1965).

The thinner and discontinuous channel-fill deposits of the Temple Butte in eastern Grand Canyon may have been intertidal deposits accumulated in tidal channels.

The regional paleogeography of the Frasnian (early Late Devonian) appears to have been shallow (but generally open-circulation) marine conditions in northwestern and central Arizona; intertidal-to-very-shallow subtidal conditions in the Grand Canyon shelf area of central and eastern Grand Canyon; and shallow, marine, restricted-circulation conditions in northeastern Arizona (Beus 1980).

SUMMARY

The Temple Butte Formation records minor deposition of a thin carbonate sequence that gradually thickens westward across the Grand Canyon region. The easternmost outcrops and some of the basal strata to the west probably were deposited in narrow tidal channels under intertidal conditions. The more laterally extensive dolomite beds in the central and western portions of Grand Canyon accumulated in more subtidal but probably very shallow-marine conditions across a gently submerged continental shelf.

CHAPTER 8

REDWALL LIMESTONE AND SURPRISE CANYON FORMATION

Stanley S. Beus

INTRODUCTION

Two formations of Mississippian age are recognized in the Grand Canyon. The oldest of these, the Redwall Limestone of Early and early Late Mississippian age, has been known and studied for more than a century and is one of the most prominent rocks units in the canyon wall. Recently, a second unit, the Surprise Canyon Formation of latest Mississippian age, has been recognized as a result of detailed mapping in the more remote parts of western Grand Canyon. It occurs as isolated patches and lenses and occupies stream valleys, caves, and collapse structures developed in the top of the Redwall Limestone.

REDWALL LIMESTONE

Nomenclature

The Redwall Limestone consistently forms massive, vertical cliffs 500 to 800 feet (150 m to 250 m) high about midway in the canyon wall (Figs. 1 and 2). The cliff face generally is stained red

Figure 1. View of Redwall Limestone cliffs along the South Kaibab Trail in eastern Grand Canyon. Lowermost sharp "V" in trail switchbacks marks approximate Redwall Limestone-Muav Limestone contact.

Figure 2. View of Redwall Limestone cliffs in western Grand Canyon near Separation Canyon, mile 240. H- Horseshoe Mesa Member; M - Mooney Falls Member; T - Thunder Springs Member; W - Whitmore Wash Member; Dt - Temple Butte Formation

by iron oxide material washed down from the redbeds in the overlying Supai Group. The name Red Wall Limestone was applied first by Gilbert (1875, p. 177) in a report for one of the early surveys west of the 100th Meridian directed by John Wesley

GRAND CANYON GEOLOGY

Powell. A type locality was established later by Darton (1910) when he introduced the name Redwall Canyon to a Grand Canyon tributary in the Shinumo quadrangle and thus provided a geographic place name for the Redwall. Four distinct stratigraphic units within the Redwall were recognized first by Darton (1910). Later, they were described in more detail by Gutschick (1943) and given formal names by McKee (1963). These units, in ascending order, are: the Whitmore Wash, Thunder Springs, Mooney Falls, and Horseshoe Mesa members. All four members have their type localities within the Grand Canyon or its tributaries, and all four can be traced throughout the Grand Canyon and beyond (McKee and Gutschick 1969).

Distribution

The Redwall Limestone originally was deposited across virtually all of northern Arizona except the Defiance positive area of east central Arizona. The Redwall is recognized as far south as the Gold Gulch area just north of Globe in south central Arizona (Racey 1974). Within the Grand Canyon, it is exposed in almost continuous outcrop bands on both canyon walls from about mile 22 in Marble Canyon, where it first appears at river level, to the mouth of the canyon at the Grand Wash Cliffs (mile 277). Thickness of the formation gradually increases northwestward. It is just over 400 feet (120 m) thick at the Tanner trail section in easternmost Grand Canyon and about 800 feet (245 m) at Iceberg Ridge, five miles (8 km) west of the Grand Canyon's mouth (Fig. 3) (McKee and Gutschick 1969, p. 3).

PRE-REDWALL UNCONFORMITY

Throughout most of the Grand Canyon, the Redwall Limestone rests without angular discordance upon Devonian strata or, where the Temple Butte Formation is missing, upon rocks of Cambrian age. The unconformity at the base of the Redwall spans all of latest Devonian (Famennian) time and the earliest part of the Mississippian period (early Kinderhookian) through central and western Grand Canyon. The magnitude of the unconformity increases eastward because of the transgressive nature of the basal Redwall (Fig. 2). Basal beds of the Redwall Limestone are Early Mississippian (Kinderhookian) age in westernmost Grand

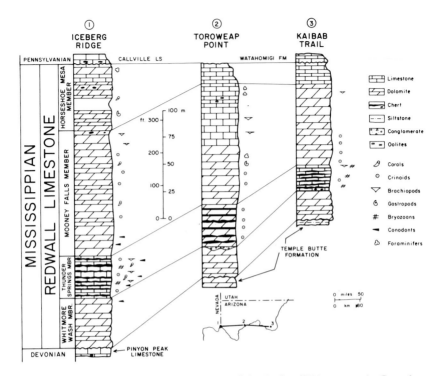

Figure 3. Selected stratigraphic sections of the Redwall Limestone in Grand Canyon

Canyon, as indicated by diagnostic fossil foraminifers, and are of late Early Mississippian (Osagian) age in eastern Grand Canyon. In western Grand Canyon, the Mississippian-Devonian contact generally is well marked by an irregular surface of erosion having up to ten feet (3 m) of relief in a lateral distance of 100 to 200 feet (30 m to 60 m).

Locally, a basal conglomerate composed of angular dolomite or limestone blocks of the Devonian Temple Butte Formation occurs at the unconformity. In eastern Grand Canyon, the Redwall-Devonian contact most commonly is a nearly horizontal surface with little or no relief. It is more difficult to recognize the unconformity where the Cambrian or Devonian strata beneath the unconformity and the basal Mississippian strata above it are both dolomite. The nature of the unconformity suggests only gentle uplift and mild, though perhaps locally prolonged, erosion of pre-Mississippian strata before the beginning of Redwall deposition.

GRAND CANYON GEOLOGY

Stratigraphy

Whitmore Wash Member. The type section for the Whitmore Wash Member is at Whitmore Wash (Colorado River mile 187.5 below Lees Ferry). Along this northern tributary to central Grand Canyon, McKee and Gutschick (1969) measured 101 feet (30 m) of thickly bedded, fine-grained dolomite. This member is composed mainly of fine-grained limestone in western Grand Canyon, but this changes to mostly dolomite in central and eastern Grand Canyon. The Whitmore Wash is nearly pure carbonate, having less than two percent insoluble residue content of minor gypsum and iron oxides (McKee and Gutschick 1969, p. 27). Common textural carbonate types are pelleted and locally skeletal or oolitic wackestones and packstones (Kent and Rawson 1980). Thickness in the Grand Canyon is from about 100 feet (30 m) in the east to nearly 200 feet (60 m) at Iceberg Ridge, five miles (8 km) beyond the western end of the Grand Canyon (Fig. 3). Bedding generally is thick, ranging from two to four feet (0.6 m to 1 m) and even thicker locally. It usually forms a resistant cliff overlying a narrow bench or series of ledges typical of the Temple Butte Formation beneath. The upper boundary with the overlying Thunder Springs Member is conformable but is easily recognized by the lowest appearance of thin, dark, chert beds alternating with the light gray limestone or dolomite beds typical of the Thunder Springs.

Fossils are rare in the Whitmore Wash Member, probably because of extensive dolomitization of the original lime mud and sand. A few brachiopods, corals, and crinoids have been recognized. The Whitmore Wash is of late Kinderhookian (early Early Mississippian) age in western Grand Canyon and early and middle Osagian (late Early Mississippian) to the east, as interpreted from brachiopod and foraminiferid fossils. Racey (1973) reported late Kinderhookian conodonts from the lower part of the member in the Salt River Canyon area south of the Grand Canyon. Ritter (1974, p. 17) noted conodonts of possible earliest Osagian age in the upper part at Iceberg Ridge.

Thunder Springs Member. The Thunder Springs type section is at the head of Thunder River about two miles (3 km) north of Colorado River mile 136 in central Grand Canyon. The Thunder Springs Member is the most distinctive member of the Redwall Limestone because of the light and dark banded appearance imparted by alternating chert and carbonate beds. It consists of thin beds of light gray limestone or dolomite alternating with thin beds of dark reddish brown or dark gray weathering beds or lenses of chert. Thickness of the member increases gradually from 100 feet (30 m) in eastern Grand Canyon to about 150 feet (46 m) in the west (McKee and Gutschick 1979, p. 41).

Most of the carbonate rock in the Thunder Springs Member is thin-bedded, crinoidal grainstone or packstone. The rock tends to be limestone in the west and dolomite in the east. Thin section analyses of the chert beds in the Thunder Springs area of central Arizona reveal them to be silicified former bryozoan wackestones and mudstones (Bremner 1986, p. 55).

Invertebrate marine fossils are especially abundant in the chert beds of the Thunder Springs Member. These include corals (particularly colonial *Syringopora*), bryozoans, brachiopods, crinoids, and a few gastropods, blastoids, and cephalopods. Similar forms are present in the carbonate beds. These forms are not so well preserved, however, probably as a result of dolomitization. Diagnostic conodonts of Osagian age are reported by Racey (1974) and Ritter (1983). The Thunder Springs is everywhere conformable with the underlying Whitmore Wash Member. It is disconformable with the overlying Mooney Falls Member except in the extreme western end of Grand Canyon (Fig. 4). Locally, the contact with the Mooney Falls is a low-angle unconformity, as at Kanab Canyon and in Marble Canyon. This indicates minor structural activity, as well as erosion between Thunder Springs and Mooney Falls deposition.

Mooney Falls Member. The type section of the Mooney Falls Member is at Mooney Falls in Havasu Canyon (mile 153), about four miles (6.5 km) south of the Colorado River (McKee 1963). It is the thickest member of the Redwall, ranging from about 200 feet (60 m) in eastern Grand Canyon to nearly 400 feet (122 m) at the western end (McKee and Gutschick 1969, p. 56). It forms the major part of the sheer wall to which the Redwall name refers.

The Mooney Falls Member is predominantly pure limestone—except locally where it is dolomitized. Insoluble residue generally is less than 0.5 percent. The rock constitutes a favorable source for lime, which is being processed at two major cement plants (one south of the Grand Canyon, near Peach Springs, and a second near Clarkdale). Carbonate grains include oolites, pel-

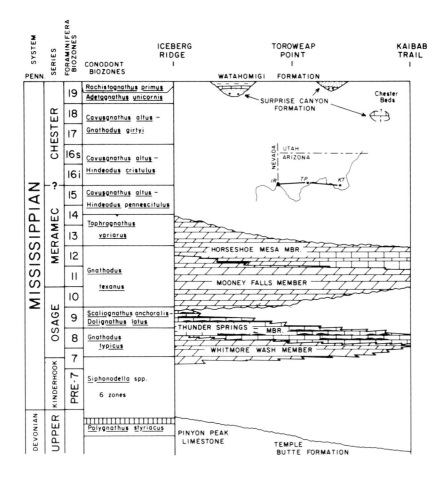

Figure 4. Diagram showing age and distribution of Mississippian rock units across northwestern Arizona. Conodont zones from Sando (1985); foraminiferal zones after Mamet (in Mamet and Skipp 1970). Modified from Skipp (1979, Fig. 76)

lets, and a variety of skeletal fragments dominated by crinoid plates. One or two zones of thin beds or lenses of chert occur in the upper part of the member—generally near the contact with the overlying Horseshoe Mesa Member. Bedding normally is thick and appears massive in outcrop. Large-scale, tabular planar cross bedding is reported in the upper third of the member at several localities in central and eastern Grand Canyon, including the North Kaibab Trail, by McKee and Gutschick (1969, p. 61).

Invertebrate marine fossils are abundant throughout the member. They include solitary and colonial corals, spiriferid brachiopods, and crinoids. Diagnostic foraminifers (Skipp 1969; Mamet and Skipp 1970, p. 338) and conodonts (Racey 1974; Ritter 1983) indicate a late Osagian and Meramecian age.

The upper contact of the Mooney Falls with the overlying Horseshoe Mesa Member is conformable and difficult to locate precisely in the field. The boundary generally is placed at the vertical change from medium-and-coarse-grained limestone and the thick or massive bedding of the Mooney Falls Member to the aphanitic and relatively thin-bedded, receding, ledge-forming limestone of the Horseshoe Mesa.

Horseshoe Mesa Member. The type section for this member is on the south rim of Grand Canyon, along the Grandview Trail north of Horseshoe Mesa. The Horseshoe Mesa is the thinnest and least extensive member of the Redwall Limestone. Its thickness in the Grand Canyon varies from 45 to 125 feet (14 m to 38 m), with the thinnest section in the east. Erosion causes the member to wedge out thirty to forty miles (50 to 65 km) south of the Grand Canyon. It also is missing from the top of the Mooney Falls Member in most of central Arizona. The Horseshoe Mesa Member is composed of thin-bedded, light gray limestone with a mudstone to wackestone texture. It typically forms weak, receding ledges in contrast to the massive cliff of Mooney Falls below. Some chert lenses occur in the lower part.

Well-preserved, invertebrate megafossils are rare but present throughout the Horseshoe Mesa. Spiriferid brachiopods, bivalves, and corals are among the most abundant forms. At least sixteen species of foraminifers are recognized in the member. These forms indicate a Meramecian (early Late Mississippian) age (Skipp 1979, p. 298). The upper boundary of the Horseshoe Mesa Member is a major unconformity overlain by Early Pennsylvanian redbeds of the Watahomigi Formation, Supai Group. In localized areas, it is overlain by the Surprise Canyon Formation of latest Mississippian age.

Paleontology and Age

Invertebrate fossils are common in certain lithofacies of the Redwall throughout the Grand Canyon (Fig. 19). Data from some

500 Redwall Limestone localities collected by McKee and Gutschick (1969) indicate that the most abundant megafossils are brachiopods and corals—followed by bryozoans, crinoids, gastropods, bivalves, and cephalopods. Additional minor elements include blastoids, trilobites, ostracods, fish teeth, and algal remains. Foraminifers are abundant and were found in half of all samples selected for thin sectioning.

Index microfossils permit relatively accurate dating of most of the Redwall Formation. As Figure 4 shows, initial Redwall deposition began with the basal Whitmore Wash Member in western Grand Canyon during latest Kinderhookian (early Early Mississippian) time. Basal Redwall deposits became successively younger as the sea transgressed eastward across northern Arizona and are no older than Osagian (late Early Mississippian) age in eastern Grand Canyon.

The Thunder Springs Member was formed by a regressing sea during middle Osagian time. Beginning in late Osagian time and extending into the early Meramecian, a second marine transgression deposited the Mooney Falls Member of the Redwall Formation. The Horseshoe Mesa Member was formed during middle Meramecian (early Late Mississippian) time as a regressive deposit.

A single, 6.5-foot (2-m) limestone outcrop of Chesterian (late Late Mississippian) age at the top of the Redwall Formation on the Bright Angel Trail was reported by McKee and Gutschick (1969, p. 74). The age assignment was based on rare brachiopod and foraminiferid fossils. This exposure, if it is dated correctly, is considerably younger than any other Redwall outcrop. It is treated here as part of the Surprise Canyon Formation, which locally overlies the Redwall and is of Chesterian age (Billingsley and Beus 1985, p. 27).

The age assignment of the Redwall places it as a correlative of the Escabrosa Limestone of southeastern Arizona, the Leadville Limestone of southwestern Colorado, and the Monte Cristo Group of southeastern Nevada. Four of the five formations in the Monte Cristo Group (excluding only the Arrowhead Limestone) are nearly identical lithologically and in stratigraphic position with the four members of the Redwall in Grand Canyon (McKee and Gutschick 1969, p. 14). It is likely that Redwall Limestone deposits were laterally continuous with all the above units at the end of the Mississippian Period.

Depositional setting

Deposition of the Redwall Limestone sediments occurred in a shallow, epeiric sea that produced a submerged continental shelf across northern Arizona. Deposits formed during two major transgressive-regressive pulses, as demonstrated by McKee and Gutschick (1969). Detailed facies analysis by Kent and Rawson (1980) and Bremner (1986) have confirmed and refined this interpretation.

The basal part of Whitmore Wash Member records initial deposition during the first transgression under nearshore, shallow, subtidal conditions where high-energy currents produced oolitic shoals. As the transgression proceeded, more offshore deposits of skeletal grainstone and packstone accumulated under quieter water and more open-marine conditions.

The Thunder Springs Member accumulated in increasingly shallow conditions as the sea regressed westward. The abundant chert layers (which exhibit a lack of sorting, an original lime mud texture, and a high proportion of delicate bryozoan fossils) are considered by Bremner (1986, p. 62) to be preferentially silicified blue-green algal mats, which locally may have baffled marine currents. Alternating with the chert beds are abraded, sorted, crinoidal grainstone and packstone deposits that record more vigorous current-washing of skeletal sand.

The Mooney Falls Member of the Redwall formed during a second marine transgression as crinoidal packstone and grainstone sediments developed widely across northern Arizona under generally open-marine, offshore conditions. The Horseshoe Mesa Member formed under conditions of increasingly shallow and more restricted circulation during a final, slow regression of the sea.

SURPRISE CANYON FORMATION

Nomenclature

The Surprise Canyon Formation is a newly recognized rock unit in the Grand Canyon. It appears as isolated, lens-shaped exposures of clastic and carbonate rocks that fill erosional valleys and locally karsted topography and caves in the top of the Redwall Limestone. McKee and Gutschick (1969, p. 76) recognized conglomerate and gnarly mudstone beds filling channels in the top of the Redwall Limestone but considered them part of the basal Supai Group. Subsequently, these strata were recognized by Billingsley (1969) as a separate unit belonging neither to the Supai nor the Redwall and were referred to as pre-Supai buried valley deposits by Billingsley and McKee (1982).

The name Surprise Canyon Formation was applied formally by Billingsley and Beus (1985, p. 27) and is taken from a large, northern tributary canyon in western Grand Canyon at mile 248. The type section is on the east-facing slope of a narrow ridge near the Bat Tower viewpoint in western Grand Canyon—about twelve miles (20 km) northwest of the mouth of Surprise Canyon (mile 265) (Fig. 5). The Surprise Canyon Formation probably is the least visible rock unit in the Grand Canyon because of its discontinuous nature and the extreme remoteness of the larger outcrops. Current research indicates that it may be one of the best-preserved rock records of an estuarine depositional system anywhere in North America.

Figure 5. Type section of the Surprise Canyon Formation, Bat Tower section 2 is located about 1 mile west of mile 263 on the Colorado River and 1.6 miles southwest of the mouth of Tincanebits Canyon.

Distribution

The Surprise Canyon Formation is nowhere a continuous stratum. Instead, it crops out as isolated, lens-shaped patches throughout much of the Grand Canyon and in parts of Marble Canyon to the east (Fig. 6). The valleys in which the formation occurs commonly are 150 to 200 feet (45 m-60 m) deep and up to 1/2-mile (1 km) wide in western Grand Canyon. They become shallower and relatively wider in central and eastern Grand Canyon (Fig. 7). Thickness of the formation corresponds to the depth of the valleys in which it occurs. The thickest section observed thus far is at Quartermaster Canyon (Fig. 8), a southern tributary canyon in western Grand Canyon, where the formation is about 400 feet (122 m) thick (Billingsley and Beus 1985, p. 27). Outcrops in central Grand canyon are up to 150 feet (45 m) thick, whereas in eastern Grand Canyon and Marble Canyon they rarely are more than 60 or 70 feet (18 m or 20 m) thick.

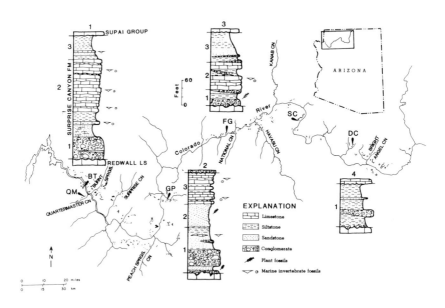

Figure 6. Index map showing major outcrops of the Surprise Canyon Formation (black patches) and selected stratigraphic sections in Grand Canyon. Section 1 is the type section near the Bat Tower (BT); section 2 is at Granite Park (GP); section 3 is at Fern Glen (FG); and section 4 is at Dragon Creek (DC); the outcrop at Quartermaster Canyon (QM) is illustrated in Figure 8.

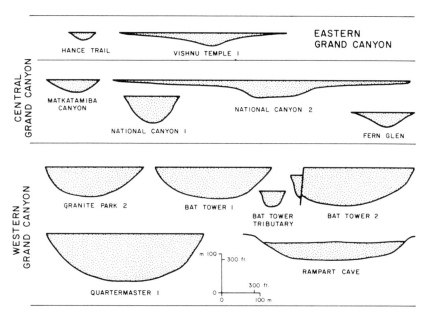

Figure 7. Cross sections of selected channel-fill outcrops of the Surprise Canyon Formation, illustrating a general increase in thickness from eastern to western Grand Canyon

Figure 8. Thickest known exposure of the Surprise Canyon Formation on the west wall of Quartermaster Canyon (QM in Fig. 6). Note distinct curved surface of the pre-Surprise Canyon valley wall cut into the top of the Redwall Limestone.

REDWALL LIMESTONE & SURPRISE CANYON   131

Redwall-Surprise Canyon Unconformity

The erosion surface that developed on the Redwall Limestone during the Late Mississippian period, and on which the Surprise Canyon was deposited, must have been a relatively flat, resistant, limestone platform but with considerable local relief. Most of the Surprise Canyon outcrops occupy gently U-shaped or V-shaped notches cut into the top of the Redwall. By their nature and distribution, these notches appear to have been part of a major dendritic drainage system that flowed generally from east to west. They also appear to have been incised up to 400 feet deep (122 m) into the top of the Redwall Limestone near the western end. A preliminary reconstruction of the drainage pattern (Grover 1987) illustrates several major valleys that merge westward (Fig. 9). In addition, solution depressions and caves in the upper Redwall Limestone are filled locally with red mudstone of the Surprise Canyon Formation, indicating the development of a karst topography prior to Surprise Canyon deposition (Fig. 10). The time available for the development of this eroded and karsted topography on top of the Redwall is just a few million years—the interval between youngest Redwall (of middle Meramecian age) and oldest Surprise Canyon (of approximately middle or late Chesterian age). The depth of the stream valleys eroded into the top of the Redwall indicates an uplift or a several hundred foot (120 m) drop in sea level.

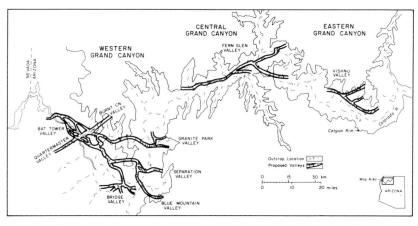

Figure 9. Hypothetical reconstruction of segments of the ancient valley system eroded into the Redwall Limestone in Late Mississippian time and into which the Surprise Canyon Formation was deposited

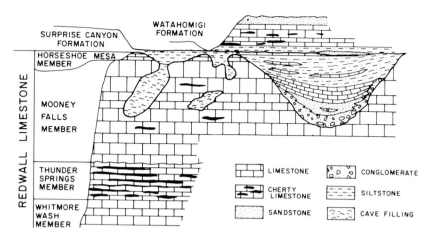

Figure 10. Cross section showing the stratigraphic relationship of the Surprise Canyon Formation to the underlying Redwall Limestone and overlying Watahomigi Formation of the Supai Group. Modified from Billingsley and Beus (1985, Fig. 3)

Stratigraphy and Lithology

The thicker sections of the Surprise Canyon Formation can be divided into three major rock units: 1) a lower conglomerate and sandstone, mainly of terrestrial origin; 2) a middle unit of skeletal limestone of marine origin; and 3) an upper, mainly marine unit of siltstone and silty, sandy, or algal limestone (Fig. 6). Since several lithofacies are included in each of the three units, the Surprise Canyon Formation exhibits a greater variety of sedimentary rock types than almost any other Paleozoic unit in the Grand Canyon.

The basal part of unit 1 in most sections consists of a ferruginous, pebble-to-cobble and local boulder conglomerate. Clasts are predominantly chert with minor limestone derived from the underlying Redwall Limestone. Some cobbles and boulders contain typical Redwall fossils. The clasts commonly are grain-supported and enclosed in a sandy matrix of nearly pure quartz grains and some hematite. Locally, the cobbles are sufficiently imbricated to indicate water current directions at the time of deposition (Fig. 11). In most sections, the conglomerate grades upward into a yellow or dark reddish brown or purple quartz sandstone or siltstone or, in some sections, a dark gray, carbonaceous shale. The sandstone beds commonly are flat bedded, but some

Figure 11. Imbricated cobbles in conglomerate of the Surprise Canyon Formation, Dragon Creek section

beds exhibit trough cross-strata or ripple laminations. *Lepidodendron* log impressions occur in the sandstone at numerous localities between Burnt Springs Canyon (mile 259.5) in western Grand Canyon and Cove Canyon (mile 169) in central Grand Canyon. A variety of plant fossils also occurs in carbonaceous shale at Granite Park near river mile 209. In eastern Grand Canyon, red-brown to purple mudstone beds, together with subordinate chert pebble-congomerate lenses typical of this lower unit, make up the entire formation. Trace fossils in sandstone beds of this unit are simple, vertical burrows and rare *Conostichus*, suggestive of the sandy-shore *Skolithos* ichnofacies of Crimes (1975).

Unit 2 is a coarse-grained, skeletal limestone that typically has a grainstone texture and is composed of whole or fragmented shells. Quartz sandstone beds up to 1.5 inches (3 or 4 cm) thick, alternating with skeletal limestone beds up to 4 inches (10 cm) thick, are a common occurrence. The base of the limestone commonly truncates the unit 1 sandstone or siltstone beds occurring below on an erosion surface. The limestone unit typically forms resistant cliffs or ledges and weathers to a yellowish brown, rusty, or purple-gray color. Small-scale trough cross strata exhibiting bimodal current directions are common in this unit. Marine invertebrate fossils are abundant and include crinoids and other echinoderms, brachiopods, bryozoans, corals, mollusks, and tri-lobites. Trace fossils typical of the shallow marine *Cruziana* ich-

GRAND CANYON GEOLOGY

nofacies of Crimes (1975) occur in sandy limestone beds. This unit is the thickest, and topographically the most prominent, feature in many sections of central and western Grand Canyon. It is absent, however, in eastern Grand Canyon east of Fossil Bay (about mile 130).

In western and central Grand Canyon, unit 3, the upper unit, is typically a dark red-brown to purple, ripple-laminated to flat-bedded calcareous siltstone or sandstone that forms weak slopes or receding ledges. Linguoid ripples are common. Resistant ledges of algal or ostracodal limestone also are common within this unit in western Grand Canyon sections. In at least three localities—Burnt Springs Canyon, Quartermaster Canyon, and National Canyon—nearly spherical algal stromatolites (oncolites) occur near the top of the unit (Fig. 12).

The boundary between the Surprise Canyon Formation and the overlying Watahomigi Formation of the Supai Group commonly is obscured by limestone rubble from above or by a covered slope developed on weak mudstone. Where well exposed, the basal Watahomigi consists of: 1) a thin, widespread, but locally discontinuous limestone pebble conglomerate that contains minor chert clasts; or 2) where the conglomerate is absent, a

Figure 12. Algal stromatolites (oncolites) from near the top of the Surprise Canyon Formation at Quartermaster Canyon section 4. Centimeter scale

purplish red calcareous siltstone and mudstone overlain generally by resistant gray limestone beds containing pale red-to-orange chert nodules (Billingsley and Beus 1985, p. 29). In a few localities, a low-angle unconformity is recognizable at the contact (Fig. 13).

Paleontology and Age

Marine invertebrate fossils are abundant in the middle limestone unit of the Surprise Canyon Formation throughout western and central Grand Canyon (Fig. 20). A preliminary list of macro-fossils includes the following:

Corals
> (identified by W. J. Sando 1985, written communication):
> *Barytchisma* spp.   *Palaeacis* sp.
> *Michelinia* sp.   *Amplexus* sp.

Brachiopods
> (identified by MacKenzie Gordon, Jr. 1985
> [written communication] and this author):
> *Orthotetes* sp.   *Cleiothyridina* sp.
> *Rhipidomella nevadensis*   *Anthracospirifer* spp.
> (Meek)
> *Linoproductus* sp.   Punctate spirifers
> *Inflatia* n.spp.
> *Flexaria* sp.   *Eumetria* vera (Hall)
> *Ovatia* sp.   *Composita* spp.
> *Antiquatonia* sp.   *Dialasma* sp.
> *Leiorhynchoidea* sp.   *Schizophoria* sp.
> *"Camarotoechia"* sp.   *Torynifer setiger* (Hall)?
> *Tetracamera* sp.

Echinoderms
> *Pentremites* n. sp.
> Crinoids
> Rare starfish

Additional forms reported include bivalves, gastropods, cephalopods, trilobites, and shark teeth (Billingsley and McKee 1982, p. 143). A reconstruction of a fossil community dominated by *Composita* and punctate spiriferid brachiopods with associated

Figure 13. Low-angle unconformity at the Surprise Canyon- Watahomigi contact (between the two arrows in the center of the photograph) about 0.3 mile due west and on the opposite side of the ridge from the type section southeast of the Bat Tower

echinoderms, bryozoans, corals, and mollusks typical of the unit 2 limestone beds in the Bat Tower area is shown in Figure 14.

Microfossil invertebrates are moderately abundant in the middle limestone unit, and some are present in the uppermost limestones of unit 3. Seven species of foraminifers were identified by Betty Skipp (Billingsley and McKee 1982, p. 144). Eight forms of conodonts have been identified tentatively from limestone beds in both the middle and upper units of the formation in the Bat Tower and Quartermaster Canyon area by Gary Webster (1985, written communication). Additional collections of all the above currently are under study.

The Surprise Canyon also has yielded a moderate amount of plant fossil material, mainly from sandstone and siltstone or shale beds of the lower unit in western Grand Canyon. Palynomorphs (spores) representing 22 species were identified from the Granite Park area by R. M. Kosanke (Billingsley and McKee 1982, p. 144). Macroscopic plant fossil material from the Granite Park area currently under study by Richard Hevly (1985, written communication) includes three forms of *Lepidodendron* logs (Fig. 19), *Calamites*, and seed ferns.

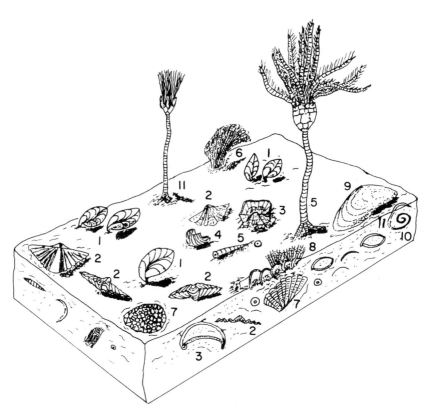

Figure 14. Reconstruction of a bottom-dwelling, marine community typical of the middle limestone unit in the Bat Tower area as it may have appeared in Late Mississippian time. Brachiopods are 1) *Composita*; 2) punctate spirifer; 3) *Inflatia*; and 4) *Ovatia*; 5) a crinoid; 6) fenestrate bryozoan; 7) the colonial coral *Michelinia*; 8) the bryozoan *Archimedes*; 9) the bivalve *Septimyalina*; 10) rare bellerophontid gastropod; 11) the blastoid *Pentremites*.

Fossil evidence from the spores, foraminifers, conodonts, brachiopods, and corals documents a latest Mississippian (Chesterian) age for the Surprise Canyon Formation (Fig. 4). Conodonts indicative of the *Adetognathus unicornis* Zone (late Chesterian) occur in the upper part of the middle limestone unit at the Bat Tower 1 section and also at Quartermaster Canyon (G. Webster 1985, written communication). In addition, thin limestone beds in the upper part of unit 3 (near the top of Bat Tower 3 section) have yielded conodonts—including the nominal species—that are indicative of the *Rhachistognathus primus* Zone of possible earliest Pennsylvanian age. There still is some question, however, whether the lowest appearance of *R. primus* actually is at the top of the Mississippian or the base of the Pennsylvanian system.

The brachiopod *Rhipidomella nevadensis* occurs through some 35 feet (10 m) of the middle limestone unit at Blue Mountain Canyon and marks the late Chesterian *Rhipidomella nevadensis* assemblage Zone that Gordon (1984 p. 76) considers a direct equivalent of foraminifer Zone 19. It appears that the Surprise Canyon Formation is the age equivalent of the following: at least part of the Indian Springs Formation (Brenckle 1973) of southern Nevada, the upper Chainman Shale and lower Ely Limestone in western Utah, and part of the Manning Canyon Formation of northern and central Utah (Webster 1984, p. 79). It probably is slightly younger than the Paradise Formation of Meramecian and early Chesterian age (Armstrong and Repetski 1980) in southeastern Arizona but may be the equivalent of the Log Springs Formation (Armstrong and Mamet 1974) of northern New Mexico.

Depositional Setting

The distribution and nature of the Surprise Canyon strata suggest that the entire formation was deposited within the confines of a broadly dendritic stream valley system and in the associated caves and collapsed depressions formed on a limestone

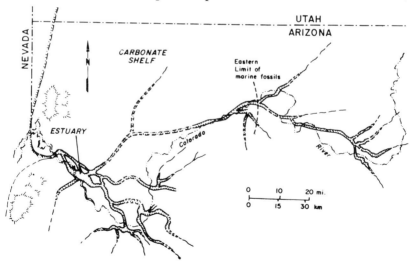

Figure 15. Hypothetical paleogeography of northern Arizona during Surprise Canyon Formation deposition in latest Mississippian time

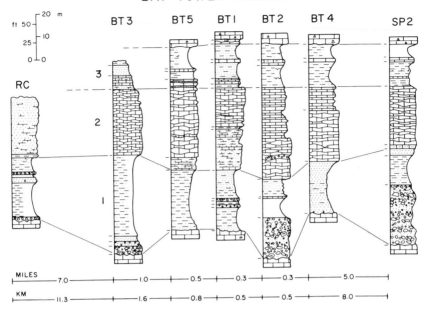

Figure 16. Stratigraphic sections of the Surprise Canyon Formation along the Bat Tower valley portion of the estuary, showing maximum development of the unit 2 limestone (from Grover 1987)

platform (Fig. 14). The major marine shoreline must have been somewhere between the present western edge of the Colorado Plateau and Frenchman Mountain, which is located east of Las Vegas, Nevada. Here, outcrops of the Indian Springs Formation record an extensive shallow-marine environment (Webster 1969). The marked lateral and vertical facies changes in the Surprise Canyon Formation are believed to record deposition in a major estuary system extending for at least eighty miles (130 km) east-west across northwestern Arizona to the eastern limit of marine fossils (Beus 1986). For the western and central Grand Canyon outcrops, the depositional environment appears to have been fluvial for the basal unit. It becomes progressively more marine-dominated up section, however, into the middle limestone and sandstone unit and ends with a restricted-marine environment for the upper unit.

The basal conglomerate beds of unit 1 exhibit local imbrication that indicates strong, unidirectional, fluvial currents. The sandstone beds above the conglomerate commonly contain cut and fill structures and locally abundant plant fossil remains. These beds grade upward into ripple-laminated or flat-bedded sandstones or siltstones. Grover (1987) has interpreted this sequence (especially well displayed in the Bat Tower area) as a record of initial continental and fluvial conditions changing to intertidal conditions as the sea transgressed eastward into the estuary and began to trap and rework clastic sediments within the tidal range.

The skeletal lime grainstone of the middle unit is well developed in both the Bat Tower and Fern Glen paleovalley systems (Figs. 9 and 16). Abundant marine fossils attest to shallow-marine conditions during deposition. The bimodal current directions (both upstream and downstream, but predominantly upstream or east), as indicated by prominent, small-scale trough cross strata, suggest deposition in an estuary dominated by flood tides (Grover 1987) (Fig. 14).

In the Granite Park area, strata at the same stratigraphic position as the middle limestone unit are predominantly cross-stratified sandstone containing abundant plant fossils (section 2, Fig. 6; Fig. 17). Grover (1987) has interpreted this as deposition in a more fluvial- and ebb tide-dominated valley where sand supply and deposition eclipsed minor marine limestone deposits.

For the most part, the upper unit (unit 3) of the Surprise Canyon Formation is ripple-laminated sandstone or siltstone,

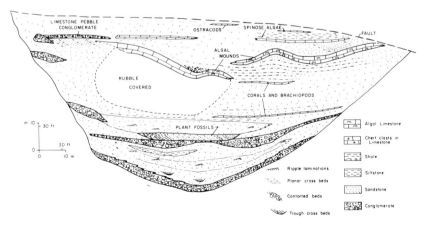

Figure 17. Sketch of a major part of the exposed lens of Surprise Canyon Formation at the Granite Park 2 section about 3 1/2 miles southeast of the mouth of Granite Park Wash and mile 209

which alternates with algal and/or ostracodal limestone beds that lack a normal, open-marine fauna. Grover (1987) has interpreted these strata as a record of restricted-marine to upper tidal-flat environments developed during final infilling of the estuary. Detailed sedimentary and paleontological analysis of sections in the Fern Glen Valley indicate that there were several fluctuations of marine and terrestrial or, at least, marginal marine conditions during deposition of the Surprise Canyon Formation (Grover 1987, Fig. 66a).

It also is possible that the original Surprise Canyon strata were part of a more widespread sheet formed in a shallow sea that covered much of northwestern Arizona. Subsequent uniform erosion then may have removed all but the lower valley-fill portion of the deposits. This would explain the great lateral extent of marine fossils. However, in the absence of convincing evidence for marine deposition outside the confines of the narrow paleovalleys that presently contain the formation, deposition confined to a narrow estuary system seems the most reasonable interpretation.

Independent environmental interpretation based upon mineralogical and geochemical analyses of clays in the generally structureless and unfossiliferous mudstones of the Surprise Canyon permits confirmation and refinement of the above interpretations. Differential flocculation of clays in the zone of sea water/ fresh water mixing within the Fern Glen Valley section of the estuary appears to have produced lateral distribution of kaolinite predominantly to the east (the landward direction) and illite to the west (seaward direction) (Shirley 1987, Figs. 23a-j), as in modern estuaries (Edzwald and O'Melia 1975). Using ratios of kaolinite/ illite clays, Shirley (1987) has demonstrated a convincing record of two marine transgression-regression cycles in the Fern Glen Valley (Fig. 18).

The depositional environment for the fine-grained, red mudstones and local conglomerates of the Surprise Canyon Formation in easternmost Grand Canyon and Marble Canyon must have been mainly fluvial—in fresh, or perhaps brackish, water conditions. Almost totally absent are limestone beds and marine fossils, and only a few plant fragments have been recovered. Imbricated cobbles in the basal conglomerate of the Dragon Creek section (Fig. 11) indicate a vigorous westward-flowing current at the time of deposition.

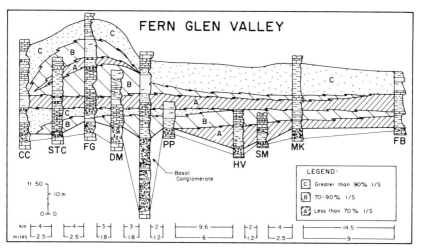

Figure 18. Selected stratigraphic sections in the Fern Glen valley portion of the estuary, showing clay mineralogy facies distribution and environmental interpretations. Correlation is attempted only on fine-grained lithologic units. Facies A, having less than 70% illite/smectite clays, is considered to record nonmarine conditions; facies B, containing 70-90% illitic clays, is considered to record transitional or intermediate conditions; facies C, greater than 90% illitic clays, is considered to mark marine conditions. Two cycles of marine transgression and regression are indicated (from Shirley 1987, Fig. 29).

ACKNOWLEDGMENTS

I am grateful for the encouragement and helpful suggestions offered during the early stages of this study by the late Edwin D. McKee, who had a continuing interest in this unit. Financial support for much of this work was provided by grants EAR 8217434 and EAR 8618691 from the National Science Foundation and faculty grants from Northern Arizona University. The cooperation of the Hualapai Tribe and the Grand Canyon National Park in allowing access to remote areas of the Grand Canyon is appreciated. Helicopter transport to critical sites was provided by The U.S. Geological Survey through the efforts of Karen Wenrich and George Billingsley. The latter also provided helpful comments and reviews of this report.

Figure 19. Invertebrate fossils from the Redwall Limestone (1-9) and plant fossils from the Surprise Canyon Formation (10, 11). 1) *Zaphrentites*, a solitary coral; 2) *Fenestella* , a bryozoan; 3) *Loxonema* ; 4) *Bellerophon*; 5) *Straparollus,* gastropods; 6) *Orophocrinus saltensis* Macurda, a blastoid; 7, 8) *Buxtonia viminolis* (White); 9) *Anthacospirifer*, brachiopods; 10, 11) two forms of *Lepido-dendron*. All figures are X 1.

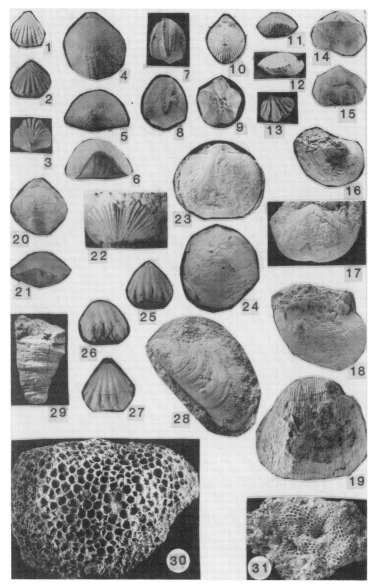

Figure 20. Invertebrate fossils from the Surprise Canyon Formation. 1-3) *Camarotoechia;* 4-6) *Leiorhynchoidea,* rhynchonellid brachiopods; 7-9) *Pentremites,* a blastoid; 10-12) *Eumetria,* 13) *Punctospirifer ?,* punctate spiriferid brachiopods; 14-16) *Inflatia,* 17-19) *Flexaria,* productid brachiopods; 20-21) *Composita;* 22) *Anthracospirifer;* 23) *Schizophoria;* 24) *Rhipidomella nevadensis* (Meek); 25-27) *Tetracamera,* Brachiopods; 28) *Septimyalina,* a bivalve; 29) *Barytchisma,* a solitary coral; 30) *Michelinia,* a colonial coral; 31) fenestellid bryozoan. All figures are X1.

CHAPTER 9

SUPAI GROUP AND HERMIT FORMATION

Ronald C. Blakey

INTRODUCTION

The brilliant red cliffs and slopes present throughout the Grand Canyon comprise strata of continental, shoreline, and shallow-marine origin. Assigned to the Supai Group and Hermit Formation of Pennsylvanian and Early Permian age, these sedimentary rocks consist of a broad variety of lithologies—including sandstone, mudstone, limestone, conglomerate, and gypsum. Geologists have divided the rocks by their red-to-tan color and their weathering characteristics into five formations recognizable throughout the region. The five formations, in ascending order, are the Watahomigi Formation and Manakacha Formation (both Lower Pennsylvanian), Wescogame Formation (Upper Pennsylvanian), Esplanade Sandstone (Lower Permian), and Hermit Formation (Lower Permian). Geologists traditionally assign the first four formations to the Supai Group.

NOMENCLATURE AND DISTRIBUTION

Although the rocks now assigned to the Supai Group and Hermit Formation have a long and complicated nomenclature history (see McKee 1982, page 3), three papers are responsible for the present terminology now in use in Grand Canyon. Darton

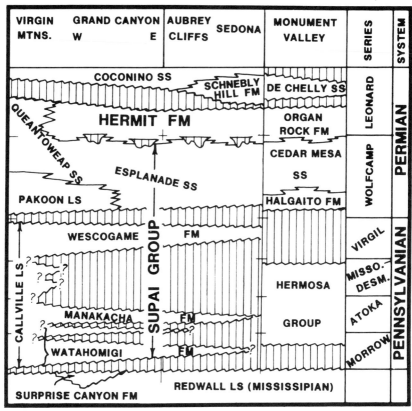

VIRGIN MTNS.	GRAND CANYON W E	AUBREY CLIFFS	SEDONA	MONUMENT VALLEY	SERIES	SYSTEM
	COCONINO SS	SCHNEBLY HILL FM	DE CHELLY SS		LEONARD	PERMIAN
QUEANTOWEAP SS	HERMIT FM		ORGAN ROCK FM			
	ESPLANADE SS		CEDAR MESA SS		WOLFCAMP	
PAKOON LS			HALGAITO FM			
CALLVILLE LS	WESCOGAME FM SUPAI GROUP			VIRGIL	PENNSYLVANIAN	
	MANAKACHA FM		HERMOSA GROUP	MISSO.-DESM.		
	WATAHOMIGI FM			ATOKA		
SURPRISE CANYON FM	REDWALL LS (MISSISSIPIAN)			MORROW		

Figure 1. Time-rock chart of Pennsylvanian and Lower Permian rocks of northern Arizona. Vertical ruled lines show time represented by unconformities.

(1910) proposed the name of Supai Formation for all the predominantly red strata between the Redwall Limestone and the Coconino Sandstone. Noble (1922) divided out the Hermit Shale from the top of Darton's Supai. McKee (1975) raised the Supai to group status and proposed four formations within the group: the Watahomigi, Manakacha, and Wescogame formations and the Esplanade Sandstone. The type section for each formation is near Supai in the Grand Canyon. Most publications within the last thirty-five years have used the term "Hermit Formation," in preference to shale, and this trend is followed here.

The Supai-Hermit terminology is sound throughout Marble Canyon and eastern and central Grand Canyon; however, several problems arise in western Grand Canyon. McNair (1951) assigned the lower part of the Supai interval to the Callville Limestone and recognized a Permian carbonate unit not present to the east, to

148   GRAND CANYON GEOLOGY

which he assigned the term "Pakoon Limestone." McKee (1982) did not recognize the Callville Limestone within the confines of Grand Canyon, and although he recognized the Pakoon Limestone, he did not include it within the Supai Group. A further complication exists in the western portions of the Grand Canyon where McNair (1951) assigned sandstone partly equivalent to the Esplanade and partly equivalent to the Hermit to the Queantoweap Sandstone.

The solution to this nomenclature problem is controversial and beyond the scope of this chapter, but the approach followed herein, pending further work, is to follow McKee's terminology in the Grand Canyon and Grand Wash Cliffs and to assign equivalent strata to the Callville and Pakoon limestones, Queantoweap Sandstone, and Hermit Formation (where applicable) in areas farther to the west, northwest, and north (Fig. 1).

To the southeast, along the southern margin of the Colorado Plateau, we can correlate the Supai-Hermit interval as far east as Sedona (Blakey 1979a, b; Blakey and Knepp 1987). Correlation to the southeast of Sedona is complicated by the addition of strata of Pennsylvanian age that is not directly time equivalent to the Supai. Termed the Naco Formation, these rocks were laid down in a basin separate from that in which the Supai was deposited. The fact that the Esplanade Sandstone is not present southeast of Sedona complicates the separation of the Supai and Hermit. Adding to the dilemma is the presence of a younger stratigraphic unit not present in the Grand Canyon, the Schnebly Hill Formation (Blakey 1980).

To the east on the Defiance Plateau, Permian redbeds have been assigned to the Supai Formation (Read and Wanek 1961). Recent, unpublished work shows that only a portion of these redbeds are equivalent to part of the Supai in the Grand Canyon (see Fig. 2).

LITHOLOGY AND STRATIGRAPHY

Watahomigi Formation

The Watahomigi Formation, the oldest formation in the Supai Group, consists chiefly of red mudstone and siltstone and gray limestone and dolomite (Fig. 3). The unit forms a broad, slightly westward-thickening sheet that ranges in thickness from 100 feet

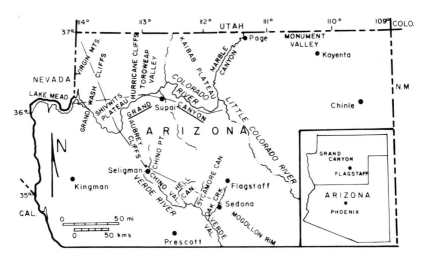

Figure 2. Index map of area of study

(30 m) in eastern Grand Canyon and at Sycamore Canyon on the east to 300 feet (90 m) in western Grand Canyon and along the Grand Wash Cliffs (Figs. 4, 5). Carbonate content, which chiefly is very fine-grained (aphanitic) limestone to the east and granular (grainstone and packstone) limestone to the west, increases dramatically to the northwest. Grains include abraded fossil fragments, accretal grains, and pellets. Mudstone consists primarily of nondescript, slope-forming, poorly exposed units with thin, intercalated, bioturbated (disturbed by organisms), bright-orange, limey sandstone. Occasional bedding-plane exposures of the redbeds reveal various tracks, trails, and burrows. Most sections contain a basal chert-pebble conglomerate in which the clasts were derived from the underlying Redwall Limestone.

Throughout the Grand Canyon and the adjacent southern Colorado Plateau, the Watahomigi Formation can be divided informally into a lower redbed slope, middle carbonate ledge, and upper redbed slope (Blakey 1980; McKee 1982). Based on included fossils and the presence of local conglomerate beneath the upper slope, McKee (1982) assigned the lower slope and the middle ledge a Morrowan age and the upper slope an Atokan age (Fig. 2).

The lower contact of the Watahomigi Formation is everywhere sharp and unconformable with the underlying Redwall Limestone. Where the Surprise Canyon Formation is present (see

GRAND CANYON GEOLOGY

Figure 3. Typical outcrops of Watahomigi Formation

Beus, this volume), geologists suspect an unconformity, but the contact is poorly exposed. The upper contact probably is conformable and was assigned by McKee to a zone of gray, jasper-bearing limestone and bright-orange sandstone and intercalated red

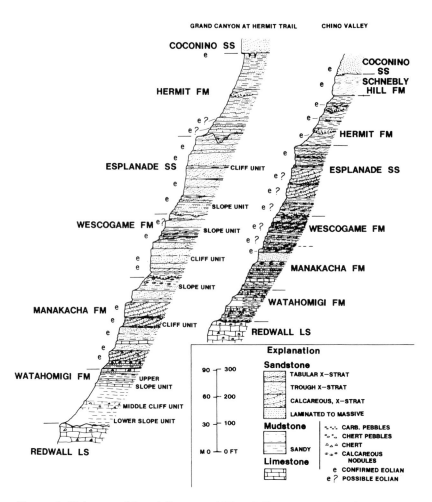

Figure 4. Columns of Supai Group and Hermit Formation and related strata showing distribution of known and suspected eolian strata

mudstone. The contact apparently occurs at the change from primarily slope below to steep slope or cliff above and, therefore, may not be a consistent stratigraphic level across the region.

The sharp increase in limestone west of a line paralleling the Hurricane Cliffs can be used to divide the Watahomigi Formation into an eastern redbed facies and a western carbonate facies (see McKee 1982, Fig. 6). The westward increase in thickness, carbonate content, and marine fossils is typical of many Paleozoic rock units in the Grand Canyon region.

GRAND CANYON GEOLOGY

Manakacha Formation

The Manakacha Formation marks an important change in the trend of Paleozoic depositional patterns in the Grand Canyon region. Following initial Cambrian sand deposition, the area was dominated by carbonates and minor mudstones from Middle Cambrian to Early Pennsylvanian time. The influx of quartz sand during the deposition of the Manakacha reflects a significant change across the western interior of the United States.

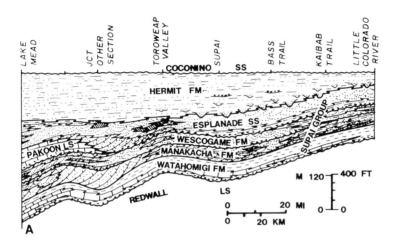

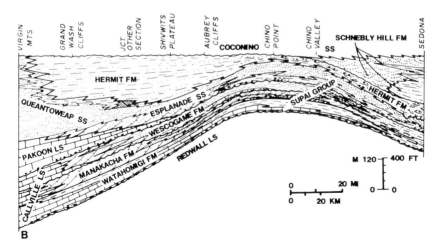

Figure 5. Restored stratigraphic cross section of Supai Group, Hermit Formation, and related strata. (a) West to east through Grand Canyon; (b) northwest to southeast from Virgin Mountains to Sedona

The Manakacha Formation consists chiefly of quartz sandstone and intercalated red mudstone (Fig. 6). Unlike most other Paleozoic rock units, the formation is thicker in central Grand Canyon than it is in western Grand Canyon (McKee 1982). The Manakacha forms a broad, sheetlike deposit across northwestern and central Arizona that averages about 300 feet (90 m) thick in Grand Canyon and 150 feet (45 m) thick across the Verde and Chino valleys.

Sandstone is the dominant lithology in the Manakacha Formation (Figs. 4, 5). The composition ranges from very fine- to medium-grained quartz grains to ooids, abraded fossils, and peloids. Most units, regardless of composition, are cemented with calcite. Jasper (red chert) is an accessory to many of the more limey units. Based on bedding types, geologists can recognize three kinds of sandstone. Cross-stratified sandstone comprises trough, planar, and compound sets that range in thickness from one foot (2.5 cm) to 30 feet (9 m). Careful examination of the strata within the sets reveals the ubiquitous presence of the climbing translatent strata described by Hunter (1977). Climbing translatent strata are thin laminae, generally less than several millimeters thick, that display reverse grading within each lamina. Each lamina displays the migration record of a single wind ripple (Hunter 1977) and, as such, is a powerful indicator of eolian deposition. The orientation of the strata reflects the nature and geometry of the surface on which they were deposited. Horizontal or very low-angle, climbing translatent strata form at the base of dunes, between dunes, and on eolian sand sheets. Climbing translatent strata in cross-stratified sets form from the migration of eolian dunes. A broad range of both types of deposition is present within sandstone units of the Manakacha Formation.

A second type of sandstone consists of horizontally laminated to very low-angle, cross-stratified, fine-grained, calcareous, and silty units up to twenty feet (6 m) thick. Generally deeper red in color than cross-stratified units, these units also show rare ripple lamination and locally abundant tracks and trails on bedding planes. The existing evidence about the sandstone's origin suggests a subaqueous origin, but few other clues are present.

The third type of sandstone typically is very fine-grained and grayish orange to bright reddish orange in color. It ranges from structureless (homogeneous) to extensively bioturbated. Actually, both the structureless and bioturbated sandstone are thought

Figure 6. Typical outcrops of Manakacha Formation

to be the result of bioturbation; however, the former is so extensively bioturbated that individual burrows or trackways cannot be distinguished. Since the latter is less thoroughly disturbed, we can see individual traces clearly.

Many sandstone units in the Manakacha Formation show additional alteration caused by diagenesis—chiefly the growth of calcite and dolomite crystals that alter or destroy the original fabric of the sediment. Such change is most prevalent in the upper portions of the sandstone units.

Mudstone and fine-grained limestone and dolomite are common throughout the Manakacha Formation. The mudstone is usually dark reddish brown in color, and it ranges from structureless to units that contain nodular carbonate concretions. Light gray carbonate units that bear jasper occur fairly commonly; they are seen with or without wispy lamination.

McKee (1982) recognized a lower cliff unit and upper slope unit throughout Grand Canyon (Fig. 6). Sandstone dominates the former and is less abundant in the latter. These subdivisions are not as apparent in the Verde and Chino valleys; units assigned to the Manakacha Formation in these areas form a lower ledge and slope unit, a middle cliff unit, and an upper ledge unit. It should be emphasized that the lower level of relief along the southern margin of the Colorado Plateau (as opposed to the Grand Canyon) undoubtedly affects the topographic expression of all units in the Supai Group.

The Manakacha Formation displays a steady increase in carbonate content to the northwest. In the Shivwits Plateau region, McKee (1982, Fig. 8) classified the entire formation as limestone.

The upper contact with the overlying Wescogame Formation marks a regional unconformity, with most of the Middle Pennsylvanian absent. Conglomerate derived from the underlying Manakacha Formation, accompanied by scouring and channeling, marks the boundary. Poorer outcrops along the southern Colorado Plateau in the western Mogollon Rim region make this contact very difficult to locate.

Wescogame Formation

The Wescogame forms a sheetlike unit that ranges from 100 to 200 feet (30 to 60 m) thick across the Grand Canyon and the western Mogollon Rim region. Lithologic components are similar to those described above for the Manakacha Formation (Figs. 4, 5,

Figure 7. Typical outcrops of Wescogame Formation

7). Sandstone dominates the Wescogame throughout much of the central Grand Canyon and in western Chino Valley; to the east, mudstone increases, whereas limestone increases west of Chino Valley (Blakey 1980; McKee 1982).

SUPAI GROUP AND HERMIT FORMATION   157

McKee (1982) informally divided the Wescogame Formation into a lower cliff unit and an upper slope unit, both traceable throughout much of the Grand Canyon. In the western part of the Mogollon Rim region, the Wescogame typically forms steep, ledgy slopes. Contrary to the implications of McKee's work, the location of the boundary between the Manakacha and Wescogame formations can be difficult to determine, both from a distance and from close range. The unconformity between the two formations varies, and even where noticeable relief is present, the boundary tends to be in a slope- or ledge-forming horizon, where debris from overlying units obscures the contact. The thickness and expression of associated conglomerate also varies and cannot be used to define the horizon in some areas. The topographic criteria provided by McKee (1982) form the most reliable method of separating the two formations. The contact lies near the top of the slope that typically forms the upper part of the Manakacha Formation, a few feet below the overlying cliff unit of the Wescogame Formation.

The upper contact of the Wescogame Formation almost certainly marks the disconformity associated with the Pennsylvanian-Permian boundary (McKee 1982). Although some of the problems discussed above also pertain to locating this contact, the top of the Wescogame Formation generally is easy to locate. Wherever exposures are clear, conglomerate and relief of up to fifty feet (15 m) mark the boundary. Based on regional patterns in the Grand Canyon and in the western Mogollon Rim region, geologists have defined three broad, north-northeast-trending facies belts. The eastern belt, the redbed facies, lies east of a line from near Hermit Basin to the western Verde Valley. It comprises subequal amounts of sandstone and red mudstone. The middle belt, the sandstone facies, extends west to a line running through the Shivwits Plateau region. The westernmost belt is chiefly cross-stratified limestone and limey sandstone. It forms the limestone facies.

Esplanade Sandstone (and Pakoon Limestone)

Sandwiched between the steep ledges and slopes of the overlying Hermit Formation, the Esplanade Sandstone forms one of the most distinctive horizons in the Grand Canyon, especially in the central and western portions. The Esplanade Sandstone contains the highest percentage of sandstone of any of the units in the

Supai Group (Figs. 4, 5, 8). The formation consists of a steadily northwestward-thickening wedge that ranges from 200 to 250 feet (60-75 m) thick in the eastern Grand Canyon and western Mogollon Rim region to over 800 feet (240 m) in the western Grand Canyon and along the Grand Wash Cliffs (Blakey 1980; McKee 1982). The thickness in the western Grand Canyon region includes the Pakoon Limestone.

Although the lithologic components are the same as those previously described in the Manakacha Formation, the dominance of cross-stratified sandstone in many sections distinguishes the Esplanade Sandstone. Most cross-stratified units are dominated by climbing translatent strata, which strongly suggests an eolian origin for the unit. The sandstone units form beds generally five to fifty feet (1.5 to 15 m) thick that are separated by thin, red mudstone; by fine-grained carbonate units; or by prominent, irregular-bedding planes. These planes cause the Esplanade to weather to an irregular or ledgy cliff. Clean outcrops range in color from pale grayish orange to pale reddish orange, but staining from the overlying redbeds in the Hermit Formation causes the unit to appear dark reddish brown.

McKee (1982) informally divided the Esplanade into a basal slope unit and main cliff unit. Typically, the upper part of the main cliff weathers back into a series of ledges, or ledges and slopes. In western portions of the Grand Canyon, the basal slope unit grades westward into the dolomite and limestone of the Pakoon Limestone. The Pakoon thickens from east to west because of increased subsidence and the replacement of siliciclastic units by carbonate units in this direction. The Pakoon Limestone is approximately 300 feet (90 m) thick along the Grand Wash Cliffs. McKee chose not to include the Pakoon formally within the Supai Group, but I would suggest that future stratigraphic work in the area consider assigning the Pakoon Limestone to the Supai Group.

The nature of the upper contact of the Esplanade Sandstone with the overlying Hermit Formation is part of a complex regional problem. McKee (1982, pp. 169-171, 202-203) examined the nature of the contact and concluded that overall evidence suggested the presence of a regional unconformity. At many locations, there is a definite transition between the Esplanade below and the Hermit above. At other locations, the Esplanade is overlain by an erosion surface with thirty to fifty feet (9 to 15 m) of local relief. The Hermit overlies that surface. Still other locations exhibit channeling into

Figure 8. Typical outcrops of Esplanade Sandstone

160   GRAND CANYON GEOLOGY

the Esplanade, but the channels originate from within the Hermit and not from the base. This suggests that the erosion surfaces formed after the deposition of the Hermit began. In a few places, (such as near Toroweap, in the Shivwits area, and around Sedona), cross-stratified sandstone typical of the Esplanade occurs as much as 100 feet (30 m) above the Hermit/Esplanade contact. This suggests that the contact is variable from place to place and that it represents a zone of transition. The erosional relief also may be related to channel cut-and-fill sequences and not to a major regional unconformity. Obviously, there is a great deal that geologists do not know about this contact zone.

The most dramatic facies change in the Esplanade Sandstone occurs at the base, where the lower slope unit grades westward into the Pakoon Limestone in the Shivwits Plateau area (McKee 1982). The upper part of the Esplanade displays a gradual change to increased carbonate content in the west. This change, however, is less sharp than that which occurs in the Manakacha and Wescogame formations. Bedded gypsum occurs near the top of the Esplanade in the Toroweap area (Fig. 5). The gypsum is intercalated with cross-stratified sandstone throughout a zone as much as 200 feet (60 m) thick.

Hermit Formation

Its overall fine-grained nature and relatively poor exposures, as well as a general lack of interest in the unit among geologists, contribute to making the Hermit Formation one of the poorest known units in Grand Canyon. White (1929) provided a comprehensive discussion of the formation, but it centered on the flora. Many other workers have provided local descriptions, but no major, recent, stratigraphic or sedimentologic study is available. This chapter will provide both sedimentologic and stratigraphic description—but primarily at the reconnaissance level.

For the most part, the Hermit Formation is composed of slope-forming, reddish brown siltstone, mudstone, and very fine-grained sandstone. The formation varies from approximately 100 feet (30 m) in thickness in the eastern portion of the Grand Canyon and near Seligman to over 900 feet (270 m) in the Toroweap and Shivwits Plateau areas (McNair 1951). Northward along the Hurricane Cliffs and into adjacent Utah, the sandstone content increases. Here, the Hermit interval generally is assigned to the Queantoweap Sandstone. In the western Mogollon Rim region east of Seligman, the Hermit can be traced eastward through discontinuous outcrops to the Sedona area, where it averages 300 feet (90 m) in thickness (Blakey 1980, Blakey and Knepp 1987).

McNair (1951) pointed out the misnomer of "shale" in the Hermit and proposed the word "formation" instead. Many subsequent workers have followed this suggestion. Overall, the Hermit Formation comprises a heterogeneous, chiefly fine-grained, siliciclastic sequence (Figs. 4, 5, 9). Neither the vertical nor the areal distribution of the various components defines any sharp trends though regional variations are apparent. Most of the lithologic components of the Hermit are present to some extent in the underlying Supai Group. Silty sandstone and sandy mudstone are the most widespread and voluminous components in the Hermit. Silty sandstone ranges from structureless to ripple laminated to trough cross stratified. Structureless units make up ledge-forming beds that average three feet (1 m) in thickness and may or may not contain limy, nodular concretions. Ripple-laminated units display subaqueous, faint- to-prominent ripple cross lamination. Trough cross-stratified sandstone consists of troughs up to several feet across (which may be organized within gently dipping master sets in which the troughs are roughly perpendicular to the inclined master sets). Termed epsilon cross stratification, this structure is associated commonly with point-bar deposition in meandering streams. Rare trough to planar-tabular sets of cross-stratified sandstone, fine grained and well sorted, with climbing translatent strata, occur near the base of the Hermit at several localities within Grand Canyon and in the Sedona area.

Sandy mudstone forms slopes in the Hermit Formation. The mudstone commonly is featureless though rare clean rock outcrops display fine ripple lamination and calcareous nodular concretions. A minor component overall, but locally abundant, is an intraformational conglomerate composed of sedimentary pebbles. Most pebbles are carbonate grains obviously derived from the adjacent carbonate concretions contained within intercalated sandstone and mudstone, but some fine-grained, limey sandstone and siltstone pebbles also are present. The conglomerate occurs both as beds and as an accessory to sandstone units. McKee (1982) reported conglomerate in the Hermit at 14 of his 35 localities in Grand Canyon; it is most abundant in the Sedona area, where 10 to 15 horizons are typical within the 300-foot-thick (90 m) section.

Figure 9. Typical outcrops of Hermit Formation

Most of the Hermit Formation consists of interbedded, silty sandstone and sandy mudstone. At most locations, sandstone is more abundant near the base of the formation, and mudstone increases upward. The two lithologies are interbedded rhythmically, typically with fifteen or more cycles present in most sections. The extent or geometry of individual sandstone bodies is uncertain because of poor exposures. Reconnaissance work suggests an increase in sandstone in the Shivwits Plateau area. This may reflect the increase in sandstone to the northwest, where the Hermit is believed to grade into the Queantoweap Sandstone (Blakey and Knepp 1987). Conglomerate is most abundant in the Sedona area. Current evidence indicates that it decreases in all directions from there.

The upper contact of the Hermit Formation is everywhere sharp and without gradation of any kind. Throughout Grand Canyon and into the Aubrey Cliffs, it is overlain by the Coconino Sandstone. Cracks twenty or more feet deep at the top of the Hermit frequently are filled with the overlying sandstone.

East of Seligman into the western Mogollon Rim region, a substantial sequence of strata not present in the Grand Canyon appears above the Hermit; these strata are defined as the Schnebly Hill Formation. The regional relations shown in Figure 5b are discussed in detail by Blakey (1980) and Blakey and Knepp (1987). Contact with the overlying Schnebly Hill Formation in this area also is sharp, but it lacks the sandstone-filled cracks. The sharp contact may be a major regional unconformity though sufficient paleontological evidence to confirm this hypothesis is not yet available.

PALEONTOLOGY

Both White (1929) and McKee (1982) have examined the faunal and floral content of the Supai Group and the Hermit Formation. Overall, both stratigraphic units are sparsely and sporadically fossiliferous, but we have recorded a broad variety of fossil types.

Trace Fossils

Trace fossils—the trackways, burrows, impressions, resting marks, and feeding marks formed by organisms in sediment and preserved in the rock record—are ubiquitous throughout the Supai and the Hermit. McKee reported their presence, but no comprehensive report on trace fossils has yet appeared. Nondescript bioturbation is the most common form of trace fossils in the Supai Group. The swirls, crinkles, and irregular patterns associated with distinct burrows attest to its abundance. Rocks showing disturbed sediment range from structureless (complete bioturbation) to distinctly burrowed. Burrowed structures are common both in the Supai and the Hermit, most frequently in silty sandstone and carbonate units. Many types of burrows are present, but those occurring most frequently are cylindrical, smooth burrows several millimeters in diameter that are parallel, perpendicular, or oblique to the bedding. McKee also described trackways, mostly attributed to vertebrates, from both the Wescogame Formation and the Esplanade Sandstone. Other miscellaneous trace fossils, including invertebrate trackways, resting marks, bioturbation caused by plant roots, and possible feeding marks, are distributed throughout the Supai Group and the Hermit Formation. Clearly, this is a field with strong future research potential.

Fossil Invertebrates

McKee (1982, Chapters E and F) described and pictured many of the invertebrates in the Supai Group. He reported a fairly rich and varied brachiopod fauna from the Watahomigi Formation. Marine fossils, including locally abundant fusulinids, occur sporadically throughout the Supai—especially in western Grand Canyon. However, the presence of abraded fossils in well-sorted, cross-stratified sandstone or limestone should not be used as a criterion for marine deposition because skeletal grains can be, and were, blown inland and incorporated into eolian deposits.

Flora

Plant remains are widespread but sparsely distributed throughout the Supai Group and the Hermit Formation. The comprehensive reports by White (1929) and McKee (1982) list the location, stratigraphic horizon, and plant taxa found throughout the units, and Blakey (1980) reported plant fossils from the Hermit Formation in the Sedona area. McKee (1982, p. 100) found that although most of the plant remains have little stratigraphic value, they do suggest the presence of broad floodplains developed during times of regressing seas and semiarid-to-arid climates.

AGE AND CORRELATION

Introduction

McKee (1975; 1982) provided us with the first detailed regional account of the age of the Supai Group. He based his age determinations primarily on brachiopods in the Watahomigi Formation and fusulinids scattered throughout the group. Using this information, he assigned to the formations of the Supai Group the following ages: Watahomigi Formation—Morrowan and Atokan; Manakacha Formation—Atokan; Wescogame Formation—Virgilian; Pakoon Limestone—Wolfcampian; Esplanade Sandstone—Wolfcampian (lower part of formation). The age of the upper part of the Esplanade Sandstone is determined by its stratigraphic position.

White (1929, p. 40) assigned the Hermit Formation an age of upper Lower Permian (Leonardian); McKee's work seems to confirm this. He used a fossil plant, *Callipteris arizonae*, found in the Bone Spring Limestone of Leonardian age in southeastern New Mexico as his point of reference. If McKee is correct, the age of the Esplanade Sandstone must be no older than Wolfcampian and no younger than Leonardian.

In the Sedona area, the Esplanade Sandstone and the Hermit Formation lie several hundred feet below the Fort Apache Member of the Schnebly Hill Formation. The Fort Apache Member has yielded conodonts of early late Leonardian age. This paleontologic evidence provides additional support for presuming an early Leonardian age for the Hermit Formation (Blakey and Knepp 1987).

Regional and Worldwide Correlation

In a 1987 study, Blakey and Knepp summarized the regional correlation of Pennsylvanian and Permian rocks of Arizona and adjacent areas. Based on age assignments presented above and using the worldwide geologic time scale of Van Eysinga (1975), they found that the Supai Group correlates with the Westphalian, Stephanian, and Sakmarian.

The Hermit Formation correlates with the Artinskian, which occurs in the standard Carboniferous and Permian units of Europe. In addition, the Esplanade Formation and the Hermit Formation both correlate with the eolian-bearing Rotliegendes of northern Europe.

SUPAI HISTORY

Introduction

Interpretation of the depositional history of Pennsylvanian and Permian redbeds in the Grand Canyon region is plagued by a variety of problems—including sparsity of fossils, the complex cyclic nature of the strata, the fine-grained nature of some units, extensive diagenesis, inaccessibility of some outcrops, and some misconceptions concerning regional stratigraphy and depositional history.

Despite these problems, recent and detailed sedimentologic work (much of it previously unpublished) suggests the need to revise our interpretation of the depositional history of the Supai Group. Earlier studies suggested a fluvial, deltaic, beach, shallow-marine, or estuarine origin for Supai sandstone bodies. New evidence indicates that an eolian origin is much more likely. McKee (1982) did not consider an eolian origin for any sections of the Supai Group. The presence of carbonate grains, including locally abundant fossil fragments, may have prevented his recognition of the eolian environment. Desert varnish and extensive diagenesis of some beds further obscure some important eolian characteristics.

Eolian Characteristics

Until fairly recently, eolian criteria consisted chiefly of the textural maturity of quartz sand, coupled with thick set size and a high-angle, cross-strata dip. Although many ancient eolian sandstones display these characteristics (the Coconino Sandstone, for example), these criteria are unreliable. Some eolian sandstones lack large-scale, high-angle sets. In some cases, sandstones with these characteristics are noneolian. Hunter (1977) found that the most reliable eolian indicators are not set size, angle of dip, or even cross stratification, but the nature of the individual laminae and strata that compose the sandstone. Migrating and climbing wind ripples, regardless of their orientation, form the inversely graded laminae (generally less than several millimeters thick) that Hunter termed "climbing translatent strata." This strata type can form and be preserved in many parts of the eolian environment—including on most parts of dunes, between dunes (interdunes), and on duneless areas of blowing sand termed sandsheets.

Many dunes contain one or more active slipfaces that are characterized by the periodic avalanching of loose sand. Hunter (1977) described deposits formed by this process as structureless strata (as much as several centimeters thick) at or near the angle of repose of dry sand (30-34°, though generally 24-26° in ancient sandstones because of compaction). He termed these "sandflow strata." Some small dunes, and most dunes without slip faces, may leave only climbing translatent strata in variously shaped sets with widely ranging angles of dip. Particularly common are trough-shaped sets several feet thick and tens of feet wide that were produced by crescentic-shaped dunes or troughlike scours (blow outs) on dunes or sandsheets. Many dunes with slip faces produce avalanche strata that fade out down the dune, where they are replaced by climbing translatent strata. This toe of the avalanche, which is termed "sandflow toe," is an easily observed eolian indicator.

Eolian sandstones typically consist of one or more facies that relate to their type and position within the eolian sand sea (erg). The central portion of most sand seas contains dunes and larger bedforms called draas. Marginal sand seas include small, possibly scattered dunes and sandsheets, and the most distal areas may consist of broad sandsheets with few, if any, dunes. Sandsheets and small dune fields may filter into adjacent environments, such as fluvial plains, dry lake beds, and coastal plains. Facies formed in central sand-sea areas typically consist of cross-stratified sandstone comprised of sandflow strata and climbing translatent strata with interdune deposits of horizontal climbing translatent strata or various types of structureless to wavy-bedded sandstone or silty sandstone. Marginal sand-sea deposits typically display smaller sets of cross-stratified sandstone and sandsheet deposits. These consist of horizontal to irregular beds of climbing translatent strata. The most complete discussion of the distribution of all known ancient eolian sandstone units and their facies can be found in Blakey and others (1988).

Controls on Deposition

Controls on deposition during Supai and Hermit time were varied and complex. They included a complicated tectonic regime, rapid rise and fall in sea level, and an influx of sand from the north.

The tectonic elements that affected sedimentation during the late Paleozoic have been addressed in several recent papers (Blakey 1980, 1988; Blakey and Knepp 1987). Areas in which the crust subsides relatively rapidly have thicker deposits than areas of less subsidence. The former tend to have more marine deposits and fewer redbeds. Areas of less subsidence are characterized by thin, red, terrigenous clastic deposits with few fossils and abundant unconformities. During the late Paleozoic and early Mesozoic in the Southwest, such deposits tended to form in fluvial, lacustrine (lake), and coastal-plain environments. Areas of greater subsidence are characterized by deposits of limestone and thin sandstone and mudstone. They typically are grayish in color and fossiliferous. Marine and shoreline deposits dominate.

Geologists classify the resulting tectonic elements by their shape and their relation to adjacent elements. Circular to oblate areas with relatively high rates of subsidence are termed basins. Areas of intermediate subsidence rates covering broad areas are shelves, and those more parabolic in shape are termed embayments. Arches are areas of little subsidence; typically, they interrupt brief periods of slight uplift. Areas dominated by uplift are referred to as upwarps or uplifts.

During the formative period of the Supai and Hermit, northern Arizona was bisected by a northeast-trending element of relatively little subsidence, the Sedona arch (Fig. 10). To the northwest, the Grand Canyon embayment dominated the Grand Canyon region. East-central Arizona was the site of the Holbrook basin and the Mogollon shelf. Northeastern Arizona was dominated by the Defiance uplift during the Pennsylvanian or the Defiance arch in the Permian. To the north, in southeastern Utah, was the Paradox basin. During periods of relatively high worldwide sea level, the seas tended to encroach on Arizona from the west into the Grand Canyon embayment, from the east and south across the Mogollon shelf, and from the northeast into the Paradox basin. Although the uplifted areas (i. e., arches) received some sediment, much of it was nonmarine. Because great fluctuations of sea level (chiefly because of glaciation in the southern hemisphere) characterized the late Paleozoic, deposits tended to be cyclical in nature. Marine-dominated cycles occurred in basins, and continental-dominated cycles occurred across the arches. Movement on arches, uplifts, and basins altered a landscape already shaped by changes in sea level. Therefore, the resulting pattern of deposition is very complex.

SUPAI GROUP AND HERMIT FORMATION   169

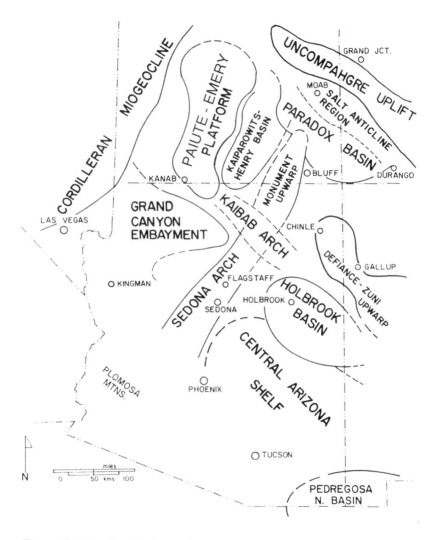

Figure 10. Map showing late Paleozoic tectonic elements of northern Arizona and vicinity

The influx of quartz sand into the Colorado Plateau region from the north also complicates the setting. The sand tended to accumulate between the arches and adjacent shelves or basins as extensive eolian deposits (Blakey 1988). When sea level was relatively low, eolian deposits became widespread, often expanding into basins and across shelves; when the sea level was high,

170   GRAND CANYON GEOLOGY

eolian deposits, if present, were confined to narrow belts along the flanks of arches (Chan and Kocurek 1988).

The Grand Canyon and the western Mogollon Rim region lay in a somewhat intermediate position between more negative areas to the northwest and higher ones to the southeast. Not unexpectedly, the resulting sedimentation is intermediate in character; cyclic sedimentation with more continental aspects is present in the east while cyclic sedimentation with more marine aspects is present in the west.

The general structural grain of northwestern Arizona tended southwest-northeast during Supai and Hermit time. Shorelines and marine trends should have paralleled this pattern, and fluvial deposits should have flowed northwesterly, perpendicular to this grain. Indeed, marine and shoreline patterns do parallel the trend, but the sandstone bodies that McKee (1982) suggested might be of fluvial origin show a paleocurrent reading to the south and southeast; this is as much as 180° from the expected trend. We have not documented the trends of Hermit Formation sandstone bodies.

Any one of the above controls on deposition was sufficient to produce a complicated depositional pattern. Together, they combined to produce a heterolithic record that varies abruptly, both vertically and laterally. The summary presented below must be considered preliminary. We have much to learn about the depositional history of the Supai Group and the Hermit Formation.

Depositional Environments

Recent studies tend to confirm the long-standing belief that the Supai Group was deposited on a broad coastal plain. Individual depositional settings ranged from shallow marine to continental. However, the recent identification of eolian-deposited sandstone units throughout the section necessitates a reinterpretation of Supai depositional history. Most of our knowledge of the Supai is based on the work of McKee, who provided a general description of sandstone bodies in the Supai Group (1982, pp. 260, 261). He interpreted the large foresets planes in the Manakacha Formation as the fronts of large, subaqueous sand sheets or small Gilbert deltas in areas strongly affected by tides. Traditionally, geologists have interpreted sandstone bodies of the Wescogame Formation as having formed in a high-energy, fluvial environment. Marine bioclastic debris was thought to have been trapped in an estuarine setting.

McKee found the Esplanade the most difficult unit to interpret and offered a compromise of mixed fluvial, estuarine, and shallow-marine origins. It should be pointed out that although McKee's Supai publication appeared in 1982, the work actually started in the late 1930s and continued into the 1970s. Delays in publication created a large gap between the time when the work was written and the time it was published. Therefore, much of the state-of-the-art sedimentology of the late 1970s and the 1980s is not reflected in the publication; this is especially true of eolian sedimentology. McKee's monograph will continue to be a valuable source of information concerning the Grand Canyon Supai, but some of the interpretations, particularly concerning depositional environments, must be considered out of date.

Given the eolian criteria discussed earlier in this chapter, most cross-stratified sandstone units in the Manakacha and Wescogame formations and the Esplanade Sandstone now are regarded as eolian in origin. Evidence for this includes the fact that wind-ripple laminae (climbing translatent strata) are ubiquitous throughout these units in the eastern and central Grand Canyon and the western Mogollon Rim regions (Fig. 11). Most sets of cross strata are composed chiefly or entirely of climbing translatent strata. There is little or no evidence, however, of avalanche-formed sand-flow strata. Studies of the eolian deposits in the Supai have not progressed enough to explain this, and several theories are possible. Kocurek (1986) reported that dunes with broad aprons (plinths) commonly do not preserve slipface deposits in the sedimentological record; longitudinal dunes, star dunes, and oblique dunes commonly have large aprons. Reversing dunes also tend to lack slipface deposits. Other late Paleozoic eolian deposits contain a high percentage of wind-ripple strata. They also contain at least locally abundant sand-flow strata (Blakey and others 1988). But only in the Esplanade Sandstone of the Oak Creek Canyon area are sand-flow strata developed extensively in the Supai Group. Some cross-stratified sets (perhaps forty percent in Grand Canyon and sixty percent in the western Mogollon Rim) have been so affected by diagenesis that the stratification type cannot be determined. A portion of these could contain sand-flow strata, and some sets could be noneolian in origin; however, close association with, and similarity to, sets with wind-ripple laminae strongly suggest that many or most unidentifiable sets formed as climbing wind-ripple deposits.

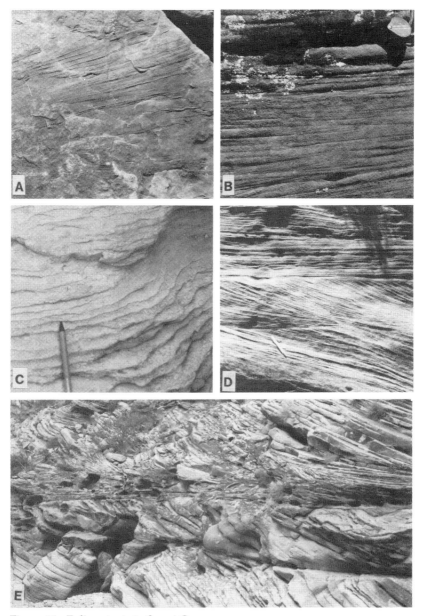

Figure 11. Eolian strata in Supai Group

The geometry of eolian sandstone bodies in the Supai Group is somewhat unusual. Many well-known eolian sandstones, such as the Coconino Sandstone and the Navajo Sandstone, are mostly or entirely cross-stratified sandstone. Redbeds, conglomerate, and limestone are rare or absent. Blakey and others (1988) have pointed out the problems of stereotyping eolian deposits as simple, pure sandstone bodies. Most are not. Individual eolian bodies range from irregular sheets to more common broad lenses, to wedges, and even small pods (Fig. 12). Bodies are separated from each other by red siltstone and mudstone, noneolian sandstone, limestone, and, rarely, conglomerate. Undoubtedly, this heterogeneous assemblage of lithologies and complex geometry is the chief reason that the Supai has not been interpreted previously as eolian in origin. Coupled with the high carbonate content and locally abundant sand-sized marine fossil grains in some eolian sandstone, the Supai eolian bodies are far removed from the "typical" eolian sandstone.

The origin of some noneolian deposits in the Supai Group is not clear. Bioturbated sandstone can form in any number of environments—including eolian, fluvial, shoreline, and shallow marine. It is possible that this material has been reworked by organisms in a marine or continental setting. Horizontally laminated to low-angle, cross-stratified sandstone may have formed as plane-bed laminae in shallow streams, in beach swash zones, or as shallow-marine storm deposits. Fine-grained redbeds are particularly enigmatic. Their potential origin can range from fluvial flood plain to lacustrine and sabkha; through low-energy, shoreline (supratidal mudflats, lagoons); and into nearshore, shallow marine.

Because the lower contacts of eolian sandstone bodies are sharp and probably disconformable (and upper contacts commonly are the same), stratigraphic position is an unreliable criteria for diagnosing the environment of noneolian deposits. At most locations, a few sequences display the following cyclic pattern: sharp basal contact abruptly succeeded by cross-stratified, eolian sandstone; the eolian sandstone grades upward into bioturbated, micritic limestone that is overlain conformably by red mudstone, horizontally stratified sandstone, or a combination of these. Unfortunately, it cannot be determined whether these cycles are transgressive, regressive, or totally continental in origin. If transgressive, the most logical interpretation would be eolian, succeeded by low-energy shoreline, to local beach deposits. If regressive, the eolian likely would be succeeded by fluvial and lacustrine or sabkha deposits. If the cycle is continental (and not necessarily related to advance or retreat of the sea), the succession

Figure 12. Eolian sandstone bodies in Supai Group

likely represents ingress and egress of fluvial and lacustrine environments into the eolian dune complex. Clearly, more work is needed on this topic.

Evaporite deposits in the Esplanade Sandstone, chiefly gypsum, could have formed either in coastal or continental sabkha environments. The evaporites are interbedded with redbed and eolian sandstone in the Toroweap area of western Grand Canyon.

We know now that not all carbonate units in the Supai Group are of marine origin; some arenaceous, cross-stratified limestones are of eolian origin. However, many Supai carbonates clearly are of shallow-marine origin. Micrite-rich, fossiliferous carbonates were deposited in open-marine, relatively low-energy environments (McKee 1982, p. 345). Carbonate grainstone and packstone formed in higher energy enviroments, such as on carbonate shoals and in tidal channels. Unfortunately, the detailed relations between marine limestone and eolian sandstone and arenaceous limestone are yet to be determined. McKee (1982) has documented the abundance of marine carbonate in western facies of the Manakacha and Wescogame formations as well as Pakoon Limestone, but we have no complete study of their presumed interbedding with eolian sandstone. These relationships would be exposed best in the Shivwits Plateau region of the western Grand Canyon.

Documentation as to how far west eolian sandstone and arenaceous limestone extends also is nonexistent. Based on similarities to eolian strata that I have seen (photographs included in McKee's 1982 study), it seems likely that eolian strata exist in western Grand Canyon and northward along the Grand Wash Cliffs.

It is equally difficult to determine the depositional setting of the Hermit Formation. White (1929) used fossil flora, general lithology, and a regional setting to postulate deposition by sluggish streams on a broad, low-lying, arid coastal plain. Details of Hermit deposition in the Grand Canyon have yet to appear in the literature. Work currently in progress in the Sedona area is providing some detail of Hermit deposition for the western Mogollon Rim region. A brief examination of scattered Hermit outcrops in the Grand Canyon tends to confirm White's general conclusions.

This information, combined with more detailed work near Sedona, provides a few more specifics on the nature of Hermit streams. Several types of stream deposits have been identified. These include: narrow, channel-shaped bodies of limestone-pebble conglomerate that formed in local, arroyolike streams as well as broader sheets of conglomerate that probably formed in wider arroyos. The conglomerate was derived from caliche deposits on adjacent alluvial plains. Sheets of limy and nodular structureless-to-ripple-laminated sandstone were deposited as relatively unconfined stream deposits or as broad levee deposits. Trough, cross-stratified sandstone (with or without associated conglomerate) laid down in broad, gently dipping, irregular sheets (epsilon cross stratification) represents the deposits of larger, more perennial, meandering streams. Red mudstone was deposited in low-lying areas as overbank deposits. Light-colored, large-scale, high-angle, cross-stratified sandstone (with climbing translatent strata) represents minor eolian dune deposits.

From the Sedona area westward to an area near Seligman, the types of deposits described above are distributed irregularly throughout the Hermit. In central and eastern Grand Canyon, geologists are able to identify all of these types. However, three general facies predominate.

The lower portion of the Hermit on both the Kaibab and Hermit trails consists of ledges and slopes of structureless and trough cross-stratified sandstone. Exposures there are not as good as some in the Sedona area, but the sand bodies are similar to the meandering stream deposits and unconfined stream deposits there. The bulk of the Hermit in eastern and central portions of the Grand Canyon consists of weak, ledge-forming, silty, faintly ripple-laminated sandstone and slope-forming mudstone. Generally, ten to fifteen cyclic alternations of these units are present. The ledges probably represent sluggish, shallow, perhaps unconfined stream deposits, and the slopes primarily are overbank deposits. Whether these cycles were formed by climatic fluctuation, by the migration of stream deposits, or by tectonic cycles is unknown.

Depositional Models

The high carbonate grain content, associated marine carbonates, and regional setting strongly suggest a coastal-plain setting for the eolian dune deposits in the Supai Group. Carbonate sand was derived locally by onshore winds from adjacent, bioclastic debris. The quartz sand was transported by wind and longshore currents into the Grand Canyon region from the north (Blakey and others 1988). To the west was a broad, shallow, epicontinental sea, and to the east were low-lying, coastal plains and low uplands (Fig. 13). The repetitive Pennsylvanian and Early Permian marine transgressions and regressions moved across this broad plain, but influxes of eolian sand formed obstacles to widespread transgression. During regression, the eolian dunes spread across the broad flats. The ensuing transgression interrupted eolian deposition and apparently prevented the development of large eolian sand seas (ergs).

The battle between dune and sea was repeated numerous times, and only a small fragment of the sedimentological record has been preserved. Sluggish and probably ephemeral streams continually washed fine-grained, terrigenous, and clastic debris into the area. The mud and fine-grained sand were trapped in lagoons, on tidal flats, and in river channels. The extremely arid climate provided a reflux action that pumped salty groundwater upward. This resulted in the precipitation of calcium carbonate and other minerals as deposits on the earth's surface. Some of these nodules were reworked later into conglomeretic deposits. Such a setting clearly was the site of a great deal of reworking of sediment.

Over time, the wind altered marine and fluvial deposits into dune deposits and eolian dust (loess). Rivers and advancing seas also reworked these eolian deposits. This intermixing and reworking of environments created an array of conditions that might seem incompatible within a given environment. For example, fluvial and tidal deposits might be better sorted than expected in one area because of the activity of the wind. Eolian deposits frequently display geometric patterns that are more typical of fluvial or shallow-marine deposits. The heterogeneity of much of the Supai is directly related to this complex intermixing of environments coupled with a broad range of available depositional material.

Hermit stream deposits do not fit readily into established fluvial models. The broad, low-gradient, arid plain was crossed by several kinds of streams, probably both perennial and ephemeral. Local coarse conglomerate indicates that streams occasionally had sufficient energy to transport larger clasts. Widespread and small dune fields were present on the floodplain, but the great eolian faucet from the north had been shut off temporarily.

GRAND CANYON GEOLOGY

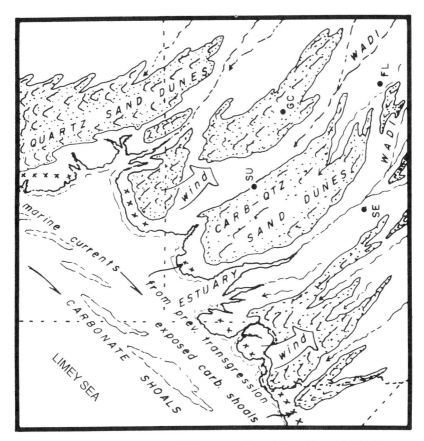

Figure 13. Map showing general depositional model and hypothetical paleogeography of eolian units in Supai Group. Map depicts region during regressive sequence. Carbonate sand is derived from exposed, older carbonate shoals; quartz sand is fed into the region from the north.

LATER HISTORY

The incredible range of Supai depositional environments and the processes within them are only generally understood. Clearly, many problems remain and some may never be solved. Patterns of transgression and regression have yet to be documented. Hermit deposition is poorly understood, primarily because of a lack of regional and local data. Despite these problems, the ideas, models, and conclusions offered here permit a preliminary analysis of Supai and Hermit depositional history.

SUPAI GROUP AND HERMIT FORMATION 179

Morrowan History

The lower two-thirds of the Watahomigi Formation was deposited in shallow-marine and low-energy shoreline environments (McKee 1982). The Morrowan portion of the Watahomigi consists generally of a lower, fine-grained, siliciclastic sequence and an upper, fossil-bearing limestone. Poor exposures and a lack of reliable criteria prevent detailed analysis of the lower slope. A basal conglomerate suggests that continental or shoreline processes reworked weathered Redwall Limestone. It is possible that the overlying mudstone accumulated in low-energy, shoreline environments or on a broad coastal plain. A rise in sea level (or subsidence in the Grand Canyon region) permitted a period of widespread, clear-water, carbonate deposition. Local intercalated mudstone suggests that there were periods of influx by fine clastics. These may have been caused by fluctuating sea levels. A widespread erosion surface at or near the top of the limestone suggests a period of erosion near the close of Morrowan time (McKee 1982, p. 160).

Atokan History

The upper Watahomigi Formation consists of thin limestone, redbeds, and minor sandstone. This suggests an advance of the sea into the Grand Canyon area. However, a strong influx of eolian material from the north initiated a trend of eolian deposition that would continue periodically for over 150 million years (Blakey and others 1988). During a major regression, quartz and carbonate sand blew across the landscape. It was derived locally from exposed, recently deposited, shallow-marine materials. Quartz sand from the north was transported across the Paiute and Emery arches and was funneled into the Grand Canyon embayment (Fig. 13). Here, the sand was deposited in coastal dunefields. We are not certain whether the eolian sand was deposited in a broad coastal erg or in a series of small dunefields. However, eolian deposits were planed off by marine transgressions and perhaps minor fluvial events. The next regression saw a return to eolian deposition. Sharp bases to eolian sandstone units suggest a rapid change to eolian conditions.

DesMoinesian and Missourian History

No rocks of these ages have been reported from the Grand Canyon or the western Mogollon Rim regions. This period of time apparently is represented by the unconformity between the Manakacha and Wescogame formations. Whether sediments of Middle Pennsylvanian age were deposited and later removed or were never deposited extensively in the region is unknown.

Virgilian History

The pattern of deposition established during Atokan time was renewed during deposition of the Wescogame Formation. Repeated transgression and regression caused numerous depositional cycles. The widespread Wescogame cliff unit, believed to be mostly eolian in origin, may represent the development of a large erg, or of several ergs, across the region (Fig. 13). Redbeds at the top of the Wescogame Formation may have formed during a period of decreased eolian influx.

Another period of erosion followed. This is indicated by the unconformity at the presumed Pennsylvania-Permian boundary. Apparently, a lowering of the sea level caused an incision of arroyolike streams into the underlying Wescogame Formation.

Wolfcampian History

During this period, the sea advanced into the region at least as far east as the central Grand Canyon. The Pakoon Limestone formed in a variety of clear-water, shallow-marine environments. Farther east, coastal-plain and minor eolian deposits are represented in the lower slope unit of the Esplanade Sandstone. Later in the Wolfcampian, a vast blanket of eolian sand spread southward across the Colorado Plateau region. Forming the Cedar Mesa Sandstone in southeastern Utah, the Queantoweap Sandstone in southwestern Utah, and the Esplanade Sandstone in northern Arizona, eolian deposits of Wolfcampain age blanketed much of the Southwest (Blakey and others 1988).

Interruptions in eolian deposition in the Esplanade Sandstone may have been caused by a rise in sea level, influxes of fluvial activity, or both. East of the Sedona arch, eolian deposits became sparse to absent, and a period of fluvial deposition dominated the landscape.

As eolian conditions waned, fluvial deposits spread westward into the Grand Canyon and the western Mogollon Rim region. Some fluvial channels were incised into the underlying eolian sandstone, but in many locations, a general transition to fluvial conditions is evident.

Leonardian History

Most likely beginning in the Late Wolfcampian and continuing into the Leonardian, low-energy, fluvial conditions developed. Based on the distribution of the Hermit and coeval deposits in Utah, Colorado, New Mexico, and eastward into the High Plains, geologists have concluded that this marked a period of extremely widespread fluvial deposition. Deposits farther east were left behind by streams that had steeper gradients and carried coarser debris. As the streams crossed the broad, low-lying coastal plain, they carried fine material and locally derived carbonate clasts. A few, scattered dune fields persisted into this period of time, but in general, the great eolian episode was brought to a temporary halt.

Eolian deposition was renewed in the Leonardian, first with the development of eolian deposits in eastern and central Arizona in the De Chelly Sandstone and the Schnebly Hill Formation and then across most of northern and central Arizona, including the Grand Canyon region.

SUMMARY

Some of the conclusions concerning the depositional history of the Supai Group and Hermit Formation are preliminary and probably will alter with further work. However, one aspect seems clear: eolian depositional processes played an important role in the formation of the Supai Group and parts of the Hermit Formation. The Atokan eolian deposits in the Manakacha Formation may represent the very earliest record of the vast eolian systems that were to dominate the Southwest until the Late Jurassic.

CHAPTER 10

COCONINO SANDSTONE

Larry T. Middleton, David K. Elliott,
and Michael Morales

INTRODUCTION

The Early Permian Coconino Sandstone is one of the most conspicuous formations in the Grand Canyon. The high-angle, sweeping sets of cross stratification, which record the southerly advance of very large dunes, are visible at great distances throughout the area. Ironically, few geologists, except for McKee (1933) and Reiche (1938), have studied the Coconino. The early work of these two, however, correctly identified the eolian (or wind-blown) origin of the Coconino. These Sahara-like dunes were part of an enormous desert that once extended north into Montana.

During the last ten years, a number of studies on modern and ancient sand seas, or ergs, have increased greatly our understanding of eolian bedform dynamics. At the same time, these studies have facilitated the development of facies models for eolian depositional systems. It is possible now to apply the results of these studies of stratification styles to the Coconino Sandstone and to discuss possible dune morphologies and migration paths. Additionally, we can describe trace fossils in the Coconino and relate them to the physical environments of deposition. This provides us with a more comprehensive reconstruction of the erg environments.

Although the cliff-forming nature of the formation can present certain logistical problems, the exceptional three-dimensional exposures in the Grand Canyon afford the sedimentologist and ichnologist the opportunity to examine the details of one of the more classic and interesting eolian deposits on the Colorado Plateau.

Since the results reported here are preliminary, it is our hope that they will spur more detailed studies of lateral and vertical facies relationships in the Coconino Sandstone, as well as initiate more effort at documenting the variety and distribution of the trace fossil assemblages.

REGIONAL STRATIGRAPHIC RELATIONSHIPS

The Middle Leonardian Coconino Sandstone crops out throughout much of the southern Colorado Plateau south of the Utah-Arizona border. The most exceptional exposures are in the Grand Canyon and the Marble Canyon areas and along the Mogollon Rim south and southeast of Flagstaff. Darton (1910) named the formation for exposures in Coconino County, Arizona. Although there have been a number of controversies regarding Permian stratigraphy in northern Arizona, recent studies have established some regional stratigraphic relationships (Fig. 1) (Cheevers and Rawson 1979; Rawson and Turner-Peterson 1980; and Blakey and Knepp, in press).

In the Grand Canyon region, the Coconino Sandstone disconformably overlies the Permian (Wolfcampian-Leonardian) Hermit Formation. Throughout the canyon, the Coconino is overlain by, or intertongues with, the Permian Toroweap Formation. In easternmost localities where the Toroweap is not present, the Kaibab Formation conformably overlies the Coconino. The Coconino Sandstone grades eastward into the Glorieta Sandstone in western New Mexico. To the south, along the Mogollon Rim, it intertongues with the Schnebly Hill Formation and is transitional into the overlying Toroweap Formation or is sharply overlain by the Kaibab Formation.

Regional structural features (Fig. 2) control the marked thickness variations in the Coconino Sandstone. In the western part of the Grand Canyon near the Grand Wash Cliffs, the formation is 65 feet (20 m) thick, but it thins progressively to the west. The

Coconino thickens in the central part of the canyon and is over 600 feet (183 m) thick near Cottonwood Creek. Eastward and northward, the Coconino thins to 57 feet (17 m) in Marble Canyon before pinching out in the vicinity of Monument Valley. The Coconino, likewise, wedges out in southwestern Utah.

McKee (1933, 1974) reported a maximum thickness of nearly 1,000 feet (305 m) near Pine, Arizona, along the Mogollon Rim (Fig. 2). This rapid southerly thickening probably is related to increased subsidence along the southern boundary of the Coconino depositional basin. The lithologies, textures, and sedimentary structures in the Toroweap Formation are similar to those of the Coconino toward the south. Hence, some of the thickening might reflect the presence of Toroweap equivalents in the upper part of the Coconino.

Blakey and Knepp (in press) proposed that the Sedona arch (Fig. 2) controlled facies distributions, as well as thickness variations, in several of the late Paleozoic units in northern Arizona. This seems likely since the Coconino thickens across and to the east of the Sedona arch, and it is in this area that the Toroweap Formation undergoes facies changes into the Coconino Sandstone.

ICHNOLOGY

Introduction and Preservation

Since body fossils have not been reported from the Coconino Sandstone, invertebrate and vertebrate trace fossils (Fig. 3) remain the only evidence of the organisms that lived in the Coconino desert. They are also the only data on which to base studies of fossil diversity, morphology, and paleoecology. In particular, trace fossils have been important evidence in the debate on the depositional environment of the Coconino Sandstone. Work by Brady (1939, 1947), McKee (1944, 1947), and Brand (1979) has shed a considerable amount of light on the formation of the traces though differences in interpretation still remain.

Brady (1939, 1947) carried out a series of experiments with small vertebrates and invertebrates and pointed out similarities between the fossil traces and those formed by modern scorpions, millipedes, and isopods. In addition, he found that in wet or slightly moist sand, no trace was left by the scorpions he was working with but that they made clear impressions in dry sand.

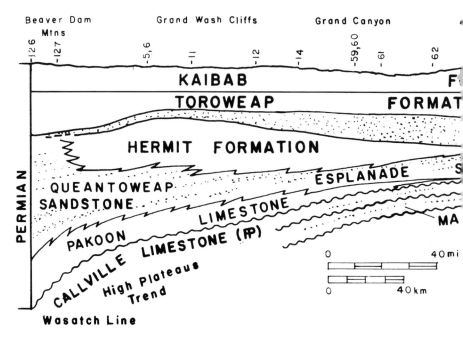

Figure 1. Regional stratigraphic relationships of Permian strata along the southern margin of the Colorado Plateau (after Blakey and Knepp)

In the 1940s, McKee conducted a series of experiments designed to duplicate the Coconino tracks. He filled a long trough with sand that formed a hill in the center. A variety of small vertebrates and invertebrates then were induced to walk along the sand and over the hill. By varying the slope of the hill and the moisture content of the sand, McKee was able to test the trace-forming capabilities of a variety of animals in a number of different environments. He tested the animals on a variety of surfaces—including dry sand, damp sand, saturated sand, and even sand that had been soaked and then allowed to dry. He found that only the largest animals tested (chuckwalla lizards) were able to make tracks in wet sand or crusted sand, and even then, the tracks were not as clear as those formed in dry sand. Smaller animals, such as millipedes and scorpions, were unable to make tracks in wet sediment and left clear traces only in dry, loose sand. He also found that at slopes below 27°, both uphill and downhill tracks were likely to be preserved. Avalanching, however, tended to destroy tracks on steep slopes.

GRAND CANYON GEOLOGY

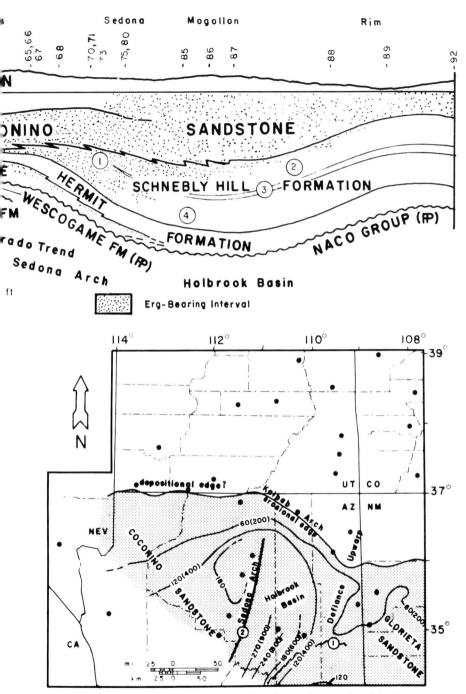

Figure 2. Isopach map of the Coconino Sandstone illustrating thinning away from the Sedona arch (after Blakey and Knepp)

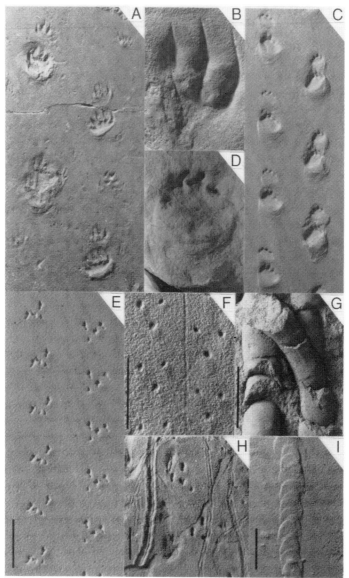

Figure 3. Vertebrate and invertebrate trace fossils from the Coconino Sandstone. (All specimen numbers refer to the Museum of Northern Arizona, Flagstaff, Geology Collection.) A. *Laoporus* sp. (V3470): notice the four toes on the forefeet prints and the five toes on the hindfeet prints; B. *Baryopus tridactylus* (V3360); C. *Laoporus* sp. (V3472): notice the sand humps behind the tracks, indicating that the animal walked uphill, pushing back loose dune sand; D. *Baropus* sp. (V3388); E. Spider track (N3089); F. *Paleohelcura dunbari* (N3666); G. *Scolecocoprus cameronensis* (N3707); H. *Diplodichnus biformis* (N3657); I. *Scolecocoprus arizonae* (N3655).

GRAND CANYON GEOLOGY

Based on previous studies of conditions necessary to preserve surface features in dune sands, McKee (1945) suggested that the tracks in the Coconino Sandstone were formed initially in loose, dry sand that was dampened subsequently before being covered. The mists and fogs that intermittently are present in areas of coastal sand dunes would provide a suitable means of dampening the surface. Recent work by Brand (1978, 1979), however, suggests that clear traces with morphologies similar to those found in the Coconino Sandstone can be generated on sandy surfaces submerged in standing water. It may be, therefore, that a study of trace morphology alone is not sufficient to distinguish between sand surfaces exposed to subaqueous or subaerial conditions. Further experimentation may help to resolve this problem.

Invertebrate Trace Fossils

In 1918, Lull described trace fossils from the Coconino Sandstone that he collected near the Hermit Trail in the Grand Canyon. Subsequently, C.W. Gilmore of the U.S. National Museum carried out more detailed work in the same area. Gilmore then published his results in a series of papers (1926, 1927, 1928), describing both vertebrate and invertebrate tracks. In these studies, he described 10 genera and 17 species, of which the following five genera and species clearly were invertebrate: *Mesichnium benjamini*, a trail consisting of two parallel lines of footprints with a median row of suboval, regularly spaced depressions; *Octopodichnus didactylus*, a trail consisting of alternating sets of impressions in groups of four; *Paleohelcura tridactyla*, a trail consisting of alternating sets of three prints with a median tail drag; and *Triavestiga niningeri*, a trail composed of paired, longitudinal grooves. Gilmore ventured little opinion as to the identity of the trail-formers beyond suggesting that *Octopodichnus* and *Paleohelcura* showed some similarity to tracks made by modern crustaceans.

Brady (in a series of publications from 1939 to 1961) continued work on these tracks. He described new traces, provided insights into the conditions necessary for their formation, and identified the trace-formers. He pointed out (1939, 1947) that the *Paleohelcura* trails were very like those formed by the modern scorpion (*Centruroides*) in dry sand when the temperature was about 60°F (15°C). He also showed, however, that with variations in temperature and surface conditions, the same animal could leave a variety of traces—sometimes impressing two, three, or four feet on each side and leaving no tail drag, an intermittent one, or a complete impression of the tail. Based on this information, he showed that a variety of trails in the Coconino Sandstone were simply variations on *Paleohelcura*. It is clear from this work that two of the ichnogenera described by Gilmore (1926), *Trivestiga* and *Mesichnium*, represent variations on *Paleohelcura* and do not merit the status of separate ichnogenera. In addition, Brady recognized one new ichnospecies of *Octopodichnus*, *O. minor*.

Brady (1947) was able to show that some two- and three-grooved trails were very similar to those formed by modern millipedes as they move up and down sand slopes. He named these *Diplopodichnus biformis* (Fig. 3h) and pointed out that the traces attributed by Gilmore (1926) to *Unisulcus* were identified wrongly. The traces attibuted to *Unisulcus* are slender, smooth, and wormlike and should be included in *Diplopodichnus*. Trails similar to those formed by the modern isopod *Oniscus* he named *Isopodichnus filiciformis*. However, he discovered subsequently that the name was already in use, and he renamed the ichnogenus *Oniscoidichnus* (Brady 1961). Traces that Brady attributed to oligochaete fecal pellets he named *Scolecocoprus cameronensis* and *S. arizonensis* (Fig. 3g and i). He described them as consisting of an overlapping series of pellets and attributed the sometimes grooved appearance of the lower surface to traces left by spines on the ventral surface of the oligochaete. It is more probable that these represent the infillings of burrows, the spreite representing what was thought to be the rounded ends of the fecal pellets. In 1961, Brady also described an additional species of *Paleohelcura*, *P. dunbari* (Fig. 3f).

Alf (1968) carried out the most recent work on invertebrate traces from the Coconino Sandstone, describing a trail composed of alternating sets of four prints (Fig. 3e). These clearly were different from *Octopodichnus*, which consists also of alternating sets of four prints, and Alf was able to show from experiments with modern tarantulas and wolf and trapdoor spiders that the most likely trace-former was a spider. He did not name the trace, however, though clearly it is a separate ichnogenus. The Coconino Sandstone's invertebrate trace fossils are listed in Table 1.

Vertebrate Trace Fossils

Lull (1918) produced the first description of tracks attributed to tetrapod vertebrates in the Coconino Sandstone. The tracks consist of forefoot prints with four toes and hindfoot impressions with five toes. Both of these tracks are relatively small, 05 - 1.5 inches (1.5 - 3.0 cm) in length, and show claw marks. Lull assigned the tracks to two ichnospecies, *Laoporus schucherti* and *L. noblei*. The rest of the vertebrate trace fossils found in the Coconino Sandstone were described by Gilmore in the late 1920s. On the basis of tracks similar to those described by Lull (though different in detail), Gilmore named the following ichnotaxa: *Agostopus matheri* and *A. medius*; *Allopus? arizonae*; *Amblyopus pachypodus*; *Baropezia eakini*; *Baropus coconinoensis* (Fig. 3d); *Barypodus palmatus, B. metszeri, and B. tridactylus* (Fig. 3b); *Dolichopodus tetradactylus*; *Nanpus merriami* and *N. maximus*; and *Palaeopus regularis*. Gilmore (1926) also redescribed *Linopus? coloradensis* and referred it to the ichnogenus *Laoporus*.

Baird (1952 and in Spamer 1984) and Haubold (1984) re-evaluated the vertebrate tracks from the Coconino Sandstone. Baird (1952) decided that *Allopus? arizonae* and *Baropus coconinoensis* are junior synonyms of *Baropezia eakini*. Later, he transferred *Nanopus maximus* to *Barypodus metszeri* and attributed the ichnospecies *N. merriami* and *Dolichopodus tetradactylus* to the ichnogenus *Laoporus*. However, Haubold (1984) maintained the generic distinction of *Dolichopodus*. He also had doubts about the taxonomic validity of *Amblyopus pachypodus* and considered *Agostopus, Barypodus, Nanopus,* and *Palaeopus* to be junior synonyms of *Laoporus* (Figs. 3a and c). This view may be a case of excessive taxonomic lumping, however, because Baird (1952 and in Spamer 1984) has not indicated such sweeping synonymies. To resolve this and related problems, a thorough taxonomic revision of the Coconino Sandstone's vertebrate tracks is needed. A list of all the

---

Table 1. Invertebrate Ichnospecies from the Coconino Sandstone

Diplopodichnus biformis Brady 1947
Mesichnium benjamini Gilmore 1926
Octopodichnus didactylus Gilmore 1926
O. minor Brady 1947
Oniscoidichnus (Isopodichnus) filiciformis Brady 1949 (1947)
Paleohelcura tridactyla Gilmore 1926
P. dunbari Brady 1961
Scolecocoprus cameronensis Brady 1947
S. arizonensis Brady 1947
Triavestiga niningeri Gilmore 1927

original names given to vertebrate trace fossils from the Coconino Sandstone is given in Table 2.

Table 2. Original Names of Vertebrate Ichnospecies from the Coconino Sandstone

Agostopus:	*A. matheri* Gilmore 1926
	A. medius Gilmore 1927
Allopus?:	*A.? arizonae* Gilmore 1926
Amblyopus:	*A. pachypodus* Gilmore 1926
Baropezia:	*B. eakini* Gilmore 1926
Baropus:	*B. coconinoensis* Gilmore 1927
Barypodus:	*B. palmatus* Gilmore 1927
	B. metszeri Gilmore 1927
	B. tridactylus Gilmore 1927
Dolichopodus:	*D. tetradactylus* Gilmore 1926
Laoporus:	*L. schucherti* Lull 1918
	L. coloradensis (Henderson 1924) Gilmore 1926
	L. noblei Lull 1918
Nanopus:	*N. maximus* Gilmore 1927
	N. merriami Gilmore 1929
Palaeopus:	*P. regularis* Gilmore 1926

Eolian Depositional Systems

Although we have known about the eolian origin of the Coconino Sandstone for a long time, there is comparatively little data on the details of dune types and distribution. Nor is there much documentation of small-scale stratification features within the larger cosets of cross stratification that characterize the formation. Within the last decade, a number of studies have examined ancient and modern eolian deposits. Geologists have developed criteria and facies models by which we can interpret these sediments. In the process, we have learned more about depositional processes and the geomorphic features of eolian sand seas, or ergs. When this information and these models are applied to the Coconino Sandstone, we get a better understanding of the genesis and evolution of the Coconino erg.

The Coconino Sandstone is composed of fine-grained, well-sorted, and rounded quartz grains and minor amounts of potassium feldspar. The cement is primarily silica in the form of quartz overgrowths. These textural and mineralogic characteristics are compatible with an eolian environment in which sediment transport involves numerous grain-to-grain collisions. These collisions result in the mechanical destruction of less stable grains and in winnowing by the wind. As McKee (1979) correctly pointed out, however, these characteristics do not substantiate conclusively a wind-blown origin. Although paleocurrent trends suggest a northern source for this sand, we cannot identify the source(s) of such a large quantity of quartz in the Coconino, as well as in correlative units to the north such as the Weber Sandstone in Utah and the Tensleep Sandstone in Wyoming and Montana.

Primary sedimentary structures in the Coconino Sandstone include small- to large-scale (up to 66-feet [20-m] thick) planar tabular and planar wedge cross stratification (Fig. 4), compound cross stratification, horizontal stratification, ripple marks (Fig. 5), and raindrop impressions (Fig. 6). Deformation features also occur and consist of small, pull-apart structures and a variety of slump-related features (Fig. 7). Small-scale stratification comprises wind-ripple laminations and sandflow strata and minor grainfall laminae. Collectively, these features, together with facies geometries, can be used to characterize bedform morphologies as well as variations in depositional and subtrate conditions.

The most obvious structures in the Coconino Sandstone are the thick sets of cross-stratified sandstone (Fig. 4). Internally,

Figure 4. Large-scale planar tabular cross stratification in the Coconino Sandstone in Hualapai Canyon. Contact with overlying Toroweap Formation is at slope above the large-scale cross strata.

COCONINO SANDSTONE 193

Figure 5. Ripple marks striking down large foreset of cross strata. This orientation of ripple crests is common throughout the Coconino and indicates wind flow and sediment transport across the surfaces of the larger bedforms. Lens cap for scale

Figure 6. Well-preserved raindrop impressions on foreset of large set of cross strata. The irregular margins of the circular pits are somewhat raised on the down dip side. Lens cap for scale

Figure 7. Pull-apart (detachment) structures on slipface of gently dipping foresets. These features probably represent small-scale avalanching of semicohesive sand. Lens cap for scale

these strata exhibit thin (less than 1 cm thick) laminae that are continuous for considerable distances along the foreset of the cross stratification (Fig. 8). These laminae conform to the shape of the foresets—planar where the foresets are straight and becoming concave upward on the tangential foresets. In most cases, these laminae contain no internal structures. Grain size, though quite uniform within the laminae, exhibits a slight inverse grading.

Recent studies, most notably that of Hunter (1977), demonstrate that these structures are the products of wind ripple migration as they climb over the upwind and downwind sides of dunes and in interdune areas. For the most part, the wind ripples in the Coconino Sandstone moved transversely across the lee side of the dunes. This cross-dip path of migration is caused by secondary air currents that parallel the strike of the lee side of simple and complex dunes (or *draas*). The low height-to-wavelength ratio of the wind ripples as measured in plan view exposures of many foresets is consistent with those recorded from modern coastal and inland dunes (McKee 1979).

Thin, down-foreset, tapering laminae occur in many of the thicker cross-stratified sets. These deposits rarely exceed three

COCONINO SANDSTONE 195

Figure 8. Thin, evenly continuous laminae that composed the bulk of the foreset deposits. These laminae represent climbing wind ripples

inches (7.5 cm) in thickness and lack any noticeable grain-size grading and internal stratification. Wind-ripple laminae or similar features surround these wedge-shaped units. McKee and others (1971) and Hunter (1977) have shown that these features form by the avalanching of loose sand on dune slipfaces. These "sandflow" strata form either by a loss of grain cohesion in a descending avalanche sheet or by scarp recession (Hunter 1977). In the latter case, sand moves down and away from the detachment area as the scarp itself migrates up and across the slipface. Whichever the mechanism, these strata represent the avalanching of loose, dry sand.

Other "detachment structures" in the Coconino Sandstone, however, suggest wetter substrate conditions. McKee and others (1971) demonstrated that the detachment and down-slope movement of moistened, somewhat cohesive sand produced steep-sided, pull-apart structures. Possible causes for the cohesiveness of the sand are difficult to verify, but wetting by rain and/or dew or early cementation are obvious possibilities. Our studies to date have demonstrated that cementation patterns generally are uniform throughout these strata, and it seems likely, therefore, that rainfall or dew caused the increase in substrate cohesiveness.

GRAND CANYON GEOLOGY

We can gain a great deal of information concerning the types of eolian bedforms in the Coconino erg by analyzing larger scale features—such as the thick sets and cosets of cross stratification and the truncation surfaces that bound or are contained within these sets. The Coconino contains both simple sets of cross strata and also complexly cross-stratified packages. Each is related to differences in dune types and/or flow processes associated with dune migration.

The most common types of cross stratification in the Coconino Sandstone are the planar tabular and planar wedge varieties (Fig.4). These sets consist of foresets with an average dip of 25° and have sharp or tangential basal contacts. McKee (1979) reported foreset lengths of up to eighty feet (24 m), but most are less than forty feet (12 m) long. In general, the foresets do not contain any smaller cross beds and are composed primarily of wind ripple and subordinate sandflow laminae. Paleocurrent trends reported by Reiche (1938) and McKee (1979) are to the southeast and southwest. Our studies corroborate these earlier findings of a unimodal path of dune migration.

In most cases, these simple cross-bedded sets are underlain and overlain by horizontal to gently dipping planar surfaces (Figs. 4, 9). These surfaces typically can be traced hundreds of meters along the outcrop. In all cases, these surfaces truncate the foresets of these thick, cross-bedded sets. Thin beds of horizontally laminated, fine-grained sandstone and coarse-grained siltstone overlie these surfaces in some localities. Typically, however, the surfaces separate large-scale cross beds.

Although the origin of these features in ancient sandstone is controversial, the fact that planar surfaces truncate the cross strata argues strongly that these are erosional features. Two major hypotheses put forth to explain the genesis of these laterally persistent surfaces involve the role of groundwater in desert basins and the phenomenon of migrating and climbing bedforms.

Stokes (1968) proposed that these extensive planes were the direct result of groundwater rise in dunes and subsequent wind erosion. According to this hypothesis, deflation of the dry, cohesionless sand above the saturation horizon resulted in the generation of a nearly planar surface. Subsequent migration of other dunes over these areas caused a vertical juxtaposition of cross strata, such as that which occurs in the Coconino Sandstone and numerous other eolian sandstones on the Colorado Plateau. The

Figure 9. Stacked sets of planar tabular cross stratification separated by horizontal to gently dipping first order bounding surfaces. Concave-upward scouring of dune surface. Hualapai Trail, western Grand Canyon

net result of a repetition of this process during periods of basin subsidence was stacked sets of thick cross strata separated by multiple truncation planes.

A major problem with the groundwater theory as applied to the Coconino Sandstone involves the nature of the water surface and the extensiveness of these surfaces. McKee and Moiola (1975) have shown that rather than being a horizontal surface, the upper level of the saturation horizon commonly is irregular and, in many instances, mimics the topography of the bedform. Deflation of the overlying dry sand in these cases would not result in a horizontal surface. It also seems rather improbable that this sand would cover a large enough area to create an extensive planation surface, notwithstanding the presence of interdune lows.

A more plausible hypothesis for the Coconino Sandstone involves the downwind migration and climb of interdunal areas and dunes. Brookfield (1977) referred to these as first-order bounding surfaces and attributed their origin to the passage of large, complex eolian bedforms or draas. The planation surface most likely is caused by the truncation of downwind dunes or draas by migrating interdune areas. Interdune areas are zones of

GRAND CANYON GEOLOGY

deflation that bevel the next downwind dune as the sand migrates downwind. Inherent in this model is the requirement that there is a net downwind climb to the bedforms. Rubin and Hunter (1982) have demonstrated the viability of bedform climb as a mechanism of generating these features. In a regional study of the Jurassic Entrada Sandstone, Kocurek (1981b) proved that first-order surfaces can develop in this manner.

Paleocurrent readings are strongly unimodal toward the south. This, coupled with the fact that the sets are internally simple, suggests that the bedforms that produced these cross strata probably were relatively straight crested and that they were oriented transverse to the prevailing winds. McKee (1979) reported that in some localities the "curved shapes of small barchan types are recognizable." The majority of the dunes of the Coconino erg appear to have been of these two morphologies.

Although the aforementioned sets are characteristic of the Coconino, in many places the formation exhibits a complex system of bounding surfaces (Fig. 9) and extremely complicated sequences of internal erosional surfaces and stratification styles (Fig. 10). The concave-upward, bounding surfaces (Fig. 9) probably formed as the result of erosion—either by wind and the subsequent infilling of the eroded surface or by scouring associated with the migration of the next upwind dune/interdune couplet. Rubin and Hunter (1983) and Blakey and Middleton (1983) have documented the occurrence of these scoured surfaces in late Paleozoic and Mesozoic sandstones on the Colorado Plateau.

Figure 10 illustrates some of the more complex geometries in the Coconino Sandstone. Low-angle surfaces truncate higher angle sets of cross stratification through this large set. These surfaces, which dip in the downwind direction, are over- and underlain by smaller sets. Brookfield (1977) has termed these low-angle, dipping surfaces second-order bounding surfaces "because they are truncated by the more extensive first-order surfaces."

These complex sets represent the deposits of large-scale, eolian draas (Wilson 1972, a, b). Draas, which are common in most modern ergs, are characterized by smaller dunes migrating on both the stoss and lee sides of the larger bedform. Although it is possible that the second-order bounding surfaces represent periods of erosion of the lee face, it is more likely that these surfaces were formed by the migration of dunes down a draa that lacked a well-developed slipface.

Draas can be oriented either transverse or parallel to the prevailing wind direction. In the absence of plan view exposures that clearly show the geomorphic shape of the bedform, it is necessary to examine the orientation of the smaller scale cross stratification relative to the dip of the major bounding surface (Rubin and Hunter 1985). In draas oriented parallel to the main wind direction (and, therefore, long-term sand transport direction), the smaller scale sets should exhibit current directions parallel to the draa crest. In all sections of the Coconino Sandstone

Figure 10. Sets of complexly cross-stratified sandstone exhibiting both first- and second-order bounding surfaces and intraset cross stratification. Hualapai Trail, western Grand Canyon

examined, the internal sets are oriented parallel to the dip of the master surfaces. Thus, it appears that flow-transverse bedforms also formed these sets.

Of minor importance in the Coconino Sandstone are sets of trough cross stratification. These sets, typically less than a meter thick, are filled by wind-ripple laminae. Winds that scoured the surface probably formed the troughs. The pit itself then was filled by wind ripples migrating into the scour during periods of reduced wind strengths. Blakey and Middleton (1983) have proposed a similar origin for these units in the underlying Schnebly Hill Formation in Oak Creek Canyon.

SUMMARY

The sediments and trace fossils of the Coconino Sandstone record the advance and passage of a major eolian sand sea. Both invertebrate and vertebrate traces indicate that the substrate was, for the most part, very dry though it is clear that light rainfall or dew periodically moistened the dune surfaces. Although we know that the erg was characterized by both simple and complex bedforms that apparently were oriented approximately transverse to the main direction of sand transport, we have much to learn about the distribution of the bedforms and the dynamics of the Coconino erg.

CHAPTER 11

TOROWEAP FORMATION

Christine E. Turner

INTRODUCTION

The Permian Toroweap Formation in the Grand Canyon region is one of the most intriguing if not the most readily noticed formation when viewed from the commonly visited overlooks in Grand Canyon National Park (Fig. 1). The Toroweap occupies the generally tree-covered interval between the overlying Kaibab Formation, which forms the rim of the Grand Canyon, and the underlying Coconino Sandstone, which forms a prominent light gray cliff that is visible from great distances. What the Toroweap Formation lacks in scenic splendor in the eastern Grand Canyon it compensates for in its geologic diversity. The Toroweap exhibits some of the most striking lateral facies changes in the Grand Canyon sequence. McKee (1938) was the first to recognize the lateral facies changes in the Toroweap in his classic monograph on the Kaibab and Toroweap formations. More recently, geologists have conducted stratigraphic and facies analyses of the Toroweap in the context of modern sedimentologic concepts. This work has permitted a re-evaluation of the depositional environments and has resulted in a fairly complete paleogeographic reconstruction of the Toroweap Formation.

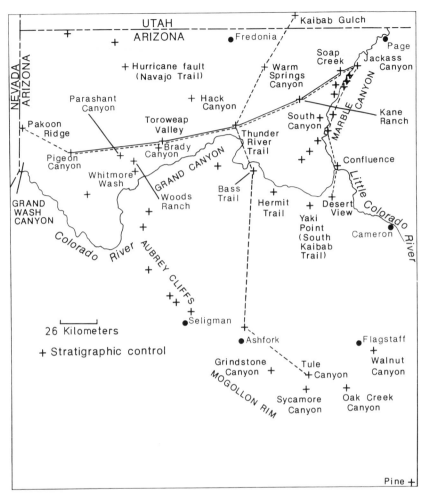

Figure 1. Index map showing location of measured sections used in this study. Dashed lines indicate lines of section used to construct panel diagram shown in Figure 3. Solid lines indicate line of section used to construct the correlation diagram shown in Figure 12.

NOMENCLATURE AND DISTRIBUTION

The Toroweap Formation in Arizona, which covers an area of approximately 25,000 square miles (65,000 km²), is best exposed in the Grand Canyon region and in outcrops along the Mogollon Rim. It pinches out in the subsurface to the east of the Grand Canyon, but extends northward in the subsurface into southern

Utah, westward into eastern Nevada, and southward in the subsurface to the outcrop belt along the Mogollon Rim. Beyond this point, erosion has removed evidence of the formation.

In 1938, McKee described and named the Toroweap Formation, which he separated from the overlying Kaibab Formation (Fig. 2). The type locality for the Toroweap is in Brady Canyon, an eastern side canyon to Tuweep (Toroweap) Valley. McKee recognized three informal members: an upper evaporite and redbed interval—the alpha member; a middle limestone unit—the beta member; and a lower sandstone and evaporite interval, which he referred to as the gamma member. Sorauf (1962) applied geographic names to McKee's informal members (Fig. 3). He named the alpha member the Woods Ranch Member; the beta, the Brady Canyon Member; and the gamma, the Seligman Member. Sorauf and Billingsley (in preparation) have proposed formally that these member names be accepted.

In addition to dividing the Toroweap Formation into members, McKee (1938) divided the formation into three lateral

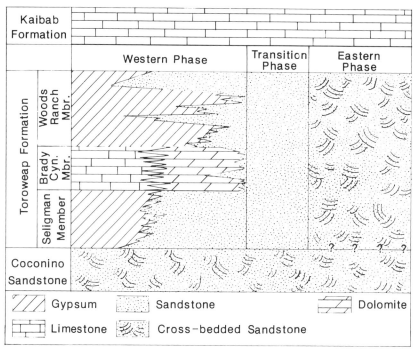

Figure 2. Schematic diagram showing members and lateral phases of the Toroweap Formation (From Rawson and Turner-Peterson 1979)

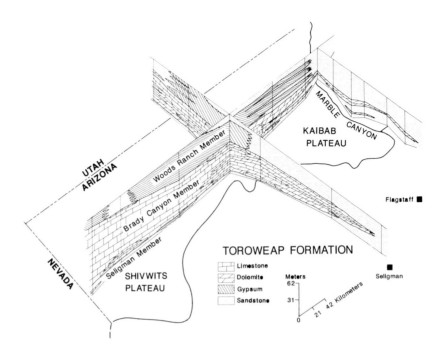

KAIBAB
PLATEAU

MARBLE CANYON

UTAH
ARIZONA

Woods Ranch Member

Brady Canyon Member

Flagstaff ■

Seligman Member

NEVADA

SHIVWITS
PLATEAU

TOROWEAP FORMATION

▦ Limestone
▨ Dolomite
▩ Gypsum
▢ Sandstone

Meters
62
31
0

21 42 Kilometers

■ Seligman

Figure 3. Panel diagram constructed chiefly in the western phase of the Toroweap Formation, along lines of section shown in Figure 1. Note thinning of Brady Canyon Member to the east and southeast, as well as an increase in dolomite in that member. Evaporites in Woods Ranch and Seligman members are confined largely to the area north of the Grand Canyon (from Rawson and Turner 1974).

phases: the western phase, where the three members are recogniz-able; the transition phase, which consists of irregular-bedded sandstone; and the eastern phase, which consists of cross-bedded sandstone (Fig. 2). The limits of the western phase are determined by the extent of the Brady Canyon Member. To the east and southeast of the pinchout of this member, the two other members (the Seligman and Woods Ranch members) cannot be distin-guished. The western phase is well developed in the Grand Can-yon region (Fig. 4). The transition phase of the Toroweap is particularly well developed in the area of Sycamore Canyon. To the south and east of this area, the sandstones are more exten-sively cross bedded. The Toroweap Formation in Oak Creek and Walnut canyons typifies McKee's eastern phase and is character-ized by cross-bedded sandstone (Fig. 5).

GRAND CANYON GEOLOGY

LITHOLOGY AND STRATIGRAPHY

Stratigraphy

Geologists can identify all three members in the Grand Canyon region (Figs. 2 and 3). The lower boundary of the Toroweap Formation is at the base of the Seligman Member. It is the first non-cross-bedded or evaporite-bearing unit above the cross-bedded Coconino Sandstone. The boundary is conformable in most locations but is not distinct in the area of Marble Canyon, where cross-bedded units of the Coconino Sandstone intertongue with the lowermost beds of the Seligman Member. The first appearance of

Figure 4. Photograph of Toroweap Formation in South Canyon, near Marble Canyon (see Fig. 1 for location). Presence of all three members is characteristic of the western phase. Pk=Kaibab Formation. Members of the Toroweap are: Ptwr=Woods Ranch Member (slope-forming unit in this region because of gypsum beds); Ptbc=Brady Canyon Member (cliff-forming carbonate units); Pts=Seligman Member (thin sandstone interval). Pc=Coconino Sandstone.

a thick carbonate unit above the noncross-bedded sandstone defines the upper contact of the Seligman Member. It is less than 45 feet (15 m) thick in the Grand Canyon, but may be as much as 450 feet (152 m) thick in the North Muddy Mountains in Nevada (Bissell 1969).

The Brady Canyon Member forms the massive cliff of carbonate above the relatively thin Seligman Member in the Grand Canyon region (Fig. 4). Limestone predominates to the west, while dolomite is abundant to the east (Figs. 2 and 3). The Brady Canyon Member has its greatest development in the western Grand Canyon, where it is up to 280 feet (93 m) thick. It thins uniformly in an easterly direction to a depositional edge just east of Marble Canyon. An aphanitic dolomite unit with prominent desiccation cracks marks the upper contact here.

The Woods Ranch Member typically extends from the top of the aphanitic dolomite that forms the uppermost unit of the Brady Canyon Member to the base of the cliff-forming limestone of the overlying Kaibab Formation. Repetitive intervals of evaporite, limestone, and sandstone form a distinctive slope in most of the Grand Canyon region. Where the Woods Ranch Member lacks evaporite, it usually forms cliffs. The Woods Ranch Member shows no consistent thickening or thinning trends throughout most of the study area. Geologists have measured a maximum thickness of about 180 feet (60 m).

Along the Mogollon Rim of central Arizona, the three members of the Toroweap disappear because of lateral facies changes. That fact, plus the presence of an additional member at the base of the Kaibab Formation, requires that different criteria be used to define the lower and upper contacts of the Toroweap Formation.

In the Sycamore Canyon area, the upper contact of the Toroweap is difficult to identify because the sandstone beds of the Fossil Mountain Member (the basal unit of the Kaibab Formation in this area) resemble the sandstone beds of the Toroweap Formation.

In the Oak Creek Canyon area, cross-bedded sandstone characterizes the entire Toroweap Formation, making it difficult to differentiate this formation from the underlying Coconino Sandstone. However, a truncation surface marked by vegetation in surface exposures separates beds contemporaneous with the Toroweap Formation from the underlying Coconino Sandstone (Fig. 5).

Figure 5. Photograph of Oak Creek Canyon showing cross-bedded sandstone typical of the eastern phase of the Toroweap Formation. Line of vegetation, indicated by arrow, marks the boundary between the Toroweap Formation and the underlying Coconino Sandstone, a unit that also contains cross-bedded sandstone. Pk=Kaibab Formation; Pt=Toroweap Formation; Pc=Coconino Sandstone

Lithofacies

Evaporite Lithofacies. Interbeds of evaporite, thin-bedded carbonate, and fine-grained sandstone characterize a significant part of the Woods Ranch Member (and to a lesser degree, the Seligman Member). The lithofacies is restricted in both members to the area north of the Grand Canyon (Figs. 6 and 7). The evaporite beds, which most often are gypsum, frequently contain laminae of limestone and dolomite (Figs. 8a and b) several millimeters thick. The evaporite in these laminated intervals usually is about one centimeter thick. Evaporite beds occur in sequences characterized by a basal carbonate unit up to 1.5 feet (0.5 m) thick, gypsum beds up to 3 feet (1 m) thick, and sandstone beds that generally are 1.5 feet (0.5 m) thick. The carbonate beds, which typically are limestone rather than dolomite, are thinly laminated and vuggy (Fig. 8c). Gypsum or anhydrite nodules fill the vugs in places. This suggests that all of the vugs originally may have been filled with evaporite minerals. Frequently, the sandstone beds are poorly cemented, and it is difficult to discern sedimentary structures.

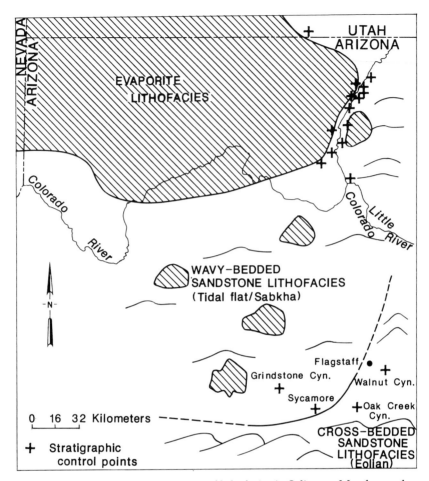

Figure 6. Map showing distribution of lithofacies in Seligman Member and equivalent strata in the Toroweap Formation. Map is based on the most abundant lithofacies present at each locality (modified from Rawson and Turner 1974).

Locally, however, geologists can identify cross bedding. With the exception of one thin limestone bed near the top of the Woods Ranch Member (a unit that contains abundant pelecypods of the genus *Schizodus*), this lithofacies is unfossiliferous. The *Schizodus*-bearing limestone bed, which also contains ooids locally, has been named informally the "Hurricane Cliffs tongue" of the Woods Ranch Member (Altany 1979).

The evaporite lithofacies probably represents deposition in a shallow, subaqueous, shelf environment in which warm re-

GRAND CANYON GEOLOGY

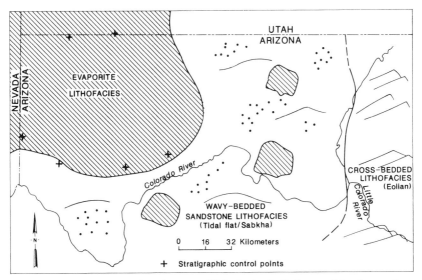

Figure 7. Map showing distribution of lithofacies in Seligman Member and equivalent strata in the Toroweap Formation. Map is based on the most abundant lithofacies present at each locality (modified from Rawson and Turner 1974).

stricted-marine waters were evaporated, promoting the precipitation of evaporite minerals. Originally, geologists thought that this lithofacies represented sabkha deposition (Turner 1974; Rawson and Turner 1974; Rawson and Turner-Peterson 1979, 1980); we now know that laminated evaporites indicate precipitation in standing water rather than formation by displacive growth within sabkha sediments (Schreiber 1986). The thinly laminated evaporite and carbonate (Fig. 8b) is similar to the anhydrite-carbonate couplets in the laminated sulfate facies in subaqueous evaporites of the Castile Formation of the Delaware Basin of southeastern New Mexico and western Texas. Some carbonate units in the evaporite lithofacies of the Toroweap Formation contain desiccation cracks, which indicate local subaerial exposure and thus shallow, rather than deep, water deposition. Most likely, this lithofacies was part of a shallow shelf sequence. The carbonate units also lack fossils, which suggests that conditions in the restricted-marine waters were not conducive to marine life. Vertical repetition of certain lithologies—a basal carbonate bed, middle evaporite bed, and upper sandstone unit—suggests that cyclic sedimentation characterized deposition of this lithofacies.

TOROWEAP FORMATION   211

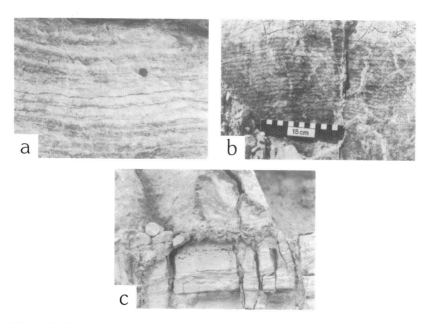

Figure 8. Features observed in the evaporite lithofacies of the Toroweap Formation. Photographs are from the Woods Ranch Member along the Thunder River Trail, Grand Canyon. (a) Gypsum bed (white bands) with thin laminae of limestone (gray), with penny for scale; (b) Laminated bed that contains alternations, on the scale of a few mm, of gypsum (white) and limestone (gray); (c) Thin-bedded limestone bed with vugs near the top (below and to the right of the quarter), overlain by bed of gypsum. Locally, similar vugs are filled with evaporite minerals, suggesting that these vugs originally contained evaporites.

Thin carbonate laminae within the evaporite beds reflect periodic freshening of the waters that were depositing the evaporite. We do not know if this freshening occurred on a seasonal basis.

Siliciclastic Lithofacies

Irregular-Bedded Sandstone. Fine- to medium-grained sandstone and siltstone characterize the intervals of the Woods Ranch and Seligman members that are laterally adjacent and equivalent to the evaporite lithofacies. This lithofacies dominates in areas just to the east and southeast of the evaporite lithofacies of the Woods Ranch and Seligman members (Fig. 6 and 7). In addition, this

212   GRAND CANYON GEOLOGY

lithofacies predominates in the entire Toroweap Formation and beyond the depositional edge of the Brady Canyon Member (Fig.2), typifying the transition facies of McKee (1938). Carbonate and evaporite beds are rare in this lithofacies, in contrast to their abundance in the evaporite lithofacies. Typical sedimentary structures include wavy lamination (Fig. 9a), flaser lamination (a discontinous form of wavy lamination), lenticular lamination, and fluid-escape structures. Minor cross-bedded, fine-grained sandstone units with sets of inversely graded laminae occur in this lithofacies. Locally, these units exhibit high index ripples whose axes trend directly down the cross-bedding laminae. Geologists also have noted wave and adhesion ripples (Fig. 9b) as well as brecciated (and rare) thin carbonate beds (Fig. 9c).

Intraformational brecciation is a common feature in this lithofacies, particularly in Marble and Sycamore canyons, within the transition facies defined by McKee (1938). The breccia most often consists of blocks of cross-bedded eolian sandstone within a matrix of flaser (or lenticular) laminated sandstone. The cross-bedded blocks typically show evidence of having collapsed downward (Fig. 9d). In the Sycamore Canyon area, laterally extensive beds of brecciated sandstone are interbedded with cross-bedded sandstone.

Most geologists interpret the irregular-bedded sandstone lithofacies as tidal-flat deposits. Flaser bedding, wavy bedding, and lenticular bedding are characteristic of intertidal environments (Reineck and Wunderlich 1968), whereas brecciated carbonate units similar to those in the Toroweap Formation generally occur in supratidal zones of tidal-flat complexes (Shinn 1983). The brecciation found in carbonate beds probably is related to the dolomitization of limestone beds in a supratidal setting. This dolomitization results in shrinkage that is caused by a loss of volume. The abundance of fluid-escape structures is consistent with a tidal flat environment. Cross-bedded units that exhibit inverse, graded bedding and high-index ripples on the slipface almost certainly are eolian dune deposits. Although dune deposits are not abundant in this lithofacies, they do indicate the occasional migration of eolian sediments across the tidal flat surface. Limited paleocurrent data from the eolian sandstones in this lithofacies show a southerly transport direction, which is consistent with the paleowind directions for the Permian in the Colorado Plateau region (Peterson 1988).

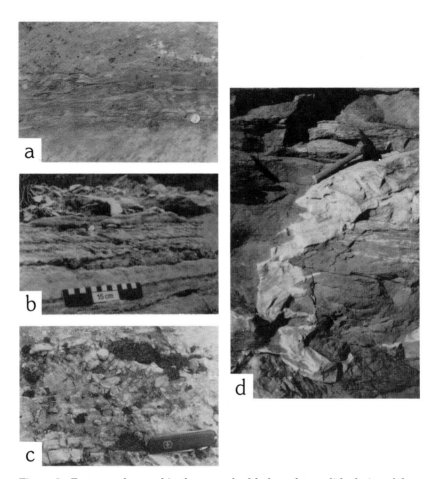

Figure 9. Features observed in the wavy-bedded sandstone lithofacies of the Toroweap Formation. Photographs a - c are from the Woods Ranch Member along the South Kaibab Trail, near Yaki Point in the Grand Canyon. (a) Wavy and lenticular bedding, with penny for scale; (b) Adhesion ripples at the top of a cross-bedded sandstone; (c) Brecciated carbonate beds, with knife for scale; (d) Cross-bedded, light-colored sandstone bed (immediately below hammer) collapsed into underlying units. This may have been caused by removal of evaporites by dissolution. From the Woods Ranch Member in Jackass Canyon, near Marble Canyon, Arizona

One possible interpretation for areas in the transition phase that contain abundant intraformational sandstone breccias is a supratidal or sabkha environment. This interpretation relies in part on the regional facies interpretation that Rawson and Turner-Peterson (1979) have proposed for the Toroweap Formation. Most of the brecciation occurs in the southeastern part of the study area, in the part of the transition facies that contains the largest number of eolian sandstone units. This is particularly true in the Sycamore Canyon area, where laterally extensive, brecciated sandstone units interbed with cross-bedded eolian sandstone. In the overall facies distribution, brecciated units occur landward of the intertidal deposits and are intercalated with eolian sandstones. Fryberger and others (1983) have suggested a supratidal or siliciclastic-sabkha environment to explain similar laterally extensive, brecciated units. It is possible that the evaporite formed by displacive growth within a siliciclastic sabkha. Subsequent removal of this evaporite by the movement of relatively fresh groundwater through the sandstones would have caused collapse and brecciation within the sabkha units.

Cross-Bedded Sandstone. The irregular-bedded and brecciated sandstone units of the transition phase of the Toroweap Formation grade laterally, in a southeasterly direction, into cross-bedded sandstone of the eastern phase of the Toroweap (Fig. 2). The eastern phase is characterized by fine- to medium-grained, cross-bedded sandstone that is indistinguishable from sandstone in the underlying Coconino Sandstone (Fig. 5). Most frequently, the cross beds are wedge and tabular planar (Fig. 10a), with some trough-shaped sets. Average set thickness is about 6 feet (2 m), and the average dip direction is S11°W, with a consistency ratio of 0.90 (Rawson and Turner-Peterson 1979). The sets of cross-bedded sandstone contain thick avalanche or sand flow toes as much as 0.5 inches (5 cm) thick (Fig. 10b) and inversely graded laminae (Fig. 10c). Slumping and deformational features are not uncommon in this lithofacies (Fig. 10d).

We base an eolian interpretation for this lithofacies on the presence of diagnostic eolian characteristics, such as large-scale cross bedding, avalanche (sand flow toe) deposits, and inversely graded ripple laminae. The deformational structures shown in Figure 10d are similar to those studied in modern environments and seem to develop exclusively in dunes that are wet or that have been wetted (McKee and Bigarella 1972). The eastern phase Toroweap Formation may have been deposited in a coastal dune environment. An eolian interpretation also is consistent with the

Figure 10. Features observed in the cross-bedded sandstone (eastern phase) of the Toroweap Formation. (a) Section composed almost entirely of tabular- and wedge-shaped, cross-bedded sandstone, Oak Creek Canyon, Arizona; (b) Avalanche (sand flow) toes observed in cross-bedded sandstone, Oak Creek Canyon, Arizona; (c) Inversely graded laminae in cross-bedded sandstone, Sycamore Canyon, Arizona. Note preservation of foreset laminations in middle part of photograph. 10X hand lens for scale; (d) Deformational structures in cross-bedded sandstone, Oak Creek Canyon, Arizona. These structures seem to develop exclusively in eolian dunes that are wet.

southeasterly transport direction for the cross-bedded sandstone lithofacies; this direction is the same as that determined for the eolian rocks of the underlying Coconino Sandstone. It also is consistent with paleowind directions during the Permian in the Colorado Plateau area (Peterson 1988).

Carbonate Lithofacies. Geologists have identified several carbonate lithofacies within the Brady Canyon Member of the Toroweap Formation. Rawson and Turner first described these lithofacies in 1974.

Skeletal Packstone. Skeletal material in the grain-supported carbonate in this lithofacies consists of disarticulated crinoid columnals; bryozoans; brachiopods; ostracods; gastropods; endothyrids; echinoid and brachiopod spines; and trilobite fragments. The matrix is micrite, which is common to all of the carbonate lithofacies. Although only a few grainstones are present, we can see rare quartz grains in thin section. The quartz, which is fine-silt-size, increases in abundance from west to east. This lithofacies is common in the western part of the Grand Canyon region.

Skeletal Wackestone. Skeletal components of this lithofacies are similar to the skeletal packstone—except that they are floating in micrite and are smaller, perhaps as a result of transportation by weak bottom currents. The micrite matrix makes up sixty to ninety percent of the rock. In outcrop, massive limestone units generally are composed of skeletal packstone. Along with the skeletal packstone, this lithofacies is common in the western part of the Grand Canyon region.

Pelletal Wackestone. This lithofacies consists of carbonate units that contain elliptical-to-spherical pellets ranging from 0.004 to 0.008 inches (0.1 to 0.2 mm) in diameter. No true ooids occur, except in the "Hurricane Cliffs tongue" of the Woods Ranch Member. The pelletal wackestone facies commonly is altered to a dolomicrite, which geologists include in this facies when they can recognize the original texture. Dolomitized pelletal wackestone is prevalent in the eastern part of the study area and also occurs in some western sections, particularly near the top of the Brady Canyon Member.

Sandy Dolomite. A sandy dolomitic lithofacies is abundant in the Marble Canyon area. It also occurs near both the base and the top of the Brady Canyon Member in the western part of the study area. The dolomite is silt-size, with abundant fine-to-coarse, sand-size quartz grains.

Aphanitic Lime Mudstone and Dolomite. Aphanitic lime mudstone and dolomite commonly occur at the base and top of the westernmost exposures of the Brady Canyon Member in the study area. This lithofacies probably is present to the east as well, but quartz grains mask the texture. The carbonate is aphanitic— (0.004 mm in diameter), and the dense rock that it forms breaks

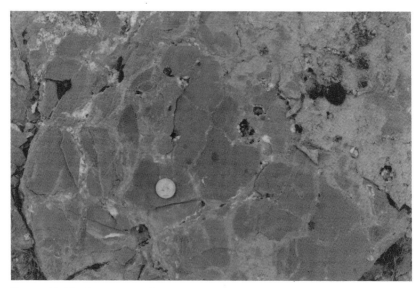

Figure 11. Desiccation cracks on upper surface of a dolomitic mudstone at the top of the Brady Canyon Member. Desiccation cracks such as these are abundant at this same interval across northern Arizona and probably indicate widespread withdrawal of the sea at the end of deposition of the Brady Canyon Member. Quarter for scale

with a conchoidal fracture. To date, geologists have not found fossils or fossil fragments in this unit. Desiccation cracks, common in this lithofacies, are readily observed at the top of the Brady Canyon Member in most localities (Fig. 11).

Distribution and Interpretation of Carbonate Lithofacies

During the formation of the Brady Canyon Member, carbonates with an open-marine fauna were deposited to the west, whereas carbonates with a restricted-marine fauna were deposited to the east (Fig. 12). This reflects an incursion of the sea from the west. The mud-supported texture that characterizes all lithofacies of the Brady Canyon Member indicates that these sediments were laid down in quiet-water conditions and that there was no significant reworking. The fauna and textural types are consistent with shallow, quiet-water conditions on a carbonate shelf. Distribution of lithofacies in the Brady Canyon Member changed through time. Figure 13 shows a representative distribution of lithofacies at one particular time (T-2 in Fig. 12). Gradation from open-marine fauna in limestone beds in the west to re-stricted-marine fauna in dolomitic mudstone in the east, as shown in Figure 13, is characteristic of the Brady Canyon Member. Of particular interest is the persistence of a dolomitic mudstone at the top of the member (R-3 in Fig. 12). This represents a westward progradation of a lithofacies deposited in the most restricted of marine conditions. Also noteworthy is the frequent occurrence of desiccation cracks on the upper surface of this dolomite unit, indicating widespread subaerial exposure at the end of deposi-tion of the Brady Canyon Member.

Current models of dolomite formation do not explain the greater abundance of dolomite in the eastern part of the Brady

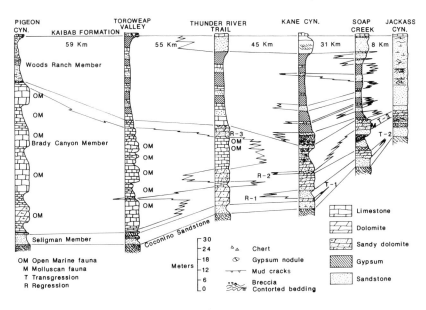

Figure 12. West to east correlation diagram of the Toroweap Formation, northern Arizona (see Fig. 1 for locations). Evaporites in the Seligman Mem-ber represent the first incursion of the sea following deposition of the underly-ing eolian Coconino Sandstone. In the Brady Canyon Member, open-marine fauna predominate to the west, and restricted-marine fauna predominate to the east. Several transgressions and regressions are apparent within the Brady Canyon Member. Regression R-3 may represent progradation westward in re-sponse to a slowing down of a relative sea level rise, and the top of the Brady Canyon Member (marked with desiccation cracks) may represent a relative drop in sea level. The evaporite lithofacies of the Woods Ranch Member is thought to reflect another relative sea level rise (modified slightly from Rawson and Turner 1974).

Canyon Member. The idea of reflux dolomitization implies a syndepositional process, where dolomitization occurs in response to the movement of hypersaline brines from a lagoonal environment into adjacent or underlying carbonate sediments (Adams and Rhodes 1960). Although evaporites are present in the Woods Ranch Member of the Toroweap Formation, the distribution of these evaporites does not coincide entirely with the distribution of dolomite in the underlying Brady Canyon Member (Fig. 3). An alternative model, the "Dorag" dolomitization model (Badiozamani 1973), implies the replacement of earlier formed limestone intervals in a "mixing zone" of seawater and fresh water.

Although some of the carbonates in the Brady Canyon may have been dolomitized in this way (e.g., some pelletal wackestone beds), replacement textures are not abundant in the carbonate beds. As Hardie (1987) points out, we do not understand the process of dolomitization very well. Distribution of dolomite in the Toroweap Formation, as in many ancient examples, clearly is greater in the more restricted-marine lithofacies. This association suggests that restricted circulation and great evaporation promote dolomite precipitation, but we do not know the mechanism by which it occurs.

PALEONTOLOGY

McKee (1938) summarized the paleontology of the Toroweap Formation in his original work on the Kaibab and Toroweap formations. Except for the "Hurricane Cliffs tongue" of the Woods Ranch Member (which contains the marine bivalve *Schizodus*), the Brady Canyon Member is the only fossiliferous member of the Toroweap Formation. McKee recognized two major faunal facies in the Brady Canyon Member: an open-marine fauna to the west and a molluscan fauna to the east. The open-marine fauna includes brachiopods, bryozoans, crinoids, and horn corals. The molluscan fauna includes bivalves and gastropods, with a few scattered scaphopods and cephalopods. Kirkland (1962) compiled a faunal listing of subsequent finds in the Toroweap Formation. Belden (1954), Mullens (1967), Miller and Breed (1964), Beus and Breed (1968), Turner (1974), and Rawson and Turner (1974) reported additional species in the Toroweap Formation.

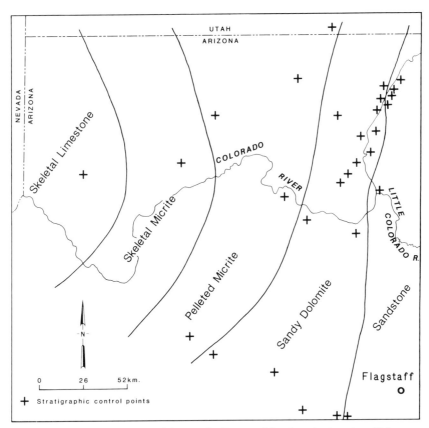

Figure 13. Lithofacies map of the Brady Canyon Member during time T-2 (see Fig. 12). From Rawson and Turner (1974)

AGE AND CORRELATION

Fossils contained in the Brady Canyon Member in northern Arizona suggest a late Leonardian age for the Toroweap Formation. Recent studies of bryozoans in Nevada (Gilmour and Vogel 1978) verify this timeframe. The late Leonardian is equivalent to the Kungurian (263 to 258 million years ago) and, possibly, late Artinskian (268 to 263 million years ago) ages on the radiometric timescale (Harland and others 1982).

As the Toroweep Formation grades southeastward into the cross-bedded sandstone lithofacies that characterizes the eastern phase, it becomes indistinguishable from eolian rocks of the underlying Coconino Sandstone. Similarly, geologists believe that the Toroweap Formation is equivalent to the White Rim Sandstone Member of the Cutler Formation in southeastern Utah. The Toroweap Formation also is equivalent to the San Andres Limestone in northwestern New Mexico.

DEPOSITIONAL HISTORY

We usually interpret the overall depositional history of the Toroweap Formation in terms of relative sea-level changes with time. Subaqueous evaporite and tidal-flat sediments in the Seligman Member, associated with a relative rise in sea level, represented the first incursion of the sea from the west. Eolian sandstone within the member indicates times when the sea withdrew from the region. The development of a thick carbonate sequence in the Brady Canyon Member indicates an incursion of the sea as far east as the Marble Canyon area.

Near the end of the Brady Canyon deposition, a progradation of carbonate lithofacies occurred. This slowing down of the relative sea-level rise is reflected in the progradation of restricted-marine lithofacies at the top of the Brady Canyon Member to the west (R-3 in Fig. 12). The period of subaerial exposure indicated by abundant desiccation cracks at the top of the dolomitic mudstone marks a relative lowering of sea level and subaerial exposure of the entire shelf.

With another incursion of the sea, probably in response to a relative rise in sea level, the shelf flooded again. Cyclic sedimentation of carbonate, evaporite, and sandstone in the Woods Ranch Member reflects conditions similar to those that persisted during the formation of the Seligman Member. Here, the evaporites and carbonates indicate periods of subaqueous deposition, while the eolian sandstones suggest times of subaerial exposure. Since no evidence for a barrier exists in the Woods Ranch Member of the Toroweap Formation, it is difficult to interpret the evaporite in terms of the traditional barred-basin model (Schreiber 1986). The evaporite lithofacies of the Woods Ranch Member extends as far as the Permian outcrops in Nevada—and without significant change. It thus appears that a shallowing of seawater across a broad shelf caused the restricted circulation that is required to generate a hypersaline brine for evaporite precipitation.

Away from the area of dominantly marine carbonate-evaporite sedimentation, tidal-flat, sabkha, and eolian depositional environments persisted throughout deposition of the Toroweap Formation. In the transition phase of the Toroweap, sediments were deposited chiefly in a tidal-flat environment. Siliciclastic sabkhas developed along the eastern and southeastern margins of the tidal flats. Farther to the east and southeast, eolian deposition that had begun during the formation of the underlying Coconino Sandstone persisted (Figs. 2 and 5). Evidence of moisture during eolian deposition in the eastern phase of the Toroweap, as suggested by the deformational features, contrasts with the scarcity of such features in the underlying Coconino Sandstone. We attribute this difference to the sea's proximity to the dune fields during the formation of the eastern phase of the Toroweap. This is in contrast to the drier, inland conditions that persisted during deposition of the Coconino Sandstone. Southwesterly transport directions in both the Coconino Sandstone and the Toroweap Formation are consistent with paleowind directions for the Permian.

SUMMARY

The Toroweep Formation, in its variety of lithofacies, reflects transgressions and regressions of an eastward-advancing and westward-retreating sea. The formation was deposited in open- and restricted-marine environments, tidal flats, sabkhas, and eolian dune fields. During times of transgression, the stable shelf was flooded by shallow-marine waters that were conducive to marine life. The climate probably was semiarid to arid, and large dune fields developed in the inland regions. The shoreline commonly was in the vicinity of the Grand Canyon, which explains the striking lateral and vertical changes in lithofacies. It is this very variety of lithofacies in a relatively small area that makes the Toroweep Formation such a fascinating unit for geologists.

CHAPTER 12

KAIBAB FORMATION

Ralph Lee Hopkins

INTRODUCTION

The Kaibab Formation comprises the caprock of the Grand Canyon and forms the surface of the Kaibab and Coconino plateaus, which is the area through which the deepest part of the canyon has been carved. When visitors view the Kaibab Formation from scenic points and trailheads within the national park, they can recognize it easily as the gray, stepped cliff directly above the vegetated slope of the Toroweap Formation (Fig.1). The Kaibab, the youngest Paleozoic rock unit on the southern Colorado Plateau, is composed of a variety of lithologic types that were deposited within a complex shallow-marine setting during the Permian. It represents the final chapter in the geologic story recorded by the sedimentary layers of the Grand Canyon.

Because it is at the top of the Grand Canyon stratigraphic section (and, therefore, is easy to see), geologists have studied the Kaibab Formation extensively. This chapter summarizes the cumulative knowledge gained through decades of observations by numerous geologists and is dedicated to Edwin D. McKee, in whose footsteps we all have followed.

Figure 1. View of precipitous Kaibab cliff along the south rim near the South Kaibab Trail (KT) showing Fossil Mountain (Kf) and Harrisburg (Kh) members. Note pinch out of sandstone unit laterally within the Fossil Mountain Member (arrow). Slope-forming Woods Ranch Member of the Toroweap Formation (Pt) everywhere underlies the Kaibab Formation in the walls of the Grand Canyon. Photograph looking east from Mather Point

NOMENCLATURE

The detailed description and naming of strata included within the Kaibab Formation has a long and interesting history. The earliest recorded observations of Permian rocks in northern Arizona were made by Jules Marcou (1856) and J. S. Newberry (1861). Originally, geologists considered the units represented by the presentday Kaibab and Toroweap formations a single formation, the Aubrey Limestone (G. K. Gilbert 1875). Walcott (1880) placed these strata within the Upper Aubrey Group. The name "Kaibab" first was applied by Darton (1910) for exposures on the Kaibab Plateau north of the Grand Canyon. Noble (1914, 1922), along with Longwell (1921), provided the initial description and correlation of Kaibab strata across the Grand Canyon region. Geologists then subdivided these strata into five basic topographic and lithologic units. Reeside and Bassler (1922) named the uppermost beds the Harrisburg Gypsiferous Member for exposures at Harrisburg Dome in Utah. Noble (1928) proposed a type section for the Kaibab Limestone in Kaibab Gulch (in Utah along the East Kaibab Monocline).

It was not until the classic work by McKee (1938) that the "Kaibab Limestone" was split into two formations. McKee proposed that the term Kaibab be applied only to the massive upper limestone and the unit directly above it, suggesting that these rocks be called the "Kaibab Formation" since they are composed of a variety of lithologic types. The lower limestone unit and adjacent slope-forming units became the Toroweap Formation (Turner, see Chapter 11). McKee's scheme recognized three members within each formation and named them "alpha," "beta," and "gamma" in descending order.

In an attempt to establish more formal rock units, Sorauf (1962) proposed a change in terminology for both the Kaibab and the Toroweap formations. He suggested that the rocks included within the alpha (or upper) member be called the Harrisburg Member. The beta (or middle) member of the Kaibab (McKee 1938) became the Fossil Mountain Member, named for Fossil Mountain along the south rim near the Bass Trail. Rocks included within the gamma (or lower) member, confined by McKee to the Mogollon Plateau south of the Grand Canyon, were not present in Sorauf's field area. Geologists interpret the gamma member as a facies within the Fossil Mountain Member (Lapinski 1976; Cheevers and Rawson 1979). Most subsequent workers have utilized the revisions in nomenclature suggested by Sorauf (1962), and the designation will be formalized in the near future (Sorauf and Billingsley, in progress).

DISTRIBUTION

Rocks of the Kaibab Formation form a continuous layer across the Grand Canyon and the surrounding region. The best exposures in cross section occur along the cliffs of the canyon and its tributaries. Although recent exposure and erosion of Kaibab rocks beneath the Permo-Triassic unconformity obscure thickness trends, we know that the formation gradually thickens to the west (Fig. 2). Along the rim of the canyon, it generally ranges between 300 and 400 feet (90-120 m) in total thickness. Geologists have measured the greatest thickness in northwestern Arizona, in an area west of Kanab Creek and northwest of the Colorado

River. Here, it exceeds 500 feet (150 m). East of the Grand Canyon, the Kaibab thins dramatically and is absent along the Defiance and Monument upwarps. Outcrops southeast of Winslow and south of Holbrook clearly show stratigraphic thinning (McKee 1938; Mather 1970; Cheevers 1980).

To the south, the Mogollon Rim, or escarpment, defines the limit of Kaibab exposure. This area represents the southern edge of the Colorado Plateau. Northward, the Kaibab Formation extends into southern and central Utah. It is well exposed along the Hurricane Cliffs, Virgin River Gorge, and Beaver Dam Mountains in the southwestern part of the state (Nielson 1981). The formation continues on into the Circle Cliffs-Waterpocket Fold and San Raphael Swell regions in central Utah (Davidson 1967) and into the subsurface across southern Utah (Irwin 1971). Some of this formation contains oil (for example, the Upper Valley Field near Escalante, Utah). Geologists have found Kaibab outcrops as far north as the Deep Creek Mountains and Confusion Range in Utah and in the isolated mountains of northeastern Nevada (Bissell 1962).

In southern Nevada, westernmost outcrops of the Kaibab Formation occur in a number of scattered mountain ranges in the Las Vegas area (Longwell 1921; Bissell 1969). The formation appears to thin westward (Fig. 2).

STRATIGRAPHY

Facies changes within the Kaibab Formation and in adjacent units complicate internal and regional stratigraphic relations.

Lower Contact

At the Grand Canyon, the Kaibab Formation is underlain everywhere by gypsum and/or contorted sandstones of the Woods Ranch Member of the Toroweap Formation (Fig.1). Originally, geologists believed that the Kaibab-Toroweap contact was unconformable. They based this impression primarily on the presence of local intraformational breccias and erosional surfaces (McKee 1938). Further study has shown that these features are related to collapse following the dissolution of evaporitic facies within the underlying Woods Ranch Member, indicating that the contact is conformable or only locally discon-

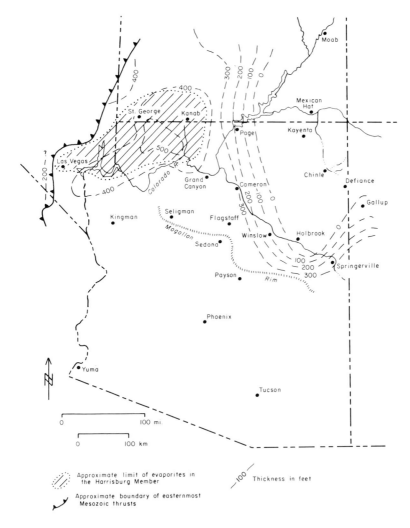

Figure 2. Isopach map of the Kaibab Formation illustrating total thickness trends across the Grand Canyon and surrounding regions. Maximum thickness of the Kaibab occurs in northwestern Arizona where the Harrisburg Member contains significant evaporites. Data are compiled from many sources.

formable. The basal Kaibab is the first cherty carbonate or sandstone unit located above the *Schizodus* bed (Hurricane Cliffs tongue) of the Woods Ranch Member.

To the south and east of the Grand Canyon, the evaporites and contorted sandstones (sabkha complex) of the underlying Woods Ranch Member are transitional and interstratified with

KAIBAB FORMATION   229 cross-bedded sandstone facies (eolian dune complex). These rocks ultimately become indistinguishable from the Coconino Sandstone (Turner, this volume). As a result, the Kaibab Formation in the Mogollon Rim region directly overlies the Coconino Sandstone. In northeastern Arizona and southeastern Utah, the White Rim Sandstone underlies the Kaibab.

Fossil Mountain-Harrisburg Contact

The contact between the Fossil Mountain and Harrisburg members of the Kaibab Formation is conformable, though different workers have placed it at slightly different stratigraphic levels. The similarity in the facies of both members at eastern localities has made it difficult to establish the contact. At the Grand Canyon, however, there are distinct textural, mineralogical, and faunal changes within the upper portion of the Kaibab sequence. Typically, a distinctive white, butterscotch, or red nodular-to-bedded chert horizon marks the base of the Harrisburg Member. This "marker-chert" most often is coincident with the disappearance of the normal-marine fauna and limestone mineralogy that is most characteristic of the cliff-forming Fossil Mountain Member (Sorauf 1962; Clark 1980; Hopkins 1986). In areas of southwestern Utah where the marker-chert is absent, geologists place the contact at the obvious change to slope-forming gypsiferous beds in the basal part of the Harrisburg Member.

East and south of the Grand Canyon, the contact between members is less obvious. It is easy to distinguish the Fossil Mountain Member in this region since it contains fauna of normal-marine affinity—particularly the productid brachiopod *Peniculauris bassi* (McKee 1938). At its eastern and southeasternmost extent, the lithology and fauna within the Fossil Mountain and Harrisburg members are similar. In this area, geologists place the contact at a distinct change from the thick-bedded sandstones and sandy carbonates of the Fossil Mountain Member to thinner-bedded units of the Harrisburg. The latter contain a more abundant molluscan fauna. Along its depositional edge, for example, southeast of Winslow and south of Holbrook, the Kaibab sequence is difficult to subdivide into members (Cheevers and Rawson 1979). Since the Harrisburg Member has disappeared from this area, only the shoreward facies of the Fossil Mountain Member have been preserved.

Upper Contact

In northern Arizona and southern Utah, the Triassic Moenkopi Formation occurs above the Kaibab Formation. Because of their less resistant nature, however, Moenkopi redbeds at the Grand Canyon have been removed almost entirely by erosion. One consequence of this erosion is that the Kaibab Formation caps many of the vast plateaus that border the Grand Canyon. Only rarely are the uppermost beds of the Kaibab preserved.

In northwestern Arizona, southeastern Nevada, and southwestern Utah, discontinuous conglomerate-filled channels and breccia deposits occur between the Kaibab Formation and the Timpoweap Member of the Moenkopi Formation. Reeside and Bassler (1922) termed these deposits the Rock Canyon conglomerate for a channel 250 feet deep (75 m) and 700 feet (210 m) wide in Rock Canyon, which is north of Antelope Spring, Arizona. At several localities (for example, in the Beaver Dam Mountains), channels of the Rock Canyon conglomerate have scoured completely through the Harrisburg and into the underlying Fossil Mountain Member (Nielson 1981). The tectonic significance of the Rock Canyon conglomerate is unresolved, though associated features may represent paleokarst depressions.

In areas of southwestern Utah and southern Nevada where carbonates of the Timpoweap Member of the Moenkopi overlie uppermost carbonates of the Harrisburg Member, the formational contact can be difficult to determine (Bissell 1969; Nielson 1981). Geologists also have trouble distinguishing the contact when gypsiferous beds of the Lower Red Member of the Moenkopi Formation occur directly above gypsiferous beds of the Harrisburg Member (Bissel 1969; Cheevers 1980). To the north, in the mountains of western Utah and eastern Nevada, the Kaibab Formation is overlain conformably by the Permian Plymptom Formation and Gerster Limestone of the Park City Group.

LITHOLOGY AND COMPOSITION

The Kaibab Formation is a complex sedimentary package composed of a variety of rock types. Due in part to mixing between carbonate siliciclastic sediment and intense post-depositional (diagenetic) changes in composition (in particular silicification [chert formation] and dolomitization), rock ledges of the Kaibab appear similar at first glance. A number of detailed studies have delineated internal facies characteristics and distribution, providing resolution with respect to major changes in lithology, mineral composition, and faunal constituents.

Fossil Mountain Member

The Fossil Mountain Member of the Kaibab Formation is a prominent cliff that weathers to form distinctive pinnacles, or "hoodoos," below the rim of the canyon. The member thickens gradually westward and typically ranges between 250 and 300 feet (75-105 m). At Fossil Mountain along the south rim, where it is over two hundred feet (60 m) thick, cherty limestones that contain abundant whole fossils characterize the member (McKee 1938; Cheevers 1980; Hopkins 1986).

As observed in outcrop along both rims of the canyon, the Fossil Mountain Member exhibits a pronounced change in lithology, mineralogy, and faunal constituents from west to east (Fig. 3). In western Grand Canyon, for example (in an area west

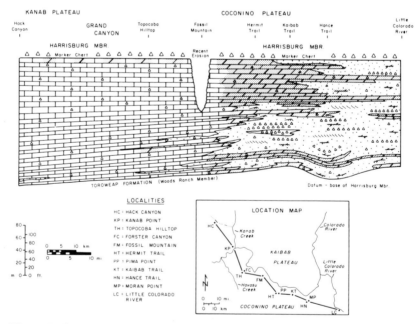

Figure 3. Diagrammatic cross section of the Fossil Mountain Member illustrating the west-to-east change in lithology and mineral composition at the Grand Canyon. See Figure 4 for list of symbols. Modified from Hopkins (1986)

GRAND CANYON GEOLOGY

of Fossil Mountain on the south rim and North Bass Trail on the north rim), the member contains a characteristic cherty, fossiliferous limestone with an abundant and diverse normal-marine fauna. This fauna includes brachiopods, bryozoans, crinoids, sponges, and solitary corals. Carbonate textures are dominated by skeletal wackestone (matrix-supported texture), with only minor amounts of admixed detrital quartz. Packstone intervals (grain-supported texture) are common locally, but typically form an insignificant percentage of the sequence. Sandstone comprises less than ten percent of total lithofacies and most commonly occurs near the base. Dolomite occurs only as scattered rhombs replacing a micrite matrix (carbonate mud).

In contrast, the Fossil Mountain Member becomes increasingly siliciclastic to the east. Here, dolomite is the predominant mineralogy, and a restricted-marine molluscan fauna is most characteristic (Fig. 3). Within the zone of most pronounced lithologic transition, at Hermit Trail, for example, sandstone and sandy carbonate comprise nearly fifty percent of the member.

Carbonate lithofacies consist of skeletal wackestone and mudstone textures that have been altered to dolostone. We can see a similar change on the north rim near Point Sublime. In the basal portion of the sequence, sandstone beds occur in close association with relatively siliciclastic-free, skeletal carbonate units, though mixing between lithologies near unit contacts is common. Scouring along basal contacts occurs locally, and in some cases, siliciclastic units pinch out laterally (Fig. 1). Preserved sedimentary structures are scarce since units typically are bioturbated intensely. We can recognize horizontal, ripple, and low-angle laminations within certain sandstone intervals. To the east along both rims, the percentage of sandstone increases significantly. Ultimately, the Fossil Mountain Member consists of approximately seventy-five percent sandstone or sandy dolostone (for example, at Desert View on the south rim and Cape Royal on the north rim).

An obvious lithologic characteristic of the Fossil Mountain Member is the amount and variety of chert (McKee 1938; Hopkins 1986). Chert is common as spherical nodules associated with siliceous sponges, which are most abundant in normal-marine carbonate facies in the western portions of the Grand Canyon. Irregular and branching chert forms also are typical within these facies. Many form by the selective replacement of burrow structures. Closely spaced nodular-to-bedded chert occurs as thin, laterally continuous intervals within sandstone facies. Petrographic study reveals that these chert horizons contain abundant relict sponge spicules and apparently formed by the recrystallization of biogenic silica during shallow burial. These horizons are most common in the eastern portions of the Grand Canyon, where they weather to form distinct recesses along cliff faces.

Another important chert type is a small, white nodule of cauliflower shape that occurs in both carbonate and sandstone lithologies, primarily where dolomitization has been pervasive within the member. These nodules represent silicified evaporites and suggest that a major portion of dolomitization in the Fossil Mountain Member was associated with the migration of hypersaline pore fluids in the shallow subsurface. The selective silicification of skeletal material represents an additional chert type. This silification results in excellent preservation of many fossils. Chert also is common as lenses and irregular nodules within sandstone units. The fundamental control on the origin of chert in the Fossil Mountain Member, and in the Kaibab Formation as a whole, is attributed to the primary distribution and abundance of siliceous sponges and spicules within the depositional environment (Hopkins 1986).

Harrisburg Member

The Harrisburg Member of the Kaibab Formation forms the uppermost cliffs and receding ledges along both rims of the canyon. This member consists of an assemblage of gypsum, dolostone, sandstone, redbeds, chert, and minor limestone. It generally is thinner than the underlying Fossil Mountain Member. It is difficult to determine its true thickness and extent, however, because of removal associated with the Permo-Triassic unconformity, evaporite dissolution, and recent erosion. Complete sections range from eighty feet (25 m) at eastern exposures along the Little Colorado River (Blakey and Middleton, unpublished data) to 300 feet (90 m) near Whitmore Wash (Sorauf 1962). The member thickens dramatically west and northwest of Kanab Creek and is thickest in northwestern Arizona, southwestern Utah, and southern Nevada, where gypsum comprises a considerable portion of the sequence (Fig. 2). In fact, gypsum currently is mined from the Harrisburg at Blue Diamond Hill, west of Las Vegas, Nevada. At its type section at Harrisburg Dome east of St. George, Utah, it is about 280 feet (85 m) thick (Reeside and Bassler 1922).

Despite considerable change in lithology and thickness, we can recognize at least six informal stratigraphic units within the Harrisburg Member (Figs. 4 and 5; Clark 1980; Blakey and Middleton, unpublished data). It is possible that a seventh unit is present in some areas. If this is the case, erosion has removed most evidence of this unit (Clark 1980).

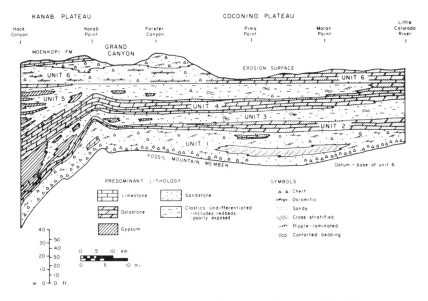

Figure 4. Diagrammatic cross section of the Harrisburg Member illustrating change in thickness and lithology laterally across the Grand Canyon. See Figure 3 for line of section. Modified from Blakey and Middleton (unpublished data)

The basal unit (#1) appears gradational with fossiliferous beds of the underlying Fossil Mountain Member. It is marked by a ragged cliff or site recess twenty to forty feet thick (6-12 m). The unit consists of bedded, nodular, and lenticular chert (marker-chert) that is gradationally overlain by sandstone and sandy dolostone. Locally, silicified evaporite nodules are abundant. Horizontal, ripple, low-angle, and hummocky cross stratification, preserved in places within the chert, occurs on a local basis.

KAIBAB FORMATION   235

Figure 5. View of receding ledges of the Harrisburg Member above cliff-forming Fossil Mountain Member. Note warping of beds (arrow) related to evaporite dissolution. Photograph by R.C. Blakey looks northwest from Kanab Point.

Petrographic study reveals the chert to contain abundant sponge spicules, varying amounts of detrital quartz, scattered peloids, but only sparse skeletal fragments.

A carbonate ledge five to twelve feet thick (1.5-3.5 m) constitutes a second unit (#2). At western localities, this unit has a limestone mineralogy and is characterized by packstone and wackestone textures containing a variety of fossil fragments that include bivalves, gastropods, crinoids, bryozoans, foraminifera (forams), and ostracods. At sections along Kanab Canyon and to the northwest, this unit has an oncolite-bearing cap and is locally brecciated. Sections to the east become increasingly dolomitic. They are dominated by mudstone textures that show cryptalgal laminations and contain intraclasts and calcite-filled vugs.

Unit three is a poorly exposed, slope-forming sequence that is lithologically variable and undergoes extreme changes in thickness. At eastern localities, this unit is about twenty feet (6 m) thick. Sandstone, sandy dolomitic mudstone, and local cryptalgal-laminated, dolomitic mudstone characterize unit three. To the west and northwest of Kanab Creek, however, portions of the sequence that contain considerable bedded gypsum and contorted red sandstone can exceed 100 feet (30 m) in thickness. In this region, this unit may show dramatic local changes in thickness and some warping of adjacent beds (Fig. 5).

A persistent medial ledge corresponds to unit four, which averages twenty feet (6 m) in thickness and ranges from ten to forty feet (3-12 m). This unit consists predominantly of sandy dolomitic mudstone that locally contains thin lenses of fossil hash, intraclastics, and cryptalgal-laminated horizons. Fossil fragments include bivalves, gastropods, and ostracods—with crinoids, bryozoans, and brachiopods occurring less commonly. The unit generally contains less sand to the west and toward the top. Local brecciation and warping is related to the dissolution of gypsum within unit three below.

Unit five, which forms the upper slope, exhibits lithologic and thickness trends that are similar to those of unit three. At eastern localities, this unit is poorly exposed (generally less than twenty feet thick [6 m]) and consists of ripple-laminated sandstone that contains chert locally. West and northwest of Kanab Creek, however, this unit approaches eighty feet in thickness (24 m) and consists of gypsum and interstratified siliciclastic redbeds. Unlike unit three, this unit contains a number of thin, laterally persistant cryptalgal-laminated dolostone beds. Apparently, it pinches out east of the Grand Canyon (Fig. 4).

Uppermost beds (unit six) form the chert-rubble erosion surface across much of the Grand Canyon region. Only locally is the unconformable contact with the Moenkopi Formation preserved. Where nearly complete, as in Robinson Wash south of Hacks Canyon (Blakey and Middleton, unpublished data), this interval approaches fifty feet in thickness (15 m) and consists of a lower ledge-forming, sandy, fossiliferous dolostone; middle slope and ledge-forming, cherty, ripple-laminated sandstone; and upper slope and ledge-forming cryptalgal-laminated and peloidal dolostone. West and northwest of Kanab Creek, the member contains a prominant molluscan fauna, which includes whole *Bellerophon* gastropods that are spectacularly jasperized locally. To the east, this unit is composed predominantly of dolostone containing increasing numbers of bivalves, gastropods, and peloids that form packstone textures.

PALEONTOLOGY

Marine invertebrate fossils are abundant in the Kaibab Formation (Fig. 6). Both members show marked changes in faunal abundance, diversity, and preservation from east to west. Since the detailed investigations by McKee (1938) and Chronic (1952), surprisingly few studies have focused specifically on paleontologic or paleoecologic aspects of Kaibab fossils (e.g., Beus 1964, 1965; McKee and Breed 1969; Mather 1970; Decourten 1976). Many of the fossil forms in the Kaibab Formation (e.g., trilobites) became extinct toward the end of the Permian.

Fossil Mountain Member

In western sections of the Grand Canyon, skeletal limestones of the Fossil Mountain Member contain a diverse faunal assemblage that includes a variety of brachiopods, fenestrate and ramose bryozoans, crinoids, siliceous sponges, and solitary corals. Fossils often are whole and unabraded—indicating little or no transport. Although fragments of trilobites, ostracods, and foraminifers can be identified in thin section, mollusks are rare within western facies.

Most characteristic of the Fossil Mountain Member at western localities are (1) large productid brachiopods (*Peniculauris bassi*), which often are silicified and found along bedding plane exposures in life-position (concave up)—sometimes with delicate spines attached—and (2) siliceous sponges [*Actinocoelia maeandrina* Finks (1960)], which commonly occur in the center of spherical chert nodules. Typically, brachiopods and sponges decrease in abundance toward the top of the member, while the number of fenestrate and ramose bryozoans and associated crinoid debris significantly increases.

Western faunal assemblages suggest unrestricted, open-marine conditions characterized by seawater of normal-marine salinity. They reflect deposition below a fair-weather wavebase. Numerous branched, horizontal, and irregularly inclined burrow traces characteristic of low-energy, offshore environments indicate a homogenization of sediment by biogenic reworking. The apparent vertical transition from abundant brachiopods and sponges to an assemblage dominated by bryozoans and crinoids suggests increasing water depths—from perhaps a few tens of meters to nearly one hundred meters.

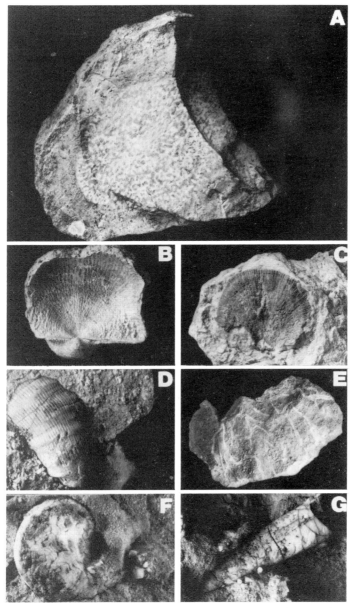

Figure 6. Fossil types representive of the Kaibab Formation at the Grand Canyon.
Examples from the Fossil Mountain Member: A. siliceous sponge (*Actincoelia
maeandrin*a) within chert nodule; B. productid brachiopod (*Peniculauris bassi*);
C. productid brachiopod (*Derbyia*); D. solitary or rugose coral; E. limestone
composed of branching or ramose bryozoans and crinoid debris. Examples from
the Harrisburg Member: F. coiled gastropod (*Bellerophontid*); G. scaphopod
(*Plagioglypta*). Photographs by Ted Melis

Associated with increasing siliciclastics eastward within the Fossil Mountain Member is a systematic decrease in skeletal types of normal-marine affinity within carbonate facies. In the eastern sections of the Grand Canyon, a molluscan fauna characterized by pelecypods (*Schizodus* most common) and poorly preserved gastropods dominates the member. These fossils typically occur as fossil molds. The most pronounced faunal transition found to date takes place in the upper portion of the member between the South Kaibab Trail and Hance Trail on the south rim and the North Kaibab Trail and Cape Royal on the north rim.

Lateral change in faunal assemblages within the Fossil Mountain Member reflects a shoreward transition to increasingly restricted-marine environments. Shallow-water, low-energy levels, limited water circulation, elevated temperature, and salinity characterize these environments. Whole productid brachiopods, particularly *Peniculauris bassi,* are present in some sandy dolomitic units to the east—suggesting that they may have been more tolerant to changing conditions or that conditions locally were transitional to normal-marine. Brachiopods that show indications of transport may have moved shoreward during storms.

Harrisburg Member

Molluscan faunal assemblages characterized by a variety of pelecypods and gastropods dominate fossils in the Harrisburg Member. In addition, scaphopods are very abundant locally. Nautiloid cephalopods and trilobites are present but vary in their distribution. Geologists can identify ostracods and foraminifers in thin section. These faunal types represent hardy individuals tolerant of a greater range in environmental conditions. Along with gypsum deposits and silicified evaporite nodules, they indicate a partially to a highly restricted, shallow-marine environment.

Brachiopods, bryozoans, crinoids, and other normal-marine organisms typically are rare, occurring as small fragments. However, they are present in increasing numbers within the carbonate beds found to the west. This suggests a possibility that sea water of normal or near normal salinity may have returned from time to time to the western Grand Canyon region during Harrisburg deposition.

AGE AND CORRELATION

Geologists have debated the age of the Kaibab Formation until relatively recently. While most agreed that the Kaibab was Leonardian in age, based on comparisons of brachiopod faunas and the distribution of the siliceous sponge *Actinocoelia maeandrina* Finks (McKee 1938; Finks and others 1961), the question remained as to whether part of the formation was slightly younger or in part Guadalupian (McKee and Breed 1969). Although fusilinids are not known from the Kaibab Formation, correlations based on brachiopod and conodont faunas have confirmed a late Leonardian age and helped to establish correlations with Permian sequences in regions beyond the Grand Canyon (Cooper and Grant 1972; Wardlaw and Collinson 1978, 1986). In general, the Kaibab Formation records widespread marine deposition about 250 million years ago, though in eastern Nevada and western Utah the Kaibab is, at least in part, older than the Kaibab in southern Utah, southern Nevada, and the Grand Canyon (Wardlaw and Collinson 1978).

The Kaibab Formation at the Grand Canyon correlates with the Concha Limestone and perhaps part of the Sherrer Formation in southeastern Arizona. To the east, the Kaibab is in part equivalent to the San Andreas Formation in New Mexico. Northward, the Kaibab correlates with part of the Grandeur Member of the Park City Formation in northern Utah and the Meade Peak Member of the Phosphoria Formation in southeastern Idaho and southwestern Wyoming. The Kaibab Formation also may be correlative with parts of the Leonard, Bone Spring, Victorio Peak Limestone, and Clearfork formations in the Permian Basin region of southeastern New Mexico and west Texas.

DEPOSITIONAL HISTORY

The close association of carbonate and siliciclastic sediments in the Kaibab Formation reflects a complex depositional history marked by major shifts of subtidal, shallow-marine environments. The overall depositional setting represents a mixed carbonate-siliciclastic ramp that existed along the southeastern margin of the Cordilleran miogeocline during the late Leonardian (Fig. 7). The Kaibab ramp extended across northern Arizona and into southern Nevada, at times exceeding 200 miles in width (125 km). In this setting, minor fluctuations in the relative position of sea level resulted in abrupt changes in depositional environments. Considering the quiescent tectonic setting of the Grand Canyon region during the Permian, it is most likely that these cycles were caused by glacial-eustatic sealevel oscillations (Kendall and Schlager 1981).

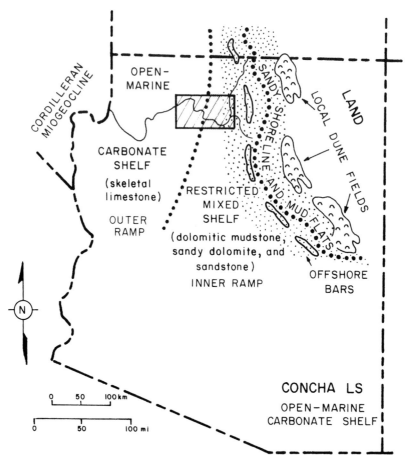

Figure 7. Hypothetical paleogeographic map of northern Arizona illustrating generalized environments and facies relationships during Fossil Mountain deposition (after Blakey and Knepp)

Fossil Mountain Member

The Fossil Mountain Member represents an overall transgressive phase of sedimentation punctuated by repeated regressive events of varying regional extent. Significant eastward shifts in carbonate facies across the Grand Canyon region record a relative rise in sea level. The westward distribution of siliciclastic facies, on the other hand, reflects seaward progradation of nearshore environments during a relative fall in sea level. The abrupt lateral and vertical lithofacies changes within the sequence (Fig. 4) reflect these lateral facies migrations.

Beyond the westward limit of siliciclastic progradation, carbonate sedimentation in the Fossil Mountain Member generally was continuous. These outer-ramp carbonate environments consisted of a diverse normal-marine faunal association that effectively baffled, trapped, and stabilized sediment, resulting in widespread limestone units dominated by wackestone textures. Discrete organic build-ups, such as sponge patch reefs, have not been documented. Local packstone textures reflect higher energy conditions that may represent localized shoals, transgressive lag deposits, or storm events. Dissipation of wave, tidal, and current energy across the broad, gently dipping seafloor was sufficient to restrict circulation in nearshore siliciclastic-dominated environments without the development of a distinct physical barrier (Irwin 1965). Since there was no exchange with seawater of normal-marine salinity, and the rate of evaporation was high, the salt content of water within inner-ramp environments increased. We see this from an eastward decrease in skeletal types of normal-marine affinity and the presence of molluscan faunas. The occurrence of silicified evaporite nodules within dolomitized facies also supports this thesis.

The sedimentary characteristics of sandstone units and their relationship with carbonate facies suggest deposition within inner-ramp, nearshore environments. Deposition may have occurred as relatively featureless sand sheets, lower shoreface deposits, or low-relief bars and sandwaves. A variety of processes may have created the currents necessary for sediment transport—including longshore processes, local wind regimes, and episodic storm events. Much of the sediment, however, experienced an intense biogenic reworking that resulted in structureless facies. The most probably source of siliciclastic detritus is coastal eolian complexes thought to border the Kaibab sea to the east (Fig 7). Unlike the underlying Toroweap Formation, however, geologists have not documented any intertonguing between marine and eolian deposits in the Kaibab Formation.

Thin, laterally continuous chert horizons, most common as interbeds within siliciclastic facies at eastern localities, could have formed by suspension settling of sponge spicules and carbonate mud within slight topographic depressions. These spicules may have moved shoreward as a suspension cloud during storm events, or they may have been winnowed by gentle fair weather and/or tidal currents.

Harrisburg Member

Sedimentary strata of the Harrisburg Member represent a distinct change from the diversely fossiliferous, skeletal carbonates characteristic of the Fossil Mountain Member. The six, informal, sedimentary units that comprise the Harrisburg sequence reflect deposition within predominantly restricted-marine environments during a cyclic westward retreat of the Kaibab sea. Repeated shifts in depositional environments are recorded by the alternation between carbonate, siliciclastic, and evaporite deposits across the Grand Canyon region (Fig. 4).

The Harrisburg sequence developed as a result of repeated transgressive-regressive cycles, with lower-order oscillations indicated by minor lithologic and textural variations within individual units. Carbonate deposition generally occurred within shallow, subtidal environments. Molluscan faunas, mudstone textures, and cryptalgal laminations suggest a low-energy, restricted setting though local fossil hash, intraclastic, and oncolite horizons indicate periodic higher energy conditions. The occurrence of brachiopods, bryozoans, and crinoids within carbonate units at western localities documents brief pulses of normal-marine conditions.

Sandstone intervals correspond to relative falls in sea level and result in a net transport of sediment derived from nearshore and coastal dune environments. Stratification is a product of the migration of small-scale bedforms, such as ripples and perhaps sand waves, generated by a combination of wave, tidal, and storm currents.

Thick accumulations of evaporites in the Harrisburg document at least two periods of extreme restriction associated with major regressive phases. The intense evaporation associated with an arid climate favored chemical sedimentation. Deposition of massive and bedded gypsum deposits probably occurred within localized hypersaline basins or lagoons bordered by a coastal mudflat/sabkha complex. The considerable thickness of the evaporite deposits suggests that sea water of normal-marine salinity replenished these restricted areas periodically.

Rapid change in the lithology and thickness west of Kanab Creek reflects periods of increased differential subsistence in this portion of the Kaibab ramp. The relationship, if any, of Harrisburg evaporite basins to recurrent movement along basement faults is uncertain. Other, more local thickness variations, along with warping and brecciation of adjacent carbonate units, are more likely related to post-depositional dissolution. Contorted beds in the upper part of the Harrisburg to the east suggest that the evaporites originally were more widely distributed and have been removed from much of this area.

SUMMARY

Rocks of the Kaibab Formation are testimony to the ancient seaway that covered most of the Grand Canyon region approximately 250 million years ago. The cyclic interbedding of carbonate and siliciclastic sediments documents a complex depositional history characterized by repeated shifts of subtidal, shallow-marine environments. Rocks of the Fossil Mountain Member record a west to east transition from fossiliferous open-marine limestone to restricted-marine sandy dolostone. The overlying Harrisburg Member reflects deposition during the cyclic retreat of the Kaibab Sea. A short walk down any of the canyon rim trails allows visitors to easily examine the great variety of rocks and fossils that comprise the Kaibab Formation.

CHAPTER 13

MESOZOIC AND CENOZOIC STRATA OF THE COLORADO PLATEAU NEAR THE GRAND CANYON

Michael Morales

INTRODUCTION

If you stand on either the north or south rim of the Grand Canyon, the soles of your shoes will rest on the cracked and weathered limestone of the Kaibab Formation. This topmost rock unit of the canyon was deposited near the end of the Paleozoic Era. As you peer into the deep chasm below, you will see a mile-thick (1.6-km) section of strata that accumulated during the Proterozoic Eon and Paleozoic Era. Now turn your gaze skyward and imagine a section of rocks extending *above* your feet for approximately another mile, about the same distance above the rim as the bottom of the canyon is below the rim. This exercise will give you an idea of the great thickness of marine and terrestrial rock layers that were deposited on top of the Kaibab Formation in several intervals during the Mesozoic Era (Hintze 1988). These sediments once covered the entire southwestern portion of the Colorado Plateau Physiographic Province (Fig. 1), an area that includes the Grand Canyon (Billingsley 1989).

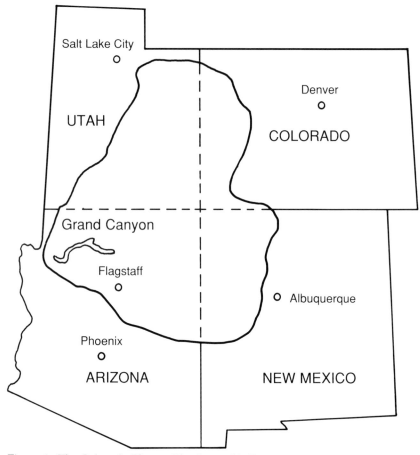

Figure 1. The Colorado Plateau Physiographic Province

In the vicinity of the canyon, denudation stripped away the Mesozoic strata during the Late Cretaceous in a major episode of erosion associated with the uplift of the southwestern Colorado Plateau (Lucchitta, chapter 15). During the early Cenozoic Era (Paleocene and Eocene epochs), one-half mile (.8 km) or more of terrestrial sediments and volcanics accumulated, only to be removed almost entirely in another episode of erosion during the Oligocene (Elston and Young 1989). Widely dispersed remnants of these deposits, primarily unnamed gravels and interbedded freshwater limestones, still crop out in the southwestern part of the Grand Canyon area (Elston and others 1989, Young 1989). Thus, nearly all Cenozoic and Mesozoic sedimentary rocks have been removed from the Grand Canyon area, leaving mainly strata of middle and late Paleozoic age (Kaibab, Toroweap, Coconino,

GRAND CANYON GEOLOGY

Redwall, Supai) as the topmost bedrock units surrounding the canyon. Scattered patches of Pleistocene volcanic rocks cover sedimentary units in and near the western Grand Canyon (Hamblin, chapter 17). South of the canyon, Tertiary and Quaternary volcanic rocks of the Mount Floyd, San Francisco, and Mormon Mountain volcanic fields (Fig. 2) cover the Paleozoic strata (Chronic 1983). These fields include a multitude of cinder cones (e.g., Sunset Crater), volcanic domes (e.g., Bill Williams Mountain), composite volcanoes (e.g., San Francisco Mountain or Peaks), and many lava flows (Holm 1987, Holm and Moore 1987).

North and east of the Grand Canyon, where less erosion has occurred, Mesozoic and Cenozoic strata of the southwestern Colorado Plateau are preserved in vast badland outcrops of cliffs, canyons, and plateaus. These rocks are especially visible in two regions: north of the canyon, in the area of northern Arizona and southern Utah that Clarence E. Dutton named the Grand Staircase, and east of the canyon to Black Mesa, Arizona. This chapter summarizes the main aspects of rock units, excluding surficial Quaternary deposits, that rest on top of the Kaibab Formation in areas of the Colorado Plateau near the Grand Canyon.

THE GRAND STAIRCASE

The topography north of the Grand Canyon is made up of alternating cliffs and flatlands that form a series of erosional steps of increasing elevation through Mesozoic and Cenozoic deposits. The Grand Staircase section of this region (Stokes 1986) includes the Uinkaret, Kanab, Kaibab, Markagunt, and Paunsaugunt plateaus; Antelope Valley; Telegraph Flat; the Little Creek, Wygaret, Kolob, and Skutumpah terraces and their equivalents to the east; the Block Mesas region; and the Moccasin Terrace (Fig. 3). The plateaus and terraces are bordered on their west and east sides by the major north/south-trending Hurricane, Toroweap-Sevier, and West Kaibab-Paunsaugunt faults (and associated monoclines and cliffs) and by the East Kaibab monocline (and related faults). In three areas, Zion and Bryce national parks and Cedar Breaks National Monument, deep canyons and extensive badlands have been carved into the rocks. Figure 4 illustrates the slightly north-dipping strata and stepped topography of the Grand Staircase.

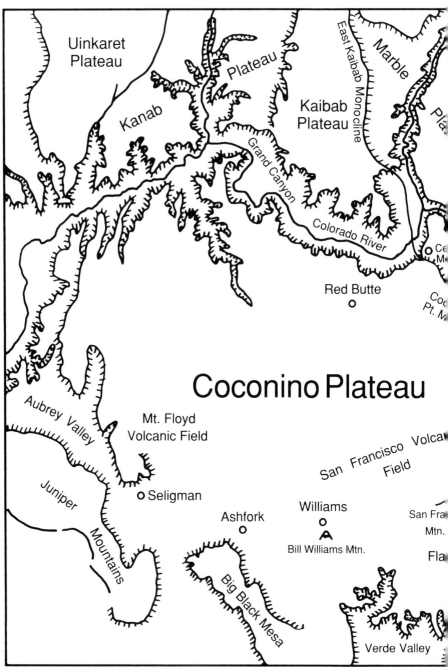

Figure 2. Physiographic map of the central and eastern Grand Canyon and surrounding areas (after Gregory 1950, Cooley and others 1969, King 1977, Billingsley and Hendricks 1989)

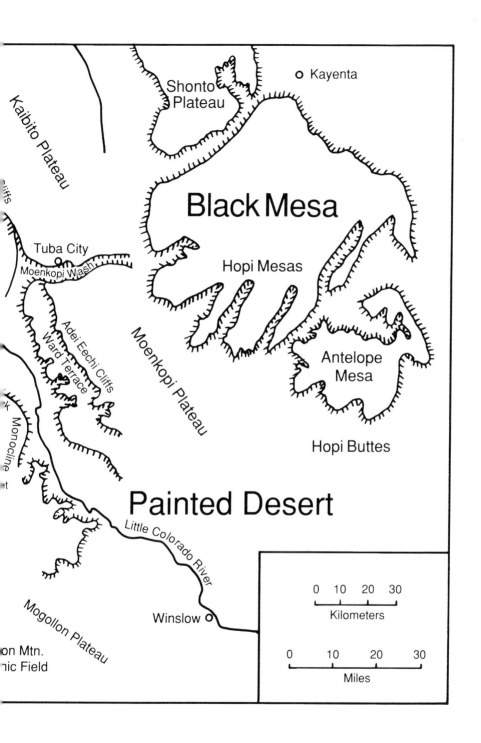

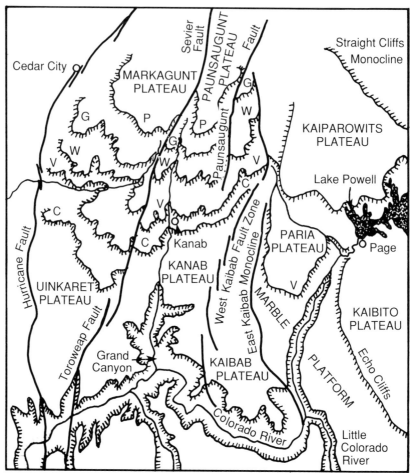

Figure 3. Physiographic map of the Grand Staircase section of the Colorado Plateau (after King 1977, Stokes 1986, Billingsley and Hendericks 1989) C - Chocolate Cliffs; V - Vermilion Cliffs; W - White Cliffs; G - Gray Cliffs; P - Pink Cliffs; T - Terrace; Cyn - Canyon.

The first line of cliffs north of the Grand Canyon is the Chocolate Cliffs. The name is derived from the red-brown mudstone of the Lower and Middle Triassic Moenkopi Formation that forms the majority of the cliff profile. This escarpment sometimes is called the Shinarump Cliffs after sandstones and conglomerates of the Shinarump Member of the Upper Triassic Chinle Formation that cap the cliffs. Although the Chocolate Cliffs officially terminate at the northern end of Kaibab Plateau, the same rock units form a cliff and slope line along the southern margin of Paria Plateau. There, however, the Moenkopi-Shinarump cliffs are not as well developed as the true Chocolate Cliffs. Erosional unconformities, which represent gaps in the rock/time record, separate the Moenkopi from the underlying Kaibab and overlying Chinle formations. The Moenkopi contains both marine and nonmarine sediments, whereas the Shinarump member of the Chinle is composed primarily of fluvial channel deposits. The flatlands and slopes above the Chocolate Cliffs, including Little Creek Terrace, Telegraph Flat, and related areas to the east, are formed by different fluvial and lacustrine members of the Chinle Formation.

The next step up the Grand Staircase is the very prominent Vermilion Cliffs. Starting at the base, red and purple deposits of the Chinle Formation, Wingate Sandstone, and the Moenave and Kayenta formations make up these cliffs, with the lower part of the Navajo Sandstone forming the caprock. The latter four formations comprise the Lower Jurassic Glen Canyon Group and are of

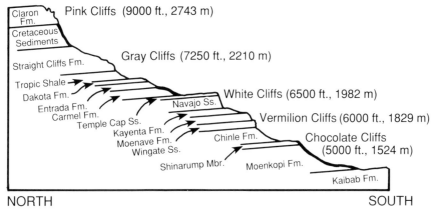

NORTH   SOUTH

Figure 4. Diagrammatic cross section of the Grand Staircase—vertical scale greatly exaggerated (after King 1977, Stokes 1986, Hintze 1988, and Clemmeson and others 1989)

primarily eolian and fluvial origin. An unconformity separates the Wingate Sandstone from the underlying Chinle Formation; however, contacts between the other units are gradational to intertonguing. Above the Vermilion Cliffs, the Navajo Sandstone forms the bench that includes Moccasin Terrace, Wygaret Terrace, and their equivalents to the east.

The next escarpment northward is called the White Cliff; it is composed primarily of the prominent buff-to-white colored Navajo Sandstone. Fluvial deposits of the Lower and Middle Jurassic Temple Cap Formation cap the cliffs, and there is an erosional unconformity between the two formations. The White Cliffs terminate on the west side of the Kaiparowits Plateau. The flatlands and slopes above the White Cliffs include the Kolob and Skutumpah terraces and their equivalents. Zion Canyon is carved into southern Kolob Terrace along the White Cliffs. The southern portions of the two terraces are made up primarily of the Dakota Formation overlying the Carmel Formation, which rests on the Navajo Sandstone. To the east, the Entrada Formation is present between the Carmel and the Dakota. The contact between each of the four formations is unconformable. Both the Carmel and Entrada are Middle Jurassic formations of the San Rafael Group. The former, however, is of marine origin, whereas the latter is predominantly tidal flat and eolian. Depositional environments of the Dakota Formation range from terrestrial to marginal marine.

The Gray Cliffs form a relatively low line that runs across the Kolob and Skutumpah terraces and extends to the eastern side of the Kaiparowits Plateau. They are formed by the gray Tropic Shale capped by the Straight Cliffs Formation. The Tropic Shale disconformably overlies and interfingers with the Dakota Formation. The Tropic Shale and the Straights Cliffs Formation have an intertonguing contact. Deposits of the Tropic Shale are marine, while those of the Straight Cliffs Formation are terrestrial to marine.

The northern portions of Kolob and Skutumpah terraces, formed by the Straight Cliffs Formation, continue northward from the Gray Cliffs to the final step up the Grand Staircase, the Pink Cliffs. These are made of Upper Cretaceous rocks of uncertain affinities (called Wahweap and Kaiparowits formations in early literature) that are capped by light pink and orange lacustrine and fluvial deposits of the lower Cenozoic (Paleocene?) Claron Formation (Hintze 1988). An unconformity separates the Claron from underlying strata. The Pink Cliffs form the southern escarpment of the Markagunt Plateau on the west and the Paunsaugunt Plateau on the east (Fig. 3). The badlands of Cedar Breaks and Bryce Canyon are carved into the Pink Cliffs on the southwestern Markagunt and southeastern Paunsaugunt plateaus, respectively.

SOUTH RIM TO BLACK MESA

The topography east of the Grand Canyon includes cliffs, terraces, plateaus, and generally north/south-trending fault zones and monoclines that cross a transect from the south rim eastward to Black Mesa. Unlike the Grand Staircase, this region of the southern Colorado Plateau has no general name. It includes the Coconino, Kaibito, and Moenkopi plateaus; Marble Platform; Ward Terrace; Black Mesa (including the Hopi Mesas); and the Little Colorado River Valley, which contains the western part of the Painted Desert (Fig. 2). The northeastern edge of the Coconino Plateau is marked by the East Kaibab, Coconino Point, and Black Point monoclines (and their associated faults). The Echo Cliffs, a result of erosion along the Echo Cliffs monocline, form the western border of Kaibito Plateau. The related Adei Eechii Cliffs comprise the western boundary of Moenkopi Plateau. Between the Kaibito and Moenkopi plateaus runs Moenkopi Wash, which extends eastward to Black Mesa. Figure 5 illustrates the stratigraphy and topography along the transect from the eastern part of the canyon's south rim to Black Mesa.

In the central and eastern portion of the Grand Canyon, the south rim represents the northern border of the Coconino Plateau. The Kaibab Formation caps the plateau here, except in rare places where remnants of Triassic rocks are preserved. At Cedar Mesa, slightly east of the south rim's Desert View overlook, the Moenkopi Formation overlies the Kaibab Formation; it, in turn, is capped by a resistant layer of Cenozoic volcanic rock. At Red Butte, south of Grand Canyon Village, the Moenkopi is covered by Shinarump deposits, which also are overlain by a younger lava flow.

On its northeastern border, the Coconino Plateau folds downward along the east to northeast-dipping East Kaibab, Coconino Point, and Black Point monoclines. This folding is especially noticeable at Gray Mountain, southeast of Desert View (Barnes 1987). Tidal-flat, fluvial, and eolian deposits of the Lower and Middle Triassic Moenkopi Formation border the edge of the plateau and extend to the middle of the Little Colorado River Valley and Marble Platform. Farther eastward, overlying red and purple deposits of the fluvial lower members of the Upper Triassic Chinle Formation make up the Painted Desert. A step up and to the east of the Painted Desert in the Little Colorado River Valley is Ward Terrace. This flat area is composed of the mainly gray lacustrine sediments of the Chinle's upper members.

East of Marble Platform and Ward Terrace, the topography includes two erosional steps upward (Fig. 5). The first is formed by the Echo and Adei Eechii cliffs, which border the Kaibito and Moenkopi plateaus, respectively. The second step includes the cliffs and top of Black Mesa. Units of the Upper Jurassic Glen Canyon Group form the Echo and Adei Eechii cliffs, which are capped by the Navajo Sandstone. The Wingate Sandstone is present only in the southern part of the Adei Eechii Cliffs; it thins northward to an erosional pinchout south of Moenkopi Wash. Strata of the Echo Cliffs dip steeply to the east because of the Echo Cliffs monocline. However, the beds of the Adei Eechii Cliffs dip only slightly in the same direction because the monocline does not extend that far south. Most of the surface of the Kaibito and

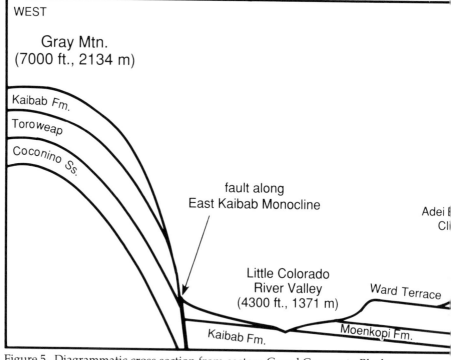

Figure 5. Diagrammatic cross section from eastern Grand Canyon to Black Mesa—vertical scale greatly exaggerated (after Brown and others 1958, Cooley and others 1969, Chronic 1983)

GRAND CANYON GEOLOGY

Moenkopi plateaus is formed by the Navajo Sandstone, except in places where less intense erosion has left younger strata, such as the unconformably overlying Carmel Formation.

The lower flatlands and slopes of Black Mesa are composed of, in ascending order, the Carmel Formation, Entrada Sandstone, Cow Springs Sandstone (west side) or Summerville Formation (east side), and Morrison Formation—all part of the Middle and Upper Jurassic San Rafael Group (Cooley and others 1969). The Entrada has conformable contacts with units below and above it, whereas the Cow Springs and Summerville, which laterally intergrade and interfinger, have an unconformable contact with the overlying Morrison. The upper slopes and cliffs of Black Mesa are made of Upper Cretaceous deposits of the Dakota Formation, Mancos Shale (correlative with the Tropic Shale), Toreva and Wepo formations, and Yale Point Sandstone. The latter three units

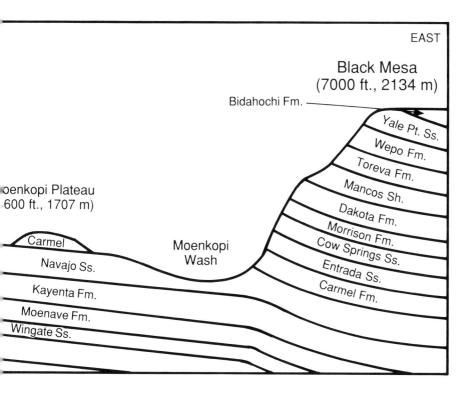

TABLE 1. Alphabetical list of formations of the southwestern Colorado Plateau north and east of the Grand Canyon. (Data compiled from many sources)

Mio-Plioc. = Miocence-Pliocene
Paleoc.? = Paleocene
Cret. = Cretaceous
Tri. = Triassic
Jur. = Jurassic

E. = Early
M. = Middle
Lt. = Late

inverts. = invertebrates
fresh. = freshwater
verts. = vertebrates
terres. = terrestrial

Formation Name	Age	Major Lithologies	Main Depositional Environment	Fossil Groups
Bidahochi	Mio-Plioc.	clay, volcanics	lacustrine, volcanic	terres. and fresh. verts; inverts; & plants
Carmel	M. Jur.	limestone, mudstone	marine	marine inverts., verts., & algae
Chinle Claron	Lt. Tri.	mudstone, sandstone	fluvial, lacustrine	terres. plants; fresh. inverts.
Cow Springs Sandstone	M. Jur.	sandstone	eolian	none?
Dakota	Lt. Cret.	sandstone, mudstone, coal	marginal marine, fluvial	terres. plants, verts., & inverts; marine inverts.
Entrada Sandstone	M. Jur.	sandstone, mudstone	fluvial, eolian	none?
Kaiparowits	Lt. Cret.	mudstone, sandstone	fluvial	terres. & fresh. verts., inverts., & plants
Kayenta	E. Jur.	siltstone, sandstone	fluvial, eolian	terres. plants, verts.; dinosaur tracks
Mancos Shale	Lt. Cret.	shale	marine	marine plants, verts., & inverts.
Moenave	E. Jur.	sandstone, mudstone	eolian, fluvial	fresh. fish, crocodiles, dinosaurs, & reptile tracks

GRAND CANYON GEOLOGY

Formation Name	Age	Major Lithologies	Main Depositional Environment	Fossil Groups
Moenkopi	E.-M. Tri.	mudstone	marine, fluvial, tidal flat	marine inverts.; terres. & fresh. verts., inverts., & plants; vert. & invert. trace fossils
Morrison	Lt. Jur.	mudstone, sandstone	fluvial	terres. & fresh. plants, verts. (esp.dinosaurs), inverts., & trace fossils
Navajo Sandstone	E. Jur.	sandstone, limestone	eolian lacustrine	terres. reptiles, plants, & invert. trace fossils; dinosaur tracks
Summer-ville	M. Jur.	sandstone	eolian	none?
Straight Cliffs	Lt. Cret.	sandstone, mudstone	fluvial, marginal	marine & fresh. inverts. fresh. marine & terres. verts.
Temple Cap	E.-M. Jur.	mudstone, sandstone	fluvial	none?
Toreva	Lt. Cret.	mudstone, sandstone,	littoral, fluvial, coal	vert. fragments?
Tropic Shale	Lr. Cret.	shale	marine	marine plants, verts., & inverts.
Wahweap	Lt. Cret.	sandstone, mudstone	fluvial	fresh. & terres. verts. & inverts.
Wepo	Lt. Cret.	sandstone, mudstone,	fluvial, paludal, coal	none?
Wingate Sandstone	E. Jur.	sandstone	eolian	dinosaur tracks
Yale Point Sandstone	Lt. Cret.	sandstone	littoral	none?

comprise the Mesaverde Group. An unconformity separates the Dakota from the underlying Morrison. The Mancos has gradational boundaries with the Dakota below and Toreva above. The lower and upper contacts of the Wepo are intertonguing. Deposits of the San Rafael Group, except the marine Carmel Formation, are terrestrial in origin, as are parts of the Dakota, Toreva, and Wepo formations. Other portions of the latter three formations represent near-shore environments (Harshbarger and others 1957, Wilson 1974). Sediments of the Mancos Shale are of marine origin.

Most of Black Mesa's top surface is formed by the Wepo Formation, but the Yale Point Sandstone caps it to the northeast. In the extreme southern portion of Black Mesa, in the area of the Hopi Mesas, the Miocene-Pliocene Bidahochi Formation unconformably overlies the Mesaverde Group (Cooley and others 1969). Lacustrine sediments dominate the Bidahochi, but fluvial and volcanic deposits also are present.

SUMMARY

Although the Proterozoic and Paleozoic rocks of the Grand Canyon record ancient environments, prehistoric life, and geologic events over hundreds of millions of years, they do not provide a complete picture of the area's geologic history. For the rest of the story, we must look to strata resting on top of the Kaibab Formation in and near the canyon.

CHAPTER 14

PHANEROZOIC STRUCTURAL GEOLOGY OF THE GRAND CANYON

Peter W. Huntoon

INTRODUCTION

The Laramide orogeny, a Late Cretaceous (Maastrichtian) through Eocene event, produced the first major episode of tectonic deformation in the Grand Canyon region since Precambrian time. The Colorado Plateau was both uplifted and horizontally shortened across reactivated basement faults that had northerly trends. Monoclines, most dipping to the east, developed in the overlying Phanerozoic rocks in response to reverse offsets along the basement faults. In addition, the intervening blocks were warped gently into generally north-trending uplifts and basins. Erosion, which accompanied the Laramide uplift of the Colorado Plateau and Mogollon Highlands to the south, beveled the generally northeasterly dipping Paleozoic and Mesozoic strata and exposed progressively older rocks toward the southwest. The resulting stripped surface, complete with incised paleocanyons, is well preserved on the plateaus surrounding the Grand Canyon. This surface, and the rocks deposited on it, have been little modified by the Pliocene incision of the Colorado River that produced the mile-deep Grand Canyon despite the fact that they are ten times older than the canyon.

The foregoing synopsis of Laramide events underscores the vast progress that has been made during the past century in deciphering the tectonic record preserved in the rocks of the Grand Canyon. Not only have we been able to map and date structures and deduce causative stresses, but we are beginning to weave these elements into the larger contexts of cordilleran tectonics, regional paleogeography, and sedimentation.

A number of important themes will be developed in this chapter. These include relating variations in the styles of deformation to different causative stress fields and documenting the overwhelming evidence for periodic reactivation of inherited fault zones by successive stress regimes. Another important theme is correlating uplift and subsidence to paleogeography, regional drainage patterns, and the movement of sediments into and out of the region.

Recent data indicate that most of the structures observed in the walls of the Grand Canyon vastly predate the Colorado River and its canyon. In fact, early Tertiary deformations are dated by sediments found in the remnants of Laramide, northward-draining canyons. Both these canyons and the sediments in them are ten times older than the Grand Canyon.

Even though late Cenozoic extensional faulting is spectacularly exposed, and exhibits an unprecedented record of recurrent activity here, the Grand Canyon still is best known in tectonic forums for its exceptional three-dimensional exposures of Laramide monoclines. The primary value of the Grand Canyon exposures is that the roots of the monoclines are unambiguously open to examination on the floor of the canyon.

The terms "Colorado Plateau region" and "Grand Canyon region" are used in this text for orientation purposes. However, recognize that the Colorado Plateau did not become fully defined until Miocene time, and erosion did not produce the Grand Canyon until Pliocene time. "Mogollon Highlands" refers to the general series of Mesozoic and Cenozoic uplifts occurring in the geographic regions south of the Colorado Plateau.

This chapter summarizes the work of many dozens of researchers spanning over a hundred years of effort. Detailed citations for each borrowed concept or fact lie beyond the scope of a work of this type. Consequently, the literature that is cited is designed to lead the interested reader to the most germane contributions or to sources that carefully develop important lines of inquiry. A reading of this text will be facilitated greatly by having the following set of geologic maps close at hand: Huntoon and others (1981, 1982, 1986) and Billingsley and Huntoon (1983).

THE COLORADO PLATEAU

As Figure 1 shows, the Grand Canyon occupies a position on the southwest corner of the Colorado Plateau, a geologic province that is underlain by a thick continental crust that became slightly separated on the east from the continental craton during the Cenozoic Era. The plateau is ringed by zones of intense deforma-

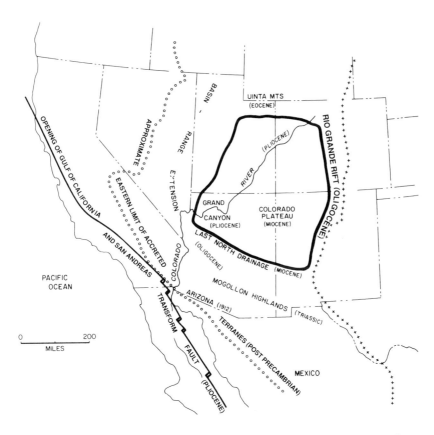

Figure 1. Selected tectonic elements in the western North American cordillera and the timing of their inception

tion, including the extending Rio Grande rift on the east, the compressional Uinta uplift to the north, and a drastically extended Basin-Range to the south and west. The Paleozoic rocks in the Grand Canyon are about a mile (1.6 km) thick. Nevertheless, they are just a thin, sensitive veneer resting on a nineteen- to twenty-five-mile (30-40 km)-thick crust comprised of an ancient basement complex.

The record of Phanerozoic deformation reveals that the Precambrian basement underlying the Colorado Plateau has enjoyed unusual stability since the close of Precambrian time. The Paleozoic rocks in the Grand Canyon region were deposited in equatorial regions of the earth as platform sediments on the southwestern part of a growing continent. At that time, the Colorado Plateau was an integral part of the larger land mass. Subsequently, the continent was rafted through plate tectonic processes some 2000 miles (3200 km) northward across the face of the earth. During this same period, the continent also was rotated a few tens of degrees in a counterclockwise direction (Elston and Breasler 1977).

Between the beginning of Cambrian and the end of Cretaceous time, a period of roughly 480 million years, net subsidence of the Precambrian surface in the Grand Canyon region amounted to between one-and-a-half and two miles (2.4 km to 3.2 km). In the last seventy million years, a period dominated by uplift, the Precambrian surface has risen in pulses a total of approximately two miles (3.2 km). Clearly, the cratonic block upon which the Grand Canyon occurs has moved great lateral distances and has both lost and gained considerable elevation. However, it has not undergone serious internal fragmentation since the close of Precambrian time.

INHERITED PRECAMBRIAN BASEMENT COMPLEX

A late Precambrian erosion surface exposed progressively older rocks in the Precambrian complex toward the west. Consequently, no younger Precambrian Supergroup sediments occur in Grand Canyon exposures west of Fishtail Canyon. Cambrian seas transgressed eastward over this erosion surface, which developed over the moderately disturbed sedimentary and volcanic Supergroup and an approximately 1.7-billion-year-old basement complex (Pasteels and Silver 1965). Relief on the erosion surface consisted of scattered, essentially isolated, small but rugged hills. These hills, some as high as twelve hundred feet (360 m), were buried by Cambrian sediments. The highest hills finally were covered by the Muav Limestone. Examples of buried Precambrian hills occur along the Colorado River immediately east of Deer Creek, in Modred Abyss and Monadnock Amphitheater, and under Isis Temple.

The younger Precambrian sedimentary-volcanic Grand Canyon Supergroup is preserved on a series of fault-bounded, northeastwardly tilted basement blocks that commonly dip between six and eight degrees. The strata occupy positions on the downturned northeastern slopes of the blocks and are beveled above by the late Precambrian erosion surface into northeastwardly thickening wedges.

Wedges to the east contain increasingly more complete, younger Precambrian sections. Consequently, we find the youngest Precambrian sediments in the easternmost part of the Grand Canyon. The maximum thicknesses of the younger Precambrian sedimentary section occur along the west side of the Butte fault, where Huntoon and others (1986) show a thickness of approximately nine thousand feet (2700 m) in the vicinity of Malgosa Canyon. Elston and Scott (1976) have reported thicknesses between twelve thousand and fourteen thousand feet (3600 to 4200 m), based on measured sections elsewhere.

The principal faults defining the boundaries of the tilted basement blocks in the eastern Grand Canyon are north- and northwest-trending Precambrian normal faults with extensive records of recurrent Precambrian displacement. These faults form zones that typically are spaced from fifteen to twenty miles (24 to 32 km) apart. The principal fault zones in the eastern Grand Canyon include the Butte, Phantom-Cremation, Crystal, and Muav faults. Synchronous Precambrian deformation consisting of parallel, parasitic folds and grabens developed in bands as wide as two and a half miles (4 km) along the eastern margins of the tilted basement blocks. Elston (1979) estimated the age of the latest Precambrian movements along these faults at 845 to 810 million years, using reset K-Ar ages from the 1.07 ± 0.07 billion-year-old whole rock Rb-Sr isochron Cardenas Lava.

The principal Precambrian faults that bound the tilted basement blocks offset an older set of northeast-trending reverse faults, the Bright Angel fault being most notable among these. The principal faults in the north-northwest and northeast sets left weaknesses in the basement that were reactivated repeatedly in Phanerozoic time throughout the eastern Grand Canyon.

The best evidence for Precambrian faulting in the western Grand Canyon occurs along the Hurricane fault. The metamorphic grades of deformed rocks between the surfaces of the Hurricane fault in exposures between Granite Spring and 224 Mile canyons represent pressures and temperatures unavailable in the region since the end of Precambrian time. These metamorphic products developed before deposition of the younger Grand Canyon Supergroup, revealing a long prehistory of activity along the fault. In addition, juxtaposition of Precambrian crystalline terrains with differing fracture-foliation fabrics and lithologies also is common along the fault in Grand Canyon exposures.

Rocks with a marked discordance in foliation trends are juxtaposed across the Meriwhitica fault, indicating a substantial but unknown magnitude of Precambrian offset. Shoemaker and others (1975) conclude that there is a good possibility that considerable right-lateral, strike-slip displacement occurred along the northeast-trending Precambrian basement faults prior to deposition of the Precambrian Grand Canyon Supergroup.

From the perspective of mechanical properties, the composition of the rocks below the Precambrian-Paleozoic contact is highly variable from west to east across the Grand Canyon. To the west, the Paleozoic rocks were deposited directly on crystalline rocks. To the east, increasingly thicker sections of younger Precambrian Supergroup rocks underlie the Paleozoic section. In some localities in the eastern Grand Canyon, Paleozoic rocks rest directly on the crystalline basement on the northeast sides of the principal Precambrian faults and on Supergroup sediments to the southwest. The structural strength of the unfaulted crystalline rocks tends to be reasonably isotropic; however, the Supergroup sediments are highly anisotropic as a result of bedding. In addition, the thick Galeros and Kwagunt formations of the Precambrian Chuar Group are very ductile even under near-surface load conditions—so ductile, in fact, that outcrops of these units in the eastern Grand Canyon commonly are obscured by landslides comprised of foundered, overlying rocks.

The geometric forms of Phanerozoic structures owing their origin to a specific stress field have been remarkably uniform throughout the region despite the wide range in the mechanical properties of the Precambrian rocks and despite the variable juxtaposition of differing Precambrian lithologies across faults in the underlying Precambrian basement. For example, the anticlinal and synclinal axes of Laramide monoclines converge downward on the coring basement fault at the Precambrian-Cambrian contact, regardless of the composition of the underlying Precambrian rocks. Consequently, it is apparent that Laramide folding does not extend downward into the underlying Precambrian Supergroup sediments significantly more than it extends into the crystalline rocks at locations where the Supergroup is missing. The implication, at least for the Grand Canyon monoclines, is that the mechanical properties of the rocks comprising the Precambrian basement were less important in dictating monoclinal fold geometry in the deforming Phanerozoic cover than were the causative Laramide stress field and the inherited geometry of the faults in the Precambrian basement.

PALEOZOIC AND MESOZOIC TECTONISM AND SEDIMENTATION

The Grand Canyon region occupied a position on a westward-sloping continental shelf during the Paleozoic Era. The huge interval of Paleozoic time, spanning 325 million years, was characterized in the Grand Canyon region by gradual net subsidence and net accumulation of sediments that now thicken westward from 3500 to 5000 feet (1100 to 1500 m) through the Grand Canyon. The top of the sedimentary pile fluctuated within several hundred feet of sea level, with roughly a third of this period spent below the ocean surface.

Mesozoic time in the Grand Canyon region was characterized by low but emergent conditions in which continental sedimentation in intracratonic basins prevailed. Large inland seas lay mostly to the north and northeast. These seas, however, did transgress southwestward over northwestern Arizona for geologically brief periods. Marine Mesozoic rocks in the Grand Canyon region include parts of the Triassic Moenkopi Formation, the Mid-Jurassic Carmel Formation, and the Late Cretaceous Mancos Shale. Although the Mesozoic section is eroded almost completely within the immediate vicinity of the canyon, upwards of

4000 feet (1200 m) of Mesozoic sediments were deposited in the region, based on thicknesses and paleotrends in Mesozoic outcrops to the east and north.

During Paleozoic and Mesozoic time, the area that was to become the Colorado Plateau was ensconced within the North American craton. Although buffeted by continental-scale tectonic events, the Grand Canyon region remained insulated by distance to the point that virtually nothing happened in the form of discrete offsets along faults in the underlying basement or local volcanism. The forces operating in the Grand Canyon region were attenuated sufficiently that deformation occurred only as broad-scale, but gentle, warping of the crust and variable rates of subsidence or uplift. The lack of angular unconformities in the Paleozoic section in the Grand Canyon reveals that, at least during Paleozoic time, we can best categorize the uplifts associated with erosional disconformities as epeirogenic in nature.

The existing major Precambrian faults, and the shear zones that preceded them, acted as structural hinges during Paleozoic and Mesozoic time by gently flexing in response to external tectonic events and to regional variations in sediment loading. Thus, the hinging faults formed internal boundaries within the mosaic of rather rigid basement blocks. Flexing across the hinges permitted minor adjustments in the dips of contiguous blocks. Such adjustments are revealed by variations in sediment thicknesses in the units being deposited over the region—and possibly by subtle facies changes across or between the hinges. Neither of these has been documented carefully in the Grand Canyon.

As the crust was thickening through sedimentation in the Grand Canyon region during Paleozoic and Mesozoic time, North America was growing rapidly in a westerly and southerly direction through the vastly different process of lateral continental accretion. The North American cordillera grew an additional thirty percent through accretion during the Mesozoic Era (Coney 1981).

Cambrian to Late Triassic Tectonics

The longest hiatus in the Paleozoic stratigraphic record is inclusive of Middle Ordovician, Silurian, and Middle Devonian time—a period of approximately 100 million years. Yet the contact between the underlying, undivided supra-Muav dolomites of Cambrian or possible Early Ordovician age and the Late

Devonian Temple Butte Formation is a disconformity throughout the Grand Canyon. Similarly, the erosion surface at the top of the Devonian rocks in the Grand Canyon, which presumably was a result of orogenic uplift associated with the Antler orogeny, is disconformable with the overlying Mississippian Redwall Limestone. Clearly, the Grand Canyon region was being subjected to large-scale cycles of uplift and subsidence; however, the crust in the region was not being internally deformed. The conclusion is that the region was distant from active structural zones.

Huge tracts of land were accreted to western North America during the Devonian-Mississippian Antler orogeny. Accretion appears to have been accomplished as shown on Figure 2a through plate convergence in an offshore arc-trench system. This mechanism involved the oceanward subduction of the oceanic lithosphere lying between an offshore trench and the continent. As the crust was consumed in the trench, both the trench and island arc migrated toward the edge of the continent. Once the trench and continent met, the island arc became sutured to the continental margin, and a new arc-trench system developed offshore to repeat the process. In this fashion, the continental margin was built progressively oceanward from central Nevada and eastern California.

Orogenic belts associated with this accretionary growth lay to the west of the Grand Canyon region and were characterized by large-scale eastward thrusting, uplift caused by crustal thickening, and deep-seated metamorphism. A north-trending depositional basin through east-central Nevada collected the clastics shed eastward off these uplifts. Marine limestones, thickening to the west (such as the Devonian Temple Butte and the Mississippian Redwall) were deposited in the Grand Canyon region on the eastern margin of the basin.

The most spectacularly preserved erosional event in the Grand Canyon Paleozoic section is the late Mississippian emergence of the Mississippian Redwall Limestone. A series of westward-draining valleys, incised as much as four hundred feet (120 m) into the limestone, occurs in the western Grand Canyon. The slightly uplifted Redwall surface was pervasively karstified so that the landscape took on the appearance of the modern Yucatan Peninsula. The Mississippian Surprise Canyon Formation fills the valleys and caves; yet it and the Supai Group lie disconformably on virtually all of the older rocks.

To date, the only offsets along faults that have been identified as dating from the Paleozoic Era in the Grand Canyon are inferred from minor angular unconformities at the top of the Mississippian

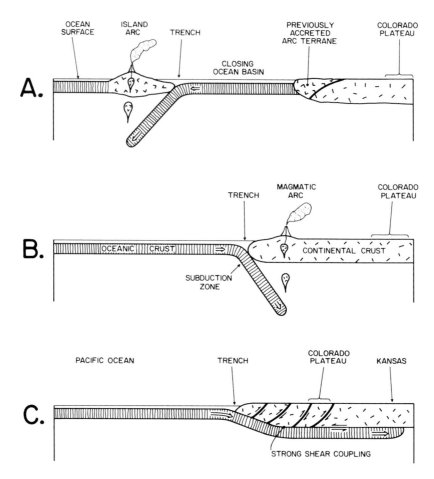

Figure 2. Convergent margin orogens along western North America. (A) Intraoceanic arc-trench orogen active periodically in post-Precambrian through Late Triassic time. Notice that the ocean basin closes, allowing island arc to accrete to continent, then another subduction zone and its island arc can form offshore and likewise eventually accrete to continent. (B) Slow landward subduction causing development of magmatic arc inboard on continent above steeply descending slab active from Late Triassic to Late Cretaceous time. (C) Rapid subduction resulting in shallow slab descent and slab underplating of continent to produce buoyant uplift and strong shear coupling with eastward telescoping of continental crust during Laramide time. Vertical scales are greatly exaggerated.

GRAND CANYON GEOLOGY

Redwall Limestone. These disturbances occurred in the interval between Mississippian Redwall Limestone and Pennsylvania Esplanade Sandstone deposition and predate deposition of the Chesterian Surprise Canyon Formation.

At least 150 feet of the Redwall section, including the entire Horseshoe Mesa Member, was truncated by erosion across the crest of a minor anticline at a site along the Tanner Trail (McKee and Gutschick 1969). This erosion surface is overlain unconformably by the Esplanade Sandstone. The fold is attributed to displacement along the underlying Precambrian Butte fault. Huntoon and Sears (1975) observed a similar truncation of the upper thirty feet (9 m) of the Horseshoe Mesa Member in a band one-quarter mile wide across the Bright Angel fault. They concluded that the truncation and an anomalous westward dip of ten degrees was caused by reverse motion along the underlying Bright Angel fault, resulting from reactivation of that Precambrian structure. This type of deformation can be attributed to horizontal, generally east-west compression in the context of present compass directions, possibly caused by compression associated with the Antler orogeny.

The Pennsylvanian and earliest Permian Ouachita orogeny developed when a terrain that now comprises the Yucatan Peninsula and South America was sutured to the North American continent along a collisional belt extending through northern Mexico, central Texas, and southeastern Oklahoma (Dickinson 1981). The resulting grand-scale internal stresses within the North American plate were sufficient to cause the intracratonic block uplifts and associated basins of the ancestral Rocky Mountains. The Uncompahgre uplift and Paradox basin developed northeast of the Grand Canyon region in response to this collision, and the Defiance uplift to the east gained additional elevation. Clastic sediments, such as those in the Supai Group, were shed from these highlands and transported into the Grand Canyon region, which remained low.

Late Triassic to Late Cretaceous Uplift

A world-wide reorganization of plate motions between Late Triassic and Late Jurassic time resulted in the opening of the Gulf of Mexico and the Atlantic Ocean through sea floor spreading. Concurrently, the Pacific Ocean crust began to subduct under North America in contrast to earlier oceanward subduction. The result was the inboard development of a Mesozoic magmatic arc south and west of the Grand Canyon region that was characterized by batholithic intrusions, arc volcanism, and thermotectonic uplift. The impact of these events was felt in the Grand Canyon region through the general emergence of the Colorado Plateau region, strong uplift of the contiguous Mogollon Highlands to the southwest, and uplift and eastward thrusting in the Sevier Highlands to the west (Woodward Clyde Consultants 1982). The uplift of the Mogollon Highlands, which caused a regression of the Moenkopi seas, was accompanied by the initiation of voluminous arc volcanism in the highlands or regions to the south of the highlands. This activity began about 230 million years ago.

Continental sedimentation became the dominant pattern in the Grand Canyon region as a result of this large-scale change. The thick, predominantly fluvial and eolian Mesozoic strata deposited across the Grand Canyon region were the result. The sources for these clastic deposits were the eroding uplifts.

The westerly Early and Middle(?) Triassic sediment dispersal patterns associated with deposition of the Moenkopi Formation in the Grand Canyon region were supplanted by Late Triassic northerly and northeasterly patterns off the Mogollon Highlands. The first of the Late Triassic units was the voluminous Chinle Formation, which, in the southwestern Colorado Plateau region, was comprised in part of volcanic detritus originating from the Mogollon Highlands and air-fall ash that probably originated from the northwest.

With the exception of the southward encroachment of the marine Mid-Jurassic Carmel Formation, continental sedimentation or erosion prevailed in the Grand Canyon throughout the rest of Mesozoic time. This lasted until the region was inundated by the southwestward-transgressing Late Cretaceous Mancos sea. Late Triassic to Miocene paleoslopes remained northeasterly off the Mogollon Highlands across the region. Debris that eroded from the Mogollon Highlands, or regions farther to the south and west, included the arc volcanics, sedimentary cover, and unroofed Precambrian basement. These constituted a primary source of sediment for many extensive Mesozoic units being deposited in the Grand Canyon region and to the northeast. Erosion in the Mogollon Highlands appears to have exposed Paleozoic and Precambrian terrains as early as Late Triassic time, based on clasts found in the Chinle Formation (Stewart and others 1972).

GRAND CANYON GEOLOGY

The last marine transgression, which occurred in Late Cretaceous time, came from the northeast. It resulted in the deposition
of both the Dakota Formation and the Mancos Shale. The surface
upon which the Dakota Formation was deposited was very
smooth and sloped gradually toward the northeast. It is doubtful,
however, if the sea extended completely to the southwest across
the entire Grand Canyon region. Substantial volumes of older
rocks had been eroded from the Grand Canyon region by this
time, especially to the southwest. Therefore, estimates of either
the total thickness of remaining Mesozoic sediments in the area or
estimates of the depth to the Precambrian basement below sea
level at the close of Cretaceous sedimentation must be speculative. Harshbarger and others (1957) report that the Dakota
Formation lies unconformably on progressively older rocks toward the south, in eastern Arizona. The unit rests directly on the
Paleozoic section at McNary.

Neither discrete offsets along reactivated Grand Canyon
basement faults nor development of local structural basins and
uplifts during pre-Laramide Mesozoic time have been documented within the confines of the Grand Canyon. The primary
reason for this is the scarcity of Mesozoic strata in the area. We do
know that Mesozoic tectonism in the region was sufficient to have
influenced facies and to have caused variations in sediment
thicknesses. Reactivation of basement faults undoubtedly occurred in response to east-southeast compression associated with
the Sevier orogeny along the northwestern margin of the Colorado Plateau in Cretaceous time.

LARAMIDE OROGENY

The Laramide orogeny, herein used to embrace latest Cretaceous through Eocene events, profoundly impacted the Grand
Canyon region. The orogeny caused (1) widespread uplift, (2)
east-northeast crustal shortening, (3) compartmentalization of
the Colorado Plateau region into subsidiary uplifts and basins,
and (4) widespread erosion. Crustal shortening was manifested
in the development of the Grand Canyon monoclines and in
regional warping of the intervening crust. The basin-uplift margins were defined for the most part by monoclines, imposing a
geography that has persisted to the present with little alteration.

Laramide erosion uncovered progressively older rocks to the south and west, including the Precambrian basement along the southwestern edge of the Colorado Plateau region. The enormous volume of detritus eroded from the Grand Canyon region and areas to the south was transported northward into the intracontinental basins of Utah and beyond. Areas in the Grand Canyon region that were monotonous lowlands at the close of Cretaceous deposition became an uplifted, partially dissected landscape characterized by north-flowing, sediment-choked streams. Developing topography included the structurally high parts of Laramide folds and the beginnings of step-bench topography.

Laramide orogenesis in western North America was characterized by widespread uplift and a significant eastward expansion of the belt of cordilleran deformation beyond the previous limits of Phanerozoic accretionary belts, orogenesis, and arc magmatism. Laramide deformation defined the cordillera as we know it in the United States. It also more than doubled the surface area encompassed within the pre-Laramide cordillera (see Fig. 1). Crustal contraction and eastward transport in a zone extending from the trenches along the west coast to the eastern limits of the Rocky Mountains characterized Laramide deformation. Types of deformation included eastward-verging thrust faulting and reverse displacements along reactivated Precambrian basement faults. The faulting was accompanied by the development of monoclines and anticlines in the covering sedimentary rocks. Arc magmatism swept eastward across the cordillera, then waned in intensity (Snyder and others 1976).

Cordilleran Tectonics

The plate tectonic explanation currently favored for Laramide orogenesis was a flattening of the angle of subduction of the Pacific oceanic plate under North America that resulted from rapid rates of subduction (Fig. 2). Dickinson (1981, p. 125) summarizes the attributes of this concept as follows. (1) The belt of magmatism moved inland as the locus of melting near the top of the subducted slab shifted away from the subduction zone. (2) Magma generation waned as slab descent became subhorizontal because the slab no longer penetrated as deeply into the asthenosphere. (3) Shallower descent of the slab increased the degree of shear and area of interaction between the descending slab and the overriding cratonic crust. As rapid subduction took place during

GRAND CANYON GEOLOGY

Laramide time, the subducted hot, buoyant, oceanic plate appears to have underplated North America as far eastward as the Great Plains, hence contributing to the uplift of the west.

The area that was to become the Colorado Plateau was caught in the eastward-compressing Laramide cordillera. The result, shown in Figure 3, was the development of generally north-striking, eastwardly verging monoclines as the underlying basement failed in response to east-northeast, horizontal compressive stresses during Late Cretaceous through Eocene time. Laramide monoclinal folding in the Grand Canyon region was accompanied by gentle regional warping of the intervening structural blocks, a process that produced uplifts such as the Kaibab Plateau and downwarps such as the Cataract Creek Basin.

Chapin and Cather (1983) present evidence for an early Eocene northeastward reorientation of Laramide compressive stresses within the Colorado Plateau region. This caused sixty

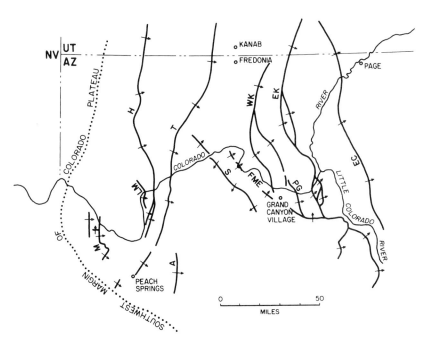

Figure 3. Locations of the Laramide monoclines in the Grand Canyon region, Arizona. From west to east: M - Meriwhitica; LM - Lone Mountain; H - Hurricane; T - Toroweap; A - Aubrey; S - Supai; FME - Fossil-Monument-Ermita; WK - West Kaibab; PG - Phantom-Grandview; EK - East Kaibab; EC - Echo Cliffs

miles (100 km) of north-northeastward translation of the Colorado Plateau along right-lateral, strike-slip faults. The movement partially decoupled the Colorado Plateau from the North American continent along the future Rio Grande rift. The early Eocene reorganization of stresses within the plateau region appears to have resulted in minor development, or reactivation, of northwest-trending monoclines in the Grand Canyon region.

Chapin (1983) proposed that the northward crowding of the Colorado Plateau into the Wyoming province was accommodated by crustal shortening manifested as thrust faulting and regional folding. The result was the production of east-and southeast-trending basins and ranges in Wyoming. The southernmost of these was the east-trending Uinta uplift bounded both to the north and south by thrust faults dipping under the range. Bernaski (1985) summarizes data that reveals that the Uinta uplift began to rise in Early Eocene time. Thus, Laramide structures outlined the eastern and northern boundaries of the Colorado Plateau as we know it today. The eastern boundary, now the Rio Grande rift, became better defined and more strikingly decoupled from the North American craton as a result of extension beginning in Late Oligocene time.

Grand Canyon Monoclines

As Figure 4 shows, Laramide monoclines formed in the Paleozoic and Mesozoic sedimentary cover throughout the Grand Canyon region in response to reverse movement along favorably oriented, preexisting faults in the Precambrian basement. The reactivated basement faults most commonly were steeply west-dipping Precambrian normal faults that served as preexisting structural discontinuities within the Precambrian basement complex. Typical east-west spacings between the monoclines in the Grand Canyon region are fifteen to twenty miles (24-32 km). These spacings gradually increase toward the east across the Colorado Plateau. The total crustal shortening that resulted from deformation within the monoclines on the western Colorado Plateau was less than one percent. The reason for this low percentage is that spacings between the monoclines are large in comparison to local shortening across them.

Figures 5 and 6 demonstrate that the Grand Canyon provides the finest cross sections through monoclines found on the Colorado Plateau. The exposures into the Precambrian basement are

A. Laramide folding over reactivated Precambrian fault; Precambrian fault was normal.

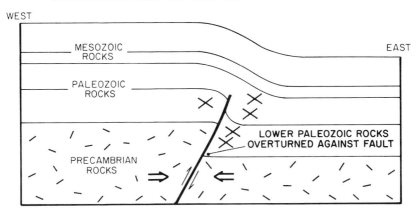

B. Late Cenozoic normal faulting.

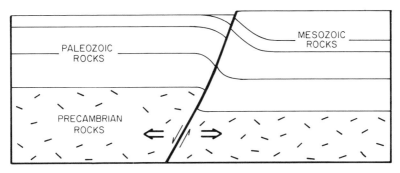

C. Late Cenozoic configuration after continued extension.

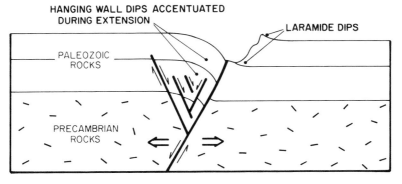

Figure 4. Stages in the development of a typical north-trending monocline-fault zone, Grand Canyon region, Arizona

Figure 5. East Kaibab monocline north of the Grand Canyon, Arizona. Dips in the upper Paleozoic section are moderate at this stratigraphic level. Plateau surfaces are comprised of the Permian Kaibab Formation. View is toward the north, Cocks Comb in center.

Figure 6. East Kaibab monocline at Chuar Butte, Grand Canyon, Arizona. Ductile thinning of Supai Group and Redwall Limestone in center is obvious. Note increasing steepness of dips with depth. View is toward the north, Kwagunt Butte to left.

particularly important. Archetypical cross sections to basement levels occur along the East Kaibab and Meriwhitica monoclines, neither of which have been downfaulted by subsequent exten-

278   GRAND CANYON GEOLOGY

sional events. The East Kaibab and Meriwhitica monoclines provide useful contrasts because the former is underlain by a thick ductile section of Precambrian Grand Canyon Supergroup sediments, whereas the latter is underlain by the crystalline basement complex. Despite these markedly different foundations, the style of folding in the two monoclines is identical.

Most segments of the Grand Canyon monoclines have developed over a single, high-angle, reverse fault, such as that shown on Figure 7. Such coring faults produced an abrupt, stepped offset at the top of the Precambrian basement surface. The dip of the basement fault typically is about sixty degrees. In profile, the anticlinal and synclinal axes in the overlying monocline converge downward on, and respectively terminate against, the underlying basement fault at or near the Precambrian-Cambrian contact. Consequently, the dips of the strata increase and the width of the fold decreases with depth in the monocline. The basement fault propagated variable distances upward into the overlying Paleozoic sediments proportional to its total displacement. The displacement on the coring fault gradually attenuated with elevation through ductile deformation of the Paleozoic sediments. The faults coring monoclines in the Grand Canyon rarely extend above the top of the Supai Group.

Shortening across the monocline at all levels is equal to the amount of horizontal override of the Precambrian-Cambrian contact across the coring reverse fault. Severe crowding (such as Fig. 8 shows) developed in the basal Paleozoic rocks in the vicinity of the synclinal axis during deformation. These rocks commonly are riven with low-angle, conjugate thrusts that mechanically thickened the strata in the syncline and operated to move material out of the syncline away from the fold. Basal Cambrian beds frequently are overturned and highly attenuated in thickness adjacent to the coring fault. The thicknesses of the Cambrian through Pennsylvanian rocks lying between the anticlinal and synclinal axis in the East Kaibab monocline are attenuated between thirty and sixty percent. As a result, the deformed Redwall Limestone has undergone low-grade metamorphism. This contrasts with comparatively gentle dips of less than fifteen degrees and virtually no attenuation at the level of the Permian strata.

The rocks occupying the anticlinal axis in the upper parts of Grand Canyon monoclines can be thinned mechanically and ductilely if displacements across the fold were sufficient to subject

Figure 7. Precambrian Cardenas Lava (left) displaced down in normal fashion against older Precambrian Dox Formation (right) along west-dipping basement fault underlying the East Kaibab monocline as viewed northward from Tanner Rapid, Grand Canyon, Arizona. In Laramide time, the fault was reactivated with approximately 600 feet of reverse (left up) offset to cause monoclinal folding of overlying Paleozoic and younger rocks at this location.

GRAND CANYON GEOLOGY

Figure 8. Mechanical crowding within the Cambrian Tapeats Sandstone at the synclinal hinge of the East Kaibab monocline, south wall of Chuar Canyon, Grand Canyon, Arizona. Notice almost vertical dips at this level in the fold. Arrow points to man for scale.

them to horizontal extension. However, because there was shortening across the monoclines, Laramide extension along the anticlinal hinges rarely resulted in brittle failure in the form of downward-propagating grabens, even at the top of the Paleozoic section.

Concurrent folding of the Precambrian rocks was relatively minor and consisted of minor horizontal shortening folds in the footwall block and drag folding adjacent to the fault surface. The stresses across the faults at the level of the Precambrian rocks were sufficiently large that nearby wall rocks frequently were deformed by numerous conjugate sets of minor, low-angle, thrust faults. In cases where the Precambrian basement is comprised of nearly horizontal, layered sediments, some horizontal shortening was accommodated in kink bands.

The Precambrian-Paleozoic contact on the footwall block along the East Kaibab monocline north of the Grand Canyon is broadly flexed over a distance of three to five miles (5-8 km) east from the fold. This flexing adds approximately 1000 feet (300 m) to the structural offset where it is best developed.

The Precambrian basement in the hanging wall of the Meriwhitica monocline in Milkweed Canyon is sharply infolded toward the coring fault in an antiform that extends a few hundred feet away from the fault. In this case, the folding of the crystalline rocks was accomplished through cataclasis, where the crystalline rocks were broken into blocks by fractures spaced a few inches apart in the tightest part of the fold. Yielding was accomplished in large part by slippage across all the fractures.

Stresses Causing Monoclines

The stress regime responsible for the development of the monoclines generally involved east-northeast, horizontally oriented maximum principal stresses and vertical minimum principal stresses. Reches (1978b) determined that the average strike of the maximum principal stress was N67E in the Paleozoic rocks along segments of the East Kaibab monocline, using stress orientations deduced from calcite twinning, minor faults, kink bands, and minor folds. Similar east-west maximum principal stresses can be demonstrated in both Precambrian basement and lower Paleozoic rocks at innumerable locations along the other Grand Canyon monoclines, using stress orientations deduced from small-scale, low-angle, conjugate thrust faults and slickenside striations on the thrust surfaces.

Hubbert (1951) used basic mechanical principals and experimental results to demonstrate that faults tend to develop in intersecting conjugate sets along planes oriented at low angles to the axis of maximum principal stress. His analysis assumes that the rock was reasonably isotropic with respect to structural strength prior to failure. Isotropy refers to the condition wherein the property being treated—in this case, strength—is the same regardless of direction about a point in space. Faults were predicted by Hubbert to develop at angles of between 27.5° and 30° as measured from the maximum principal stress axis.

The relationship between the angles of fracture and the axis of the maximum principal stress is summarized as follows. In brittle materials, the acute angle formed by intersecting shearing planes is bisected by the axis of maximum principal stress, and the obtuse angle by the axis of minimum principal stress. The intermediate principal stress is defined by the line of intersection between conjugate faults. In environments where vertical stresses dominate, the maximum principal stress is vertical, and the faults are predicted to dip between 60 and 62.5 degrees. In contrast, horizontal compression results in horizontally oriented maximum principal stresses and the development of thrust faults that dip theoretically between 27.5 and 30 degrees.

The presence of inherited zones of weakness render the rocks anisotropic with respect to strength, and such anisotropy can destroy the simple angular relationships between principal stress orientations and the conjugate fractures. There is overwhelming evidence from outcrops in the eastern Grand Canyon that the monoclines developed over high-angle, reverse faults, usually with dips of 60 degrees or more. It is tempting, therefore, to argue that the causative maximum principal stresses were vertical, the implication being that some type of block uplift mechanism was responsible for the deformation.

The fact is that the faults under the monoclines are reactivated Precambrian normal faults. What now is the up-thrown block in the basement was the down-dropped block in Precambrian time. Consequently, when Laramide deformation occurred, the Precambrian basement clearly was anisotropic as a result of the Precambrian faulting. The well-developed system of inherited Precambrian high-angle, normal faults was already in place to accommodate the Laramide strain. As a result, the predetermined dips of the reactivated faults do not bear a definitive relationship to the Laramide stress regime.

In order to use the dips of the faults to deduce causative principal stresses within the Laramide monoclines, we had to find one of two situations. (1) A monocline had to be located that did not develop over a preexisting basement fault. (2) Conjugate fractures had to be found in isotropic rocks that were far enough from the Precambrian faults so that reasonably accurate stress information could be deduced from them. Both objectives have been realized.

Huntoon (1981) found that a segment of the Meriwhitica monocline developed in a region that was not underlain by a Precambrian fault. This fortuitous circumstance occurs under a segment of the fold that developed over a six-mile-wide block of previously unfaulted crystalline basement rocks lying between two reactivating Precambrian faults. For the monocline to develop through the area, the intervening basement block had to fail along a new fault. The block did fail, and the basement exposures lying west of Milkweed Canyon, shown on Figure 9, reveal that

Figure 9. Laramide thrust fault (arrows) underlying the Meriwhitica monocline in small canyon to west of junction of Milkweed and Spencer canyons, Grand Canyon, Arizona. Thrust fault developed in previously unfaulted Precambrian crystalline basement in response to horizontally oriented Laramide maximum principal stresses.

the fault dips between 17 and 27 degrees. This dip of the fault is accompanied by numerous pairs of low-angle conjugate thrusts in the enclosing Precambrian and Cambrian rocks. East-northeast horizontal compression was the causative stress regime.

A similar basement thrust fault underlies a monocline in embryonic stages of development in the south wall of Diamond Canyon 1.5 miles (2.4 km) east of the Hurricane fault. In this case, the southwest-dipping fault crosscuts strong north-northeast striking vertical foliation, showing no regard for that preexisting basement fabric. Once again, horizontal compression is implied.

Small-scale, low-angle thrust faults, many occurring in conjugate sets (see Fig. 10), are very common in the lower Paleozoic and Precambrian rocks in close proximity to all of the Grand Canyon monoclines. The strikes of these faults parallel the strikes of nearby monoclines, but they tend to be discontinuous. Their presence reveals horizontal compression across the axes of the monoclines.

GRAND CANYON GEOLOGY

Figure 10. Minor conjugate thrust faults (arrows) in the Cambrian Bright Angel-Muav section east of the synclinal axis of the East Kaibab monocline, south wall of Kwagunt Canyon, Grand Canyon, Arizona. Maximum principal Laramide stresses were horizontal, intermediate principal stresses paralleled strike of monocline, and minimum principal stresses were vertical. Thrusting served to mechanically thicken the section parallel to the minimum principal stresses and shorten it horizontally parallel to the maximum principal stresses. View is approximately 30 feet high.

Sinuosity and Branching of Monoclines

As Figure 3 shows, the strikes of the Laramide monoclines are sinuous, and they tend to branch. Branching is well developed along the East Kaibab monocline in the eastern Grand Canyon. Here, prominent northwest-trending branches splay from the main fold, such as the Grandview-Phantom segment. The Hurricane monocline in the western Grand Canyon bifurcates southward into two parallel branches that die out within ten miles of their junction. In addition, monoclinal segments develop *en echelon* patterns, such as those observed between the various elements of the East and West Kaibab systems in the eastern Grand Canyon. Some monoclines are segmented with intervening gaps having no discernible deformation, such as the Fossil, Monument, and Eremita segments, which collectively constitute a very weakly developed branch of the East Kaibab monocline. Changes in strike and complicated branching, such as occur along the East Kaibab monocline in the eastern Grand Canyon, are linked directly on outcrops to intersecting basement fault patterns that have been reactivated. Favorable preexisting dips appear to be of paramount importance as reactivation occurs.

The lack of favorably dipping basement faults, or complexities in basement-fault geometry, resulted in anomalies in monoclinal geometry. For example, there is a nine-mile (14-km) gap in the Toroweap monocline in the vicinity of the headwaters of Diamond Canyon in the western Grand Canyon. The gap is deformed by a series of obliquely trending minor anticlines and synclines at the level of the Supai Group. Complicated post-Laramide faulting in the vicinity reveals the presence of a complex interlocking basement fault configuration that prevented the development of the monocline through this reach.

Age of Monoclines

It is difficult to establish a time for the inception of monoclinal folding in the Grand Canyon region because we have not yet discovered unconformities involving the Late Cretaceous section. However, such unconformities do exist in outcrops in the southern high plateaus of Utah and elsewhere in the Rocky Mountain region, and these establish a Maastrichtian initiation for Laramide deformation (Dickinson and others 1987).

The utilization of the Utah unconformities to reveal the onset of Laramide folding in the Grand Canyon region requires acknowledgement of the simultaneous development of (1) a regionally extensive Laramide erosion surface across the region and (2) sedimentary basins containing angular unconformities within and between Late Cretaceous, Paleocene, and Eocene rocks in southern Utah (Anderson and others 1975).

The presence of angular unconformities between the Late Cretaceous Kaiparowits, Campanian-Paleocene(?) Canaan Peak, Paleocene(?) Pine Hollow and Paleocene-Eocene Wasatch formations described by Bowers (1972) reveals that folding was occurring in the southern Utah basins during Sevier and Laramide time. Laramide sediments in the southern Utah basins were derived in part from the Grand Canyon region and Mogollon Highlands to the south, demonstrating that the southern regions were undergoing concurrent uplift.

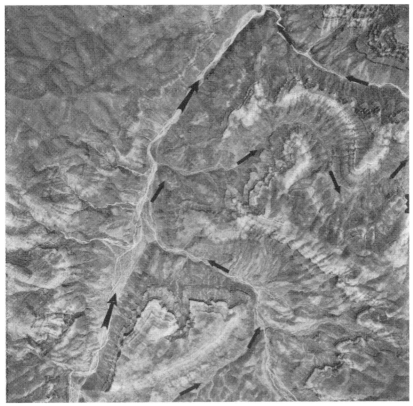

Figure 11. Deeply incised Laramide paleochannels in the Peach Springs Canyon area, Arizona. The two meander loops (small arrows) are older and less deeply incised than the Peach Springs paleocanyon (large arrows), which is colinear with the Hurricane fault. Offset Early Tertiary rocks along the floor of Peach Springs Canyon reveal that Late Cenozoic normal faulting postdated incision of both channels. Arrows show northward paleocurrent directions. U.S. Geological Survey photograph

Laramide erosion in the Grand Canyon area removed most of the remaining Mesozoic strata from the Permian surface east of the Aubrey Cliffs and removed Paleozoic strata as low as the upper part of the Muav Limestone to the west. The Meriwhitica and southern segment of the Toroweap monoclines were beveled during this erosional event prior to deposition of the Paleocene(?)-Eocene(?) rocks on the erosion surface described by Young (1982). The beveling of the monoclines by Laramide erosion indicates that they were developing concurrently with the localized folding that produced the Late Cretaceous and Paleocene unconformities in Utah. Young (1979) makes a case for recurrent Eocene(?) folding along the Meriwhitica monocline. He found lacustrine limestones that are restricted to Laramide paleo-canyons on the up-thrown block upstream from the anticlinal axes of the monoclines. He concluded that renewed folding caused the anticlinal hinges to rise enough that water ponded in the channel on the hanging wall block. The ponded lacustrine sediments associated with this fold represent the youngest epi-sode of Laramide monclinal folding found to date in the Grand Canyon region.

EOCENE TO MIOCENE STABILITY

Rapid rates of oceanic crustal subduction in trenches slowed along the west coast and attendant shallow angles of descent of the oceanic slab steepened under the North American plate about 45 million years ago (Dickinson 1981). The result of the steepened slab descent was a return of arc magmatism to the southern cordillera and a significant relaxation of east-west compressive stress across the region deformed during Laramide contraction. As a result, deformation ceased in the Grand Canyon region.

Tectonic quiescence at shallow crustal levels prevailed in the Grand Canyon region from about Late Eocene through early Miocene time, longer than in the Basin-Range to the west and south. The southwestern part of the Colorado Plateau remained undifferentiated from the Mogollon Highlands to the south, and together they constituted an uplift that early on was tilted slightly more steeply toward the north than is the case at present. The rapid early Tertiary erosion that accompanied Laramide uplift waned. This left an erosion surface that was beveled across successively older rocks toward the Mogollon Highlands. Drain-age was toward the north in canyons that originated in the Mogol-lon Highlands and terminated in basins on the Colorado Plateau.

Inherited Laramide Erosion Surface

Remnants of the Laramide drainage system are best pre-served on the southwest corner of the Colorado Plateau as hang-ing valleys west of the Toroweap monocline and south of the Colorado River. These paleovalleys bear silent testimony to the changing, tectonically driven paleogeography in the region during early Tertiary time. Also, early Tertiary sediments preserved in them have proven invaluable in dating Cenozoic structures. As Figure 11 shows, progressively younger and more deeply incised trunk channels in the drainage system are superimposed along the Hurricane fault zone. The deepest of these is incised over 1600 feet (500 m) below the Laramide surface. The youngest trunk channel is preserved as Peach Springs Canyon. It was reorganized in pre-Miocene time from a meandering pattern into a straight canyon that exited the region at some location between the Shivwits and Kaibab plateaus over what now is the north rim of the Grand Canyon.

The meandering, older paleocanyons in the vicinity of the Hurricane fault zone reveal gentle gradients prior to Late Cretaceous(?) or Early Tertiary incision. Channel filling, followed by uplift during late Laramide time, caused meander cutoffs and deep incision of the younger paleocanyons. Young and Brennan (1974) state that regional northward dips in the vicinity of the Hualapai Plateau during the period of maximum incision had to be between one half and one degree greater than at present in order for the streams to clear the present north rim of the Grand Canyon between the Shivwits and Kaibab plateaus.

Remnants of the Laramide erosion surface preserved under Miocene volcanics in the vicinity of the Meriwhitica monocline on the Hualapai Plateau reveal that the monocline was beveled to very low relief. Rocks as low as the upper Cambrian Muav Limestone were stripped from the monocline. Hindu Canyon, an Early Tertiary channel, crossed the fold with little regard for its presence despite structural offset of about 1000 feet (300 m).

The Laramide erosion surface occupies a position on successively younger strata from west to east along the south rim of the Grand Canyon. This occurred because the north-trending Hurricane, Toroweap, and Aubrey monoclines progressively stepped the Paleozoic strata down toward the east, so that the gentle Laramide surface developed across successively younger strata in this direction.

Early Tertiary Rocks

The floors of the Laramide paleocanyons, and the oldest sediments preserved in them, probably date from Paleocene(?) or possibly even Late Cretaceous(?) time. Northward flow is revealed by channel slope, imbricated pebbles, and increasing

Precambrian clast concentrations toward the south within specific sedimentary intervals. The early Tertiary rocks, which fill and partially bury the drainage system, are extremely important because they constrain the timing of the onset of post-Laramide faulting in the Grand Canyon region.

At Long Point, in Cataract Basin south of the Grand Canyon, Young (1982) has documented lacustrine deposits resting on the Moenkopi Formation that appear to date from Paleocene(?)-Eocene(?) time. This conclusion is based on viviparid gastropods. These lake deposits indicate that the Laramide erosion surface was virtually flat-lying at that time. He attributes the ponding to the development of the Cataract structural basin adjacent to the rising Kaibab uplift during the last stages of Laramide deformation.

Tertiary arkoses filled the earliest Laramide canyons in the vicinity of Peach Springs Canyon and on the Hualapai Plateau. They then spread over the adjacent plateaus. The earliest of the arkoses are characterized by abundant Precambrian clasts, rarity of volcanic clasts, deeply weathered profiles, and a deep, red appearance in many outcrops (Koons 1948). These crystalline arkoses—the Music Mountain Conglomerates—are in turn overlain by Late Oligocene volcanics in the Aquarius Mountains south of the Colorado Plateau.

The onset of Oligocene-Miocene arc volcanism to the southwest of the Colorado Plateau was heralded by increased amounts of volcanic and pyroclastic fragments in the upper part of the Oligocene Buck and Doe Conglomerate. These fragments were deposited in the aggrading paleocanyons on the Hualapai Plateau. This unit, in turn, is overlain by younger lavas and the ~17 million-year-old Miocene Peach Springs Tuff (Young and Brennan 1974). The volcanic activity was a harbinger of extensional tectonics that would impact the region strongly. It also would structurally differentiate the Colorado Plateau from the Basin-Range to the west and disrupt northward drainage onto the plateau from the Mogollon Highlands.

LATE CENOZOIC EXTENSIONAL TECTONISM

The onset of normal faulting and renewed uplift in the Grand Canyon region in Middle(?) to Late Miocene time was the outgrowth of complex plate-tectonic interactions along the west coast that began in Late Oligocene time. The sequence of events is illustrated on Figure 12, and the description that follows is summarized from Dickinson (1981, p. 129).

During Late Oligocene time, the Pacific-Farallon ridge and the Farallon-American trench obliquely converged upon each other along the continental margin near the border between what is today the United States and Mexico. The result was an annihilation of the ridge and trench at the point of contact and the mechanical substitution of right-lateral transform faulting along the area of contact. The transform fault progressively grew northward and southward as the trench-ridge system continued to converge. During this period, the transform faulting successively stepped inboard onto the continent. Near the beginning of Pliocene time, the San Andreas fault emerged as the principal transform fault along this plate boundary.

These events caused the cessation of subduction, eventual termination of arc magmatism, and widespread extensional tectonism in the region bordering the San Andreas transform fault

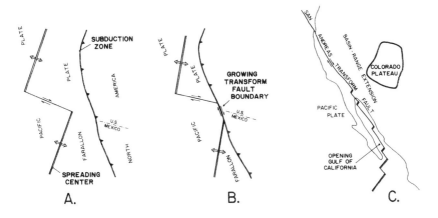

Figure 12. Stages in the Cenozoic collision of the East Pacific spreading center and Farallon subduction zone, western North America. (A) Spreading center and subduction zone converge during Early Tertiary time. (B) Spreading center begins to subduct near U.S.-Mexican border in Late Oligocene time, and interplate motion is accommodated along transform fault that grows north and south at continental margin as Farallon plate continues to subduct. (C) Remaining part of Farallon plate subducts between Late Oligocene and Late Miocene time, resulting in annihilation of both spreading center and subduction zone; interplate motion accommodated by San Andreas transform fault beginning in Pliocene time and continuing to present.

system. The result was wholesale extension within the Basin-Range to the west and south of the Grand Canyon region during Late Oligocene through Middle Miocene time and east-west opening of the Rio Grande rift beginning in Late Oligocene time. The opening of the Rio Grande rift caused between two and four degrees of clockwise rotation of the Colorado Plateau. The plateau was fully defined as a structural entity by Middle Miocene time, and its modern style of tectonic deformation commenced. In addition, more than 3000 feet (900 m) of uplift occurred along the Grand Wash Cliffs at the mouth of the Grand Canyon in Pliocene time, indicating that rates of uplift in the region have accelerated during Late Cenozoic time.

The major phase of Basin-Range extension in the vicinity of the Whipple Mountains along the lower Colorado River southwest of the Colorado Plateau commenced in Late Oligocene time, considerably earlier than the first surface manifestations of normal faulting on the Colorado Plateau. Detachment faulting in the Whipple Mountains is characterized by east-northeast dipping, low-angle normal faults in which the upper plate glided down-slope to the northeast on an unextended lower plate (Davis and others 1980). Although presently unproven, similar extension also may have been occurring during this period at middle and lower crustal levels under the southwestern part of the Colorado Plateau.

A speculative tectonic model accommodating Late Oligo-cene-Late Miocene extension at deep crustal levels under the Colorado Plateau and at shallower levels in the Basin-Range involves either a gently northeasterly dipping, crustal-penetrating normal fault or lateral extension within shear–bounded lenses, respectively illustrated in Figures 13D and 13C.

Figure 14 illustrates that northwest-striking blocks calving off the thin, trailing edge of the Colorado Plateau could account for tectonic erosion of the southwestern edge of the Colorado Plateau. These upper plate blocks now comprise the ranges in the area south and west of the plateau—including the Hualapai, Cerbat, Juniper, Aquarius, and Peacock ranges. In either scenario, the northeasterly tilted mountain blocks rest in tectonic contact on lower plate rocks, such as observed in similar environments in the Whipple Mountains. The detachment surface defining the base of the plateau appears to be spoon shaped and concaved upward, thus accounting for the southwestern bulge of the Colorado

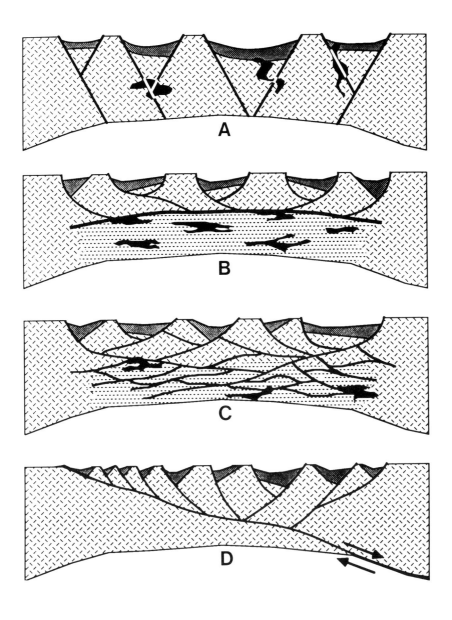

Figure 13. Summary of models used to explain extension in the Basin and Range Province southwest of the Colorado Plateau, Arizona (from Allmendinger and others 1987, Fig. 7). (A) Classic horst-graben model, (B) subhorizontal decoupling zone model, (C) shear zone bounded lense model, (D) crustal penetrating shear zone model. Either models C or D could have produced crustal thinning and subsidence along the southwestern Colorado Plateau in Late Oligocene-Early Miocene time.

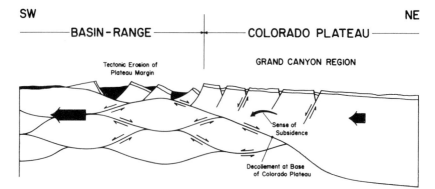

SW   NE
————————BASIN-RANGE————————+————————COLORADO PLATEAU————————

Tectonic Erosion of   GRAND CANYON REGION
Plateau Margin

Sense of
Subsidence

Decollement at Base
of Colorado Plateau

Figure 14. Cartoon illustrating tectonic erosion and subsidence along the southwestern edge of the Colorado Plateau over extending shear-bounded lenses in Late Oligocene-Early Miocene time. Lense concept from Hamilton (1982). Heavy arrows show absolute motions within the Basin-Range and Colorado Plateau provinces. Fine arrows show relative motions between shear surfaces. Notice that the relative motion between the lenses causes the crust to both thin and lengthen. Vertical scale greatly exaggerated, particularly at top

Plateau along the Hualapai Plateau over the thickened, northeasterly plunging axis of the upper plate.

Deep crustal extension probably was underway in Late Oligocene time. This allowed for (1) progressive tectonic erosion of the Colorado Plateau toward the northeast, (2) structural differentiation of the plateau from the Basin-Range in Miocene time, (3) thinning of the crust under the western part of the Colorado Plateau from about twenty-five to nineteen miles (40-30 km), and (4) nearly one degree of down to the southwest tilting of the southwestern edge of the Colorado Plateau. The latter easily accommodates sufficient subsidence of the Laramide erosion surface on the Colorado Plateau to force the abandonment of Laramide paleocanyons and to force the establishment of the west-flowing Colorado River in Late Miocene(?)-Pliocene time. Such subsidence also readily explains why early Tertiary rocks in the southern high plateaus of Utah now lie 2000 feet (600 m) or more above some of their source areas in the Grand Canyon region.

The relative motion of the Colorado Plateau above the decollement was down toward the northeast. Because such motion also corresponded to the opening of the Rio Grande rift, absolute regional motions had to include a slight clockwise rotation of the

294   GRAND CANYON GEOLOGY

plateau (Fig. 15) and southwestward extrusion of lower plate rocks from under the plateau. This model implies that the Colorado Plateau was rotating into extensional space created within the Basin-Range Province, wherein the lower plate was moving into, and westward in concert with, the Basin-Range at a rate slightly greater than the upper plate carrying the Colorado Plateau. Thus, the entire Basin-Range Province was moving away from the Colorado Plateau to the southwest at all crustal levels.

By Miocene time, the western part of the Colorado Plateau was undergoing the first significant east-west crustal extension to affect the region since late Precambrian time. The Grand Canyon

Figure 15. Postulated clockwise rotation of the Colorado Plateau into extensional space created in the Basin-Range during Late Oligocene-Early Miocene time. Arrows in the Basin-Range schematically illustrate that all rocks within the province moved away from the Colorado Plateau at rates which increased to the west as the crust within the Basin-Range simultaneously extended.

region was experiencing east-west extension that caused down to the west normal faulting. This tectonic stress regime has persisted to the present though activity probably peaked in Pliocene time. The principal Miocene and younger faults coupled with the Laramide monoclines to outline the plateaus and basins in the Colorado Plateau as we know them today.

Normal Faulting

The strain resulting from late Cenozoic east-west extension in the Grand Canyon region was accommodated mostly by normal displacements along faults in north/south-trending zones. The first failures, such as along the Hurricane fault zone (Fig. 16), reactivated the principal Precambrian basement faults in the zone. In most instances, the principal normal faults severed the Laramide monoclines, displacing the west block down, opposite to Laramide downfolding to the east (Fig. 17). This resulted in laterally continuous faulting along the strikes of most of the Laramide monoclines. Strong local northwest, north, and north-

Figure 16. View toward the northeast along the main strand of the Late Cenozoic Hurricane fault from the top of Diamond Peak, Grand Canyon, Arizona. Arrows point to the Cambrian Tapeats Sandstone on the respective sides of the fault. The east (right) dipping Laramide Hurricane monocline is not developed in this reach.

GRAND CANYON GEOLOGY

Figure 17. Two down to the west, Late Cenozoic normal faults sever the east (left) dipping Lone Mountain monocline south of Parashant Canyon (foreground). Faulted surface is the top of the Permian Esplanade Sandstone.

east trends developed along the principal faults, reflecting prominent local trends along reactivated basement faults.

New faults developed as the strain increased within specific zones. These faults are arranged planimetrically in two, or even three, intersecting sets. The sets thereby break the surface into a mosaic of fault-bounded blocks. In profile, each of the sets contains conjugate faults (faults with parallel strikes but opposing dips) that converge downward within the zone. Thus, the shallow crust was broken into blocks bounded by faults having four, or even six, different orientations. As extension progressed, the shattered Phanerozoic section deformed almost entirely by faulting. The net result was an east–west lengthening as well as an vertical thinning of this section. The presence of numerous faults with different orientations allowed the mass to deform without significant rotations of the fault-bounded blocks. This style of deformation has been treated theoretically by Reches (1978a) and is well exposed along the Hurricane fault zone. A predominance of dipslip faulting is revealed by slickenside striations such as those from the Hurricane fault illustrated in Hamblin (1965, Fig. 2).

Both displacements along individual faults and fault densities increase with depth in the Paleozoic section within the major fault zones, such as the Hurricane, Toroweap, and Eminence zones.

This finding reveals that the subsidiary faults propagated upward from shallow levels within the brittle lower Paleozoic carbonate section or, in some cases, from the upper part of the Precambrian basement. Displacements attenuate with elevation in the Paleozoic section through ductile deformation within the clastic Pennsylvanian and Permian sediments. This relationship is especially clear on the map of the Hurricane fault zone by Huntoon and others (1981).

There appears to be little indication that one or another of the prominent northeast, north, or northwest fault trends developed before the others in the western Grand Canyon region. Rather, as extension progressed, all trends were active to some degree to compensate volumetrically for both the east-west lengthening of beds and vertical thinning. In contrast, Huntoon and Sears (1975, p. 470-471) provide evidence, based on fault intersections exposed in Bright Angel Canyon, that much of the northeast-trending late Tertiary displacement along the Bright Angel Fault postdates late Tertiary faulting along northwest trends in that area.

Extensional Sag

Infolding of the Paleozoic rocks toward the fault on the downdropped, hanging wall blocks is common along many large displacement normal faults in the Grand Canyon. Dips of normal faults in the Paleozoic section commonly are greater than seventy degrees. This contrasts with sixty-degree dips on the same fault surfaces at deeper levels where reactivated basement faults are involved in the deformation. In some locations, the dips of fault surfaces tend to decrease gradually with depth. In either case, as displacement proceeded in the less steeply dipping, deeper reaches, separations tended to develop between the fault surfaces at higher elevations. The result was infolding of the upper part of the hanging wall block toward the fault surface, a phenomenon that is well documented along the Bright Angel fault, and a process that has reached advanced stages along the Hurricane fault zone.

In zones where a Laramide monocline was downfaulted opposite in sense to the dip of the fold, the Laramide dips in the hanging wall became progressively more accentuated as extension continued, yet dips in the footwall remained unaffected (see Fig. 4). Hamblin (1965) was able to document this fact by comparing the dips in successive Cenozoic lava flows that crossed the Hurricane monocline north of the Grand Canyon. He found that the younger flows in the hanging wall block were less steeply infolded toward the fault than the older flows. Also, the offsets across the younger flows were less than those across the older flows. It is clear, therefore, that the degree of infolding of the hanging wall is related to the amount of displacement along the fault.

Extensional Basins

The area encompassed within the Toroweap and Hurricane fault zones along the Colorado River is characterized by modest, late Cenozoic structural subsidence resulting from east-west crustal extension across the zones. Extension across numerous northwest-trending faults in the Hurricane fault zone in the area centered on Parashant Canyon has progressed sufficiently to create a ten-mile-(16-km)-diameter structural basin with approximately six hundred feet (180 m) of structural closure.

The finest example of a late Cenozoic extensional basin in an embryonic stage of development occurs on the Coconino Plateau in the Rose Well area between the headwaters of National Canyon and Chino Wash. Numerous normal faults, most downthrown to the west, comprise a north-trending zone that deforms the Kaibab surface into an elongated basin measuring five by fifteen miles (8 by 24 km). Sedimentation in the basin has not kept pace with subsidence. Consequently, there is somewhat more than thirty feet (10 m) of topographic closure at the north end of the basin at Sink Tank. Total structural closure is approximately five hundred feet (150 m).

A larger, north-trending extensional basin with about 1000 feet (300 m) of closure is centered around Blue Springs in the canyon of the Little Colorado River. The high-angle, normal faults allowing for the sag are exceptionally well exposed in three dimensions in the Little Colorado gorge.

Age of Faulting

Definitive timing for the inception of late Cenozoic faulting in the Grand Canyon has eluded researchers. Tertiary rocks that could be used to bracket the onset of faulting are missing from the eastern Grand Canyon, and those in the western Grand Canyon, such as the rocks shown in Figure 19, have so far provided only

Figure 18. Two prominent Late Cenozoic normal faults, which comprise the southwestern boundary of a graben, offset the Permian Esplanade Sandstone down to the northeast, four miles southeast of the mouth of Whitmore Canyon, Grand Canyon, Arizona.

minimum ages. It is likely that the first faults to become active in the region were the Grand Wash and Hurricane faults, and this activity probably dates from Middle Miocene time. It may have begun as far back as Early Miocene(?) time (see Fig. 20).

The Grand Wash fault, which comprises the western boundary of the Colorado Plateau in Arizona, is downthrown to the west and is buried by the upper part of the Miocene(?)-Pliocene Muddy Creek Formation. The offset along the fault at the mouth of the Grand Canyon has been estimated by Lucchitta (1967, Fig. 13) to be as great as 16,000 feet (4900 m). Downdropping, accompanied by down to the east rotation of the western block, created the Grand Wash trough, which then filled with Muddy Creek sediments. It appears that the initial faulting along this major extensional structure was synchronous with the latter part of the major phase of Basin and Range extension to the southwest, which Hamilton (1982) assigns to Late Oligocene through Late Miocene time. Young and Brennan (1974) observed that eruption of the ~17 million-year-old Peach Springs Tuff predated most of the offset along the southwestern part of the Grand Wash fault. They based their findings on slight pre-tuff, southwestern-facing scarp de-

Figure 19. Early Tertiary Roberts Roost Fanglomerate offset down to the east by Late Cenozoic normal faulting (arrows) at the head of Diamond Canyon, Grand Canyon, Arizona. View is toward the north.

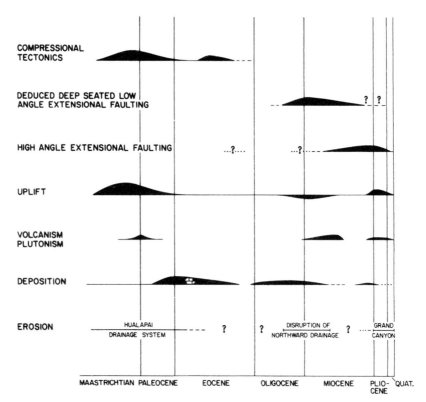

Figure 20. Temporal relationships between tectonics, sedimentation, and erosion in the Grand Canyon region, Arizona, during Laramide and post-Laramide time

velopment along the fault and the fact that the tuff is tilted on ranges that subsequently were partially buried by the Muddy Creek Formation. Therefore, major movement occurred along the Grand Wash fault after deposition of the Peach Springs Tuff, but prior to deposition of the upper part of the Muddy Creek Formation. These relationships bracket the major offsets across the fault between earliest Middle Miocene and latest Late Miocene time.

Some of the offsets along the Grand Wash fault probably were synchronous with deposition of the lower Muddy Creek Formation in the deepening Grand Wash trough, but relationships proving this remain buried. Minor post-Muddy Creek displacements along the Grand Wash fault appear to account for the folding of the Muddy Creek Formation in the vicinity of Lake Mead and for the faulting, by a few feet, of a gravel deposit located three and one-half miles (5.6 km) south of the Diamond Bar Ranch

(Lucchitta 1967). Clearly, however, most of the activity along the Grand Wash fault in Arizona was finished by the beginning of Pliocene time.

The Hurricane fault is the southern, waning extension of the Wasatch fault zone that, in Utah, comprises the physiographic boundary between the Colorado Plateau and the Basin-Range Province. The province boundary steps westward to the Grand Wash fault in Arizona. Maximum late Tertiary offset across the Hurricane fault zone in the Grand Canyon is in excess of 3000 feet (900 m) at Granite Park.

The timing of the onset of late Cenozoic normal faulting along the Hurricane fault is difficult to determine. Young and Brennan (1974) point out that Truxton Valley and upper Peach Springs Canyon, which are aligned along the Hurricane fault, are unusually wide and filled with sediments that predate the Peach Springs Tuff. They conclude that the fault existed prior to eruption of the tuff, so erosion progressed more rapidly along the trend. Thus, they push the inception of faulting back in time. An alternative explanation is that erosion followed the coeval trend of a southern, now eroded, segment of the Laramide Hurricane monocline. This interpretation does not require such early faulting.

The Peach Springs Tuff and underlying Tertiary rocks are offset by the Hurricane fault near the head of Peach Springs Canyon. Unfortunately, the closest exposed offset Paleozoic rocks lie three miles (5 km) to the north, precluding accurate differentiation of pre- and post-Peach Springs Tuff faulting. The offset across the tuff is between 200 and 300 feet (60 and 90 m) as is the offset associated with the Paleozoic rocks to the north. Consequently, it is doubtful if there was pre-Peach Springs Tuff displacement in the area. However, the uncertainties in correlating the similar offsets between the two sites does not justify discounting the possibility of small pre-Peach Springs Tuff displacement.

If we assume that there was no faulting along the Hurricane fault prior to the eruption of the tuff, the inception of late Tertiary faulting is post ~17 m.y., which is post-earliest Middle Miocene in age. Gardner (1941) concluded from outcrops in southern Utah that 75 percent or more of the total offset across the fault occurred in Late-Miocene(?) and Pliocene time, an observation that commonly is cited to support a peaking in the rate of normal faulting in the region during Pliocene time.

Recurrent Late Cenozoic Faulting

The most spectacularly exposed and complete records of late Cenozoic faulting in the Grand Canyon region occur along the Hurricane and Toroweap faults. The quality of this record is made possible by deep cross sections and the presence of successive late Cenozoic lava flows, each of which exhibits an offset proportional to its age. In addition, as shown in Figure 21, there are Quaternary cinder cones and alluvium that record the latest offsets. The resolution of the number and timing of late Cenozoic displacements is constrained only by the number of late Cenozoic rocks deposited across the two faults.

Highlights from outcrops near Whitmore Wash along the Hurricane fault serve to illustrate the quality of the record present in that single area within the Grand Canyon. The Paleozoic rocks are offset approximately 1000 feet (300 m), whereas older Pleistocene basalt flows that fill an old channel of Whitmore Wash to a depth of about 600 feet (180 m) are displaced only 75 feet (23 m). A slightly eroded scarp in Whitmore Wash extends eight miles (13 km) north of the Colorado River and offsets by fifty feet (15 m) a

Figure 21. West down-faulted Quaternary basalt (right of cone) and alluvium along the Toroweap fault in Prospect Valley south of the Colorado River, Grand Canyon, Arizona. Notice that the offset of the basalt is greater than that of the alluvium. View is toward the southeast.

　　　　　　　　　　　GRAND CANYON GEOLOGY

stage IV younger Pleistocene basalt flow that cascaded into the canyon from the east. It also offsets alluvium by about fifteen feet (5 m). With only four distinctly different units, we are able to bracket a minimum of four faulting events. Certainly, there were others during this interval of time. The offsets of the basalts were Quaternary events, but probably not Holocene.

Scarps in Pleistocene(?) alluvium along the Aubrey, Toroweap, and Hurricane faults also demonstrate Quaternary activity in the western Grand Canyon region. However, Quaternary offsets have not been documented in the eastern Grand Canyon region, indicating that there was a lack of Quaternary activity or that evidence for such offsets has been obliterated by erosion. No Quaternary basalts cross faults in the eastern Grand Canyon though alluvium is common along many faults, especially on the plateaus. Holm (1987) summarizes evidence for Quaternary faulting at the northeastern edge of the San Francisco volcanic field immediately to the southeast of the Grand Canyon. Although Holocene offsets have not been proven in that area, Akers and others (1962) found a fresh scarp in a Quaternary gravel that they suspected might be Holocene in age.

Near Surface Structural Localization of Volcanism

Volcanic rocks erupted through vents on the plateaus immediately adjacent to the Grand Canyon range in age from Late Miocene through Holocene. Best and Brimhall (1970) note the following relationships between volcanism and structure in the area of the Uinkaret Plateau. (1) Although the inception of high-angle, normal faulting predates basaltic volcanism, normal faulting and volcanism have operated simultaneously throughout most of Late Cenozoic time. (2) A shift in fault activity from the Grand Wash to the Hurricane-Toroweap zones has been paralleled in time by an eastward shift in volcanism. (3) With but a few exceptions, vents throughout the region lie between fault lines and are independent of them. (4) Many of the basalts on the Uinkaret Plateau carry mantle-derived peridotite inclusions, indicating that the magma was derived from the upper mantle.

Dutton (1882) and Koons (1945) observed the tendency for cones on the Uinkaret Plateau to occur in areas between major faults rather than along them. While there are several hundred basalt cones present in the San Francisco volcanic field south of the eastern Grand Canyon, only a few are related in an obvious way to major fault systems. In contrast, half the silicic-to-interme-diate volcanics lie along the Mesa Butte fault zone, where it passes under the San Francisco volcanic field.

Although the correlation between basalt vents and surface expressions of faults is tenuous, many vents are aligned in trends that parallel the local strikes of structures in the Grand Canyon region. For example, the Hancock Knolls are comprised of eleven centers in a nine-mile (14-km)-long alignment six miles (10 km) west of, and parallel to, the Supai monocline and fault. Three basalt plugs form a second parallel alignment two miles west of the Supai monocline in exposures on the Esplanade bench next to the Colorado River. The Colorado River gorge exposes a small dike swarm in lower Paleozoic rocks between two of the cones along this second alignment. The dikes at lower depths of expo-sure reveal that laterally continuous fissures deep in the Paleozoic section served as feeders to the discontinuously spaced plugs higher in the section.

Numerous generally north-trending alignments of cones that lie parallel to nearby segments of the Toroweap and Hurricane faults occur in the Uinkaret volcanic field. Deeply eroded out-crops of associated dike swarms on or below the Esplanade surface in the western Grand Canyon reveal that the dikes are localized along fractures and minor faults that parallel nearby normal faults.

Examples include dike swarms within a mile west of the Hurricane fault and south of the mouth of Whitmore Canyon and another swarm found in the walls of Tincanebitts and Dry can-yons. These relationships reveal that upward movement of magma probably was facilitated by the presence of extensional fractures in the upper part of the basement complex and in the Paleozoic section. The fact that vents do not occur preferentially along the principal late Cenozoic faults in the region indicates that, in general, upward movement of magma in the crust was in-dependent of these faults at great depths. Rather, dikes and vents tend to localize on extended fractures only in close proximity to the surface.

Disruption of Drainage

Tectonic differentiation of the Colorado Plateau from the extending Basin-Range to the south and west had progressed sufficiently to sever the northward flow of streams across the plateau margin in the western Grand canyon region by late Early Miocene time. Similar severing of the northward-flowing streams took place as early as Oligocene time in the region south and east of the eastern Grand Canyon (Peirce and others 1979). Two other processes operated to further disrupt these streams on the Colorado Plateau itself: (1) partial burial by Miocene volcanics and (2) southwestward tilting of the plateau surface of probable Late Oligocene to Early Miocene age.

The modern Colorado River became integrated from east to west across the region following the close of Miocene(?)-Pliocene Muddy Creek deposition in the Grand Wash trough. The river cut into the formerly northward-sloping Laramide erosion surface and carved the Grand Canyon to within fifty feet (15 m) of its present depth by early Pleistocene time.

The western margin of the plateau is demarcated by the west-facing Grand Wash fault scarp from the area north of the Utah state line southward around the entire Hualapai Plateau to a terminus south of the Cottonwood Mountains. The rate of fault-scarp development that accommodated the downdropping of the terrane to the south and west in this region was sufficiently rapid to sever the streams that formerly flowed onto the plateau. Once blocked, the streams drained into growing extensional basins such as the Grand Wash trough and Red Lake basin. Young and Brennan (1974) place the timing of final drainage disruption along the Hualapai Plateau as occurring at about the time the Peach Springs Tuff was erupted in late Early Miocene time. Their interpretation is based on the fact that sediments preserved in the early Tertiary canyons on the Hualapai Plateau above the Peach Springs Tuff do not contain clasts derived from the region south of the plateau.

MODERN GEOPHYSICAL SETTING AND TECTONISM

Although the Grand Canyon is situated on the western edge of the Colorado Plateau physiographic province, the modern geophysical properties of the crust under it appear to be transitional in character between those of the stable interior of the plateau and the extending Basin-Range Province to the west. For example, the thickness of the crust is 25 miles (40 km) or more under most of the plateau, but the crust tapers to about 19 miles

(30 km) thick from the easternmost Grand Canyon to the Grand Wash fault (Smith 1978). The crust is only 15 miles (24 km) thick in southern Nevada and western Utah. Heat flow through the plateau interior typically is 1.5 heat flow units (1 heat flow unit = 10^{-6} calorie/centimeter-second), but from east to west across the Grand Canyon it rises to two units (Blackwell 1978). Increased heat flow in the western Grand Canyon appears to be substantiated by a warming of groundwater discharged from springs west of Kanab Canyon. Temperatures of waters from springs in the lower Paleozoic section in the eastern part of the Grand Canyon range from 50 to 73 degrees F, whereas those sampled from the same rocks to the west range from 64 to 86 degrees F (Loughlin and Huntoon 1983).

The intermountain seismic belt coincides with the Basin and Range/Colorado Plateau crustal transition zone. The seismic belt is characterized by high seismicity and tectonic extension. Seismicity associated with it extends as far eastward in the Grand Canyon as the West Kaibab fault zone (Smith and Sbar 1974). Fault plane solutions for earthquakes in the intermountain seismic belt in southern Utah reveal that east-west extension is occurring and that focal depths are shallow, most at depths of less than ten miles (16 km). The faults have steep dips, and motion on them tends to be near vertical. Smith and Sbar (1974) report that an earthquake on November 11, 1971, near the Hurricane fault zone north of Cedar City, Utah, produced three north-trending fractures in alluvium. The longest fracture exhibited horizontal east-west extension and could be traced for half a mile. Focal depths during the quake were from near surface to slightly over a mile (1.6 km) deep.

Citing a combination of refraction, reflection, low resistivity, and high Pn velocity data, Keller and others (1975) postulate the presence of a mantle upwarp that occupies a band at least fifty miles (80 km) wide under the transition zone. The upwarp extends approximately thirty miles (50 km) eastward under the Colorado Plateau. The eastern limit of the upwarp boundary appears to correlate with a lateral change in crustal magnetization reported by Shuey and others (1973).

The coincidence between the postulated mantle upwarp and the intermountain seismic belt led Keller and others (1975) to speculate:

The presence of the upper crustal low-velocity layer may be related to the mechanism of Cenozoic faulting and seismicity. Thus the presence of (a low velocity layer) east of the Wasatch front could provide an explanation for the presence of Cenozoic block faulting east of the province boundary. The shallow seismicity characteristic of the intermountain seismic belt also extends east of the Wasatch front and is roughly coincident with the easternmost zone of Cenozoic normal faulting.

Late Cenozoic basaltic volcanism also characterizes much of this same region, a finding that is consistent with extensional tectonics.

SUMMARY

Late Precambrian time in the Grand Canyon region was characterized by erosional beveling across an uplifted, block-faulted mountainous terrane produced by extensional tectonism. That same stress regime again has been impressed on this region. The outcome once again is fairly well assured—the plateaus surrounding the Grand Canyon will erode; the canyon, in all likelihood, will gradually disappear. Over 800 million years ago, the Precambrian normal faulting was just the latest in a series of earlier convulsions that deformed the crust here. Are we seeing tectonic cyclicity here? Is tectonism in the Grand Canyon record a good example of the geologic doctrine of uniformitarianism?

On a trivial level, the answer to the former appears to be "yes," but if we carefully honor our semantics, the answer is "no." Cyclicity requires the repeated return to a starting point. We are not seeing true tectonic cycles here, such as rotations through crustal extension and compression, or through passive-margin and active-margin plate tectonics.

However, the tectonic processes do adhere to uniformitarianism. Similar stress regimes, regardless of their origin, repeatedly produce predictable styles of deformation. Uplift, regardless of the cause, does lead to erosion. When it comes to tectonic processes, perhaps Ecclesiastes (1:9) stated it best: "The thing that hath been, it is that which shall be; and that which is done is that which shall be done: and there is no new thing under the sun."

CHAPTER 15

HISTORY OF THE GRAND CANYON AND OF THE COLORADO RIVER IN ARIZONA

Ivo Lucchitta

INTRODUCTION

Many desert rivers, including the Nile River of Africa, the Tigris and Euphrates rivers of Asia, and the Colorado River of the western United States, obtain their waters from mountain ranges far removed from the warm and parched lands that border most of their courses. This combination of abundant water and warm climate has made parts of these rivers lifelines, allowing the cultivation of large areas that otherwise are desert.

The Tigris, Euphrates, and Nile led to the establishment of ancient civilizations such as those of Mesopotamia and Egypt—and probably to the development of settled urban living as we know it today. It is no coincidence that cities such as Ur, Babylon, Thebes, and Memphis were sited on the banks of these rivers.

The Colorado River cannot claim to be the cradle of civilization; its virtue, instead, is to support modern cities such as Los Angeles, Las Vegas, and Phoenix that could not flourish in its absence. Many of these cities are far removed from the Colorado, whose waters are brought to them by a system of artificial impoundments and aqueducts. The Colorado, therefore, has enabled us to carry out a great experiment, based on contemporary technology, in settling areas that are in themselves inhospitable.

Great demands are made on the waters of the Colorado River, yet the amount of available water is finite. This has resulted in a host of social, political, and engineering problems that have centered chiefly on how many reservoirs should be built and who should get the water. Many of these problems are fertile areas of endeavor for geologists concerned with the practical application of their science.

For other geologists, however, the river has been the source of quite different interests and controversies. These interests are theoretical in nature and have to do with the general problem of how rivers are born and develop. When and how did the Colorado River come into being? How do rivers like the Colorado evolve? When did canyon cutting and correlative uplift occur? How and why has the Colorado cut across the many belts of high ground astride its course? How quickly was the Grand Canyon cut? Might the answers to these questions give us an insight into how rivers in general establish their courses?

This chapter considers the history of the Colorado River and its Grand Canyon (Fig. 1). To understand what follows, it is important to remember two contrasting views regarding the history of the Colorado River. The first is that the river from birth has been part of an integrated drainage system with a course approximating the present one.

According to the first view, the river came into being pretty much as it is today and at some well-defined time, such as the Eocene. A statement that is made about any part of the river applies to the river as a whole; the entire river is young, or old, as the case may be.

The second view is that most rivers are continually changing entities that have evolved from various ancestors and will continue evolving into progeny whose configuration depends on factors such as tectonism and climate. According to this view, the answer to the question "When was the Colorado River born?" can only be another question: "How much departure from the present configuration is one willing to tolerate and still speak of the Colorado?"

It is important to remember that the Colorado traverses two contrasting terrains in Arizona. The first is the canyon country, typified by the Grand Canyon. This is highly dissected terrain, commonly with substantial topographic relief. The second is the plateau country, which is typified by most of the Navajo and

GRAND CANYON GEOLOGY

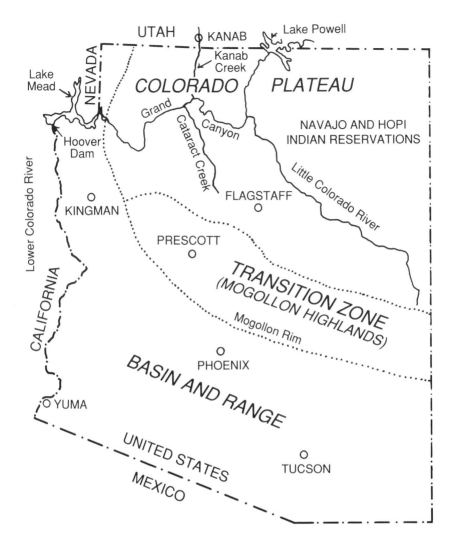

Figure 1. Location map, Arizona

Hopi reservations (Fig. 1). This landscape is characterized by low relief, wide mature valleys, and scarps that develop on beds of contrasting resistance and retreat down structural slopes. The plateau country is older and more widespread than the canyon country, which is encroaching on it.

HISTORY OF THE GRAND CANYON   313

A HISTORY OF IDEAS

For the first sixty years or so after John Wesley Powell's 1875 journey of discovery, geologists subscribed to the idea of a river with a simple history; it was born with the same course that it has now. The questions of paramount importance were: when was it born and when did the uplift of the region (which was considered responsible for the cutting of the canyons) occur? Because erosion of the plateau country is pervasive, these early geologists inferred that the erosion was also deep—the "Great Denudation" of Dutton (1882). Consequently, the denudation, the canyon cutting, and the uplift ultimately responsible for both must have occurred a long time ago, presumably shortly after retreat of the great inland seas at the beginning of the Tertiary Period. According to this view, the Colorado River, the uplift, the canyon cutting, and the Great Denudation all began in Eocene time—and perhaps even earlier in the Tertiary.

The origin of the river and the Grand Canyon seemed safely established. Attention, therefore, was focused on geomorphic problems highlighted by the textbooklike character of the Grand Canyon region, where sparse vegetation and simple structure make it possible to see landforms clearly and to trace them for great distances. These characteristics led to the development of several concepts of fundamental importance in geomorphology, among which are the principles of antecedence, superposition, consequence, and anteposition, all having to do with the relations between drainage systems, structure, and topography (Davis 1901, 1903; Babenroth and Strahler 1945; Strahler 1948).

Storm warnings signaling danger for the view that the Colorado River is old were hoisted in the 1930s and 1940s by geologists studying the Basin and Range country (Fig. 1). These geologists found that interior-basin deposits of late Miocene and Pliocene age are common along the course of the Colorado River. They also could find no evidence for an older drainage system that could be called the Colorado. In conformity with the concept of a monophase history for the river, these geologists concluded that the entire river, and thus the Grand Canyon as well, was no older than late Tertiary (Blackwelder 1934; Longwell 1936, 1946).

The next development occurred in the plateau country of Arizona, Utah, and Colorado. Here, widespread evidence, eventually summarized by Hunt (1969), showed that drainage

GRAND CANYON GEOLOGY

systems, locally departing from the present course of the Colorado River but arguably ancestral to it, existed certainly in the Miocene and very probably as early as the Oligocene. They might have existed even earlier, but if so, the evidence is gone. There was now a major paradox; the same river seemed to be at least as old as Miocene-Oligocene in its upper reaches, but no older than latest Miocene or Pliocene in its lower ones.

In an attempt to shed light on the paradox, E. D. McKee and the Museum of Northern Arizona sponsored studies on critical areas at and near the mouth of the Grand Canyon. Results by Lucchitta (1966) and Young (1966) showed no stratigraphic or morphologic evidence of a through-flowing drainage system during the deposition of Miocene interior-basin materials related to Basin-Range deformation. Nor could the lack of evidence for through-flowing drainage be bypassed by looking elsewhere along the course of the lower Colorado River or the southwest margin of the Colorado Plateau in Arizona (Fig. 1). In this area, interior-basin deposits are ubiquitous, and deposits older than Basin-Range rifting indicate drainage northeastward, from what now is the Basin and Range Province onto what now is the Colorado Plateau.

The northeast drainage existed as recently as the emplacement of the Peach Springs Tuff, as 18-million-year-old ignimbrite that flowed onto the Colorado Plateau. Before rifting of the Basin-Range, therefore, drainage was not to the west or southwest (as would be required for a river with a course similar to that of the present Colorado) but in the opposite direction—to the northeast.

The next step in the conceptual journey was the idea of a polyphase history for the river. Hunt (1969) contributed to it by proposing drainage systems initially departing markedly from the present Colorado River, but gradually evolving into this configuration. However, Hunt, as well as Lovejoy (1980), still postulated an ancestral Colorado River flowing westward from the Colorado Plateau even before Basin-Range deformation, a concept not supported by the evidence.

The concept of a polyphase history for the Colorado River was developed fully for the first time by McKee and others (1967). These authors accepted the antiquity of the upper part of the drainage system, as documented by Hunt, but could not accept a continuation of this drainage westward through the

Grand Canyon into the Basin and Range Province. Instead, they proposed that the ancestral Colorado followed its present course as far as the eastern end of the Grand Canyon, but then continued southeastward (not westward) along the course of the present Little Colorado (Fig. 1) and Rio Grande rivers into the Gulf of Mexico. In Pliocene time, a youthful stream, emptying into the newly formed Gulf of California and invigorated by a steep gradient, eroded headward and captured the sluggish ancestral river somewhere in the eastern Grand Canyon area. It was then that the river became established in its present course, and the carving of the Grand Canyon began.

The concept is pivotal because it introduces the idea (even though the point is not made explicitly) that drainage systems evolve continually—and do so chiefly through headward erosion and capture and in response to tectonic movements. During this process, the configuration and course of a drainage system may change so much that it becomes difficult and rather arbitrary to continue calling the ancestral drainage system by its present name.

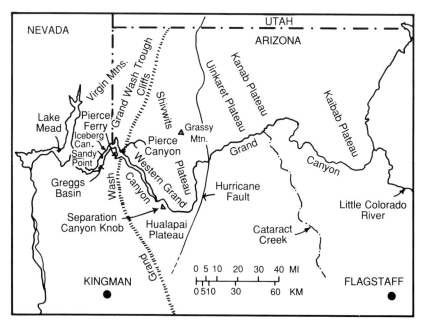

Figure 2. Map showing selected geologic and geographic features in northern Arizona

GRAND CANYON GEOLOGY

Evidence accumulated since 1967 argues against drainage southeastward along the Little Colorado and Rio Grande rivers, as proposed by McKee and others (1967). On the other hand, evidence has continued to grow that an ancient river could not have flowed through the western Grand Canyon region (Fig. 2) into the nearby Basin and Range Province (Lucchitta 1972, 1975; Young 1975; Young and Brennan 1974).

Analysis of deposits along the course of the lower Colorado River in the Basin and Range Province has confirmed that this part of the river is no older than latest Miocene. It also has shown that the capture of the ancestral Colorado River is documented by the appearance within river deposits (Imperial Formation of Miocene and Pliocene age) in California's Salton trough of coccoliths otherwise found only in the Cretaceous Mancos Shale of the Colorado Plateau (Lucchitta 1972).

An attempt to synthesize current information led Lucchitta (1975, 1984) to postulate that the ancestral Colorado did not flow to the southeast along the valley of the Little Colorado River, as proposed by McKee and others (1967). Instead, the river crossed the Kaibab Plateau along the present course of the Grand Canyon, then continued northwestward along a strike valley in the area of the Kanab, Uinkaret, or Shivwits plateaus (Fig. 2) to an as yet unknown destination. After the opening of the Gulf of California, this ancestral drainage was captured west of the Kaibab Plateau by the lower Colorado drainage. According to this concept, the upper part of the Grand Canyon in the Kaibab Plateau area is old and related to the ancestral river, whereas the lower part of the canyon in this area and in all of the western Grand Canyon postdates the capture and was carved in a few million years, a process aided by nearly 0.6 miles(0.9 km) of regional uplift since the inception of the lower river (Lucchitta 1979).

This hypothesis is based on (1) the occurrence of gravels of probable river origin in the area of the Kanab, Uinkaret, and Shivwits plateaus and (2) the observation that northwest-trending drainages along strike valleys were common and persistent before canyon cutting, as evidenced by fossil valleys preserved under Miocene lavas in many places in the southwestern Colorado Plateau and by ancient valleys in the plateau country. Examples of such valleys are those of Cataract Creek (Fig. 2) and the Little Colorado River (Fig. 1), which predate canyon cutting and have not yet been affected appreciably by it.

The old problem of how the Colorado could have crossed the Kaibab Plateau can be analyzed by going backwards in time and restoring rocks removed from this upwarp in the past few million years (Fig. 3). This shows that a river such as the ancestral Colorado could have flowed readily across the Kaibab (then lower topographically than its surroundings) in an arcuate racetrack corresponding to the present configuration of the eastern Grand Canyon. The racetrack was localized by north-facing monoclinal flexures that cross the Kaibab and interrupt the general southward plunge of this dome.

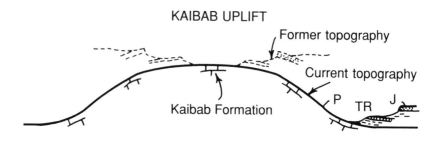

Figure 3. Cross section showing uplift of the Kaibab Plateau both a few million years ago and currently. P - Permian; TR - Triassic; J - Jurassic

THE EVIDENCE

The Colorado River and its tributaries have been cutting down vigorously during the last several million years of their history. Such erosion—or at least nondeposition—may have been typical of this river system through most of its life. This characteristic creates great difficulties for the geologist intent on reconstructing the history of the river because such a history can only be pieced together from evidence left behind by the river.

The evidence is of two kinds—river deposits such as gravel, sand, and silt and landforms such as river valleys and canyons. Of the two, the deposits are by far the more useful because most can be attributed unequivocally to a specific river, on whose provenance, direction of flow, and age they provide valuable information. In contrast, a landform such as a valley can result from a river other than the one of interest or even from the action of an entirely different agent, such as a glacier.

Any river (even one that is cutting down) leaves behind deposits. For rivers that are cutting down, such as the Colorado, most of the deposits are removed soon after deposition. This means that deposits with which the geologist can work are few and scattered, especially in the Grand Canyon. Furthermore, the deposits preserved tend to reflect only the most recent part of the river's history.

Because of these factors, study of the Colorado River consists largely of a detectivelike piecing together of circumstantial evidence, most of which is negative. In other words, the evidence is more likely to show that the Colorado River did *not* go through some area at a specified time than to document its existence in some specific place at a specific time. When circumstantial evidence is not negative, it typically attributes to an inferred Colorado River the properties known to be widespread at the time in question. For example, if at some time in the past most drainages followed northwest-trending strike valleys, one can reasonably infer that the Colorado also followed such a valley.

These points may seem too obvious to be worth repeating, but they need to be made once more because many people are taken aback by the lack of hard data pertaining to the history of the Colorado River. It is true that we do not have much direct evidence, given the problems mentioned above, but the circumstantial information is of many different kinds and from different places. Collectively taken, it enables us to construct a solid history that has a good chance of being correct, at least in its major aspects.

The history of the Colorado River and the Grand Canyon is best subdivided into three periods of tectonism that profoundly affected the drainage patterns of their time. These periods are:

1. *Pre-rifting.* The interval between the beginning and the middle of the Tertiary, at which time basin-range rifting got underway along the present course of the Colorado River, and the Colorado Plateau became distinct structurally and morphologically from the adjacent Basin and Range Province.

2. *Rifting.* The time of basin-range extension, with intensity tapering off toward the end of the interval—five to eight million years ago, at most. This was a time of widespread interior drainage in the Basin and Range Province.

3. *Post-rifting.* During this time—between five to eight million years ago and the present—rifting ceased, the Gulf of

California opened, and through-flowing drainage became established.

Pre-rifting

Before the onset of rifting in late Oligocene to Miocene time, the terrain south and southwest of the Colorado Plateau (Fig. 1) was higher than the nearby plateau, both topographically and structurally (Fig. 4). It remains high structurally today, even after the rifting. This belt of uplift, often referred to as the Mogollon Highlands, presumably was formed during the orogenic events at the end of the Mesozoic Era and the beginning of the Tertiary Period. It caused the gentle northeastward tilting of strata near the southwest margin of the plateau, as witnessed by the widespread "rim gravels" (Finnell 1962, 1966; McKee and

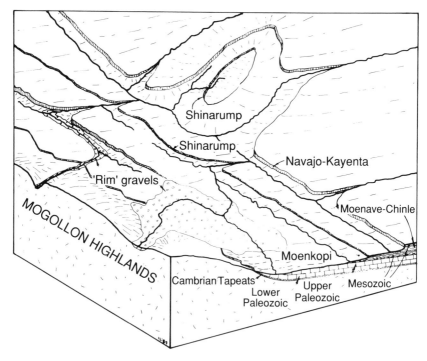

Figure 4. Block diagram showing schematically the relation of Mogollon Highlands to the Colorado Plateau before Miocene rifting. Diagram shows scarps composed of hard-over-soft couplets retreating down structural slope, the inferred topography on the Kaibab uplift, and the trellis drainage network. Looking about northwest, Hualapai Plateau would be at left side of diagram, the area of the Little Colorado River at the right side.

GRAND CANYON GEOLOGY

McKee 1972; Peirce 1984; Peirce and others 1979; Peirce and Nations 1986). Along much of the rim, these gravels were deposited on rocks high in the Paleozoic section. They occur locally on Mesozoic rocks as well, yet they contain clasts of Precambrian igneous and metamorphic rocks that could have come only from the belt of uplift south of the rim, where such rocks were exposed.

Along the southwestern margin of the Colorado Plateau, in the area of the Hualapai Plateau, the rim gravels are present within ancient canyons trending northeastward, down the structural slope (Fig. 4). Clast provenance, imbrication, and gradient of the canyon floors all indicate derivation from the southwest (Young 1966, 1970). A similar drainage direction is indicated by the 18 million-year-old Peach Springs Tuff, an ignimbrite that flowed northeastward onto what now is the Colorado Plateau from a source area in the Basin and Range Province (Young 1966; Young and Brennan 1974; Glazner and others 1986). In the area southwest of the present plateau rim, erosion already had cut down to the Precambrian basement by Miocene time, and Phanerozoic strata were retreating northward from a structural and topographic high near Kingman (Lucchitta 1966, 1972).

At the southern margin of the Colorado Plateau, "rim gravels" are widely distributed (Fig. 4) along the Mogollon Rim (Fig. 1), which is the physiographic southern edge of the Colorado Plateau. Even though gravel-filled channels are not as prominent there as they are on the Hualapai Plateau, gravel-filled channels do occur, notably along the Little Colorado River. These gravels delineate the probable drainage pattern near the southern margin of the Colorado Plateau for much of Tertiary time (Lucchitta 1984).

For most of its course, the Little Colorado River flows in a mature and subdued valley that trends northwestward—parallel to the regional strike of Mesozoic and Paleozoic units (Figs. 1, 5). The valley is at the erosional featheredge of the Triassic Moenkopi Formation on the Permian Kaibab Formation. The northeast side of the valley is in the Moenkopi, capped by the resistant Shinarump Member of the Triassic Chinle Formation, which dips gently to the northeast and is very resistant to erosion. Because of this resistance, the topographic surface is near the top of the Kaibab over wide areas of the Colorado Plateau.

Figure 5. High-altitude (U-2) aerial photograph looking north to northeast. The Little Colorado River is in the foreground. Dark mass in middle distance is Black Mesa. Wide valley of the Little Colorado River and scarps forming its northeast side are clearly visible.

The hard-over-soft couplet represented by the Shinarump over the Moenkopi is the lowest, stratigraphically, of several such couplets within the Mesozoic section (Fig. 6). The scarps formed by the couplets face southwest and with time migrate northeastward down the structural slope. The couplets that are stratigraphically highest and youngest have migrated the farthest to the northeast. Those that are lowest and oldest, on the other hand, are found to the south or southwest. These couplets are closest to the belt of uplift (Mogollon Highlands) from which they started. The Grand Staircase described by Dutton (1882) is composed of these couplets.

A northeastward migration can be documented for the valley of the Little Colorado River, where the ancient gravel-filled channels are cut into the top of the Kaibab Formation southwest of the present channel of the river. Both the northwest-trending valleys at the foot of the Shinarump-Moenkopi couplet and the

322 GRAND CANYON GEOLOGY

northeastward migration of such valleys with time can be documented on the Shivwits Plateau (Figs. 2, 6), where various basaltic lavas of late Miocene age have flowed down successive stands of the valleys (Lucchitta 1975).

Most tributaries of the Grand Canyon are short, steep, and immature (Fig. 7). The Little Colorado River, Cataract Creek, and Kanab Creek (Figs. 1, 7) have the length and appearance of conventional, mature river valleys. The Little Colorado and the Cataract Creek flow northwestward—parallel to the strike of beds; Kanab Creek also shows the influence of structure on its course by flowing south around the western flank of the Kaibab uplift, which has overprinted and modified the regional northwestern strike.

These three streams illustrate features characteristic of the pre-Grand Canyon drainage. One feature is the control of drainage by structure, which is represented by the gentle dip of beds modified locally by folds and faults. Another is the marked effect of lithology, as represented by couplets within the Mesozoic section that differ in resistance to erosion and form strike valleys and scarps. Together, these features have given the region a trellis drainage pattern that consists of northwesterly segments parallel to strike and northeasterly segments trending down the structural slope.

The streams bringing rim gravels onto the Colorado Plateau and ancient drainages such as the Little Colorado must have

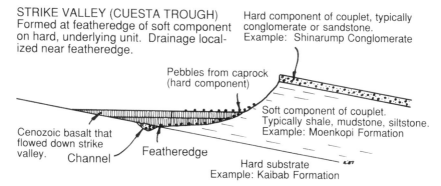

STRIKE VALLEY (CUESTA TROUGH) Formed at featheredge of soft component on hard, underlying unit. Drainage localized near featheredge.

Hard component of couplet, typically conglomerate or sandstone. Example: Shinarump Conglomerate

Pebbles from caprock (hard component)

Soft component of couplet. Typically shale, mudstone, siltstone. Example: Moenkopi Formation

Cenozoic basalt that flowed down strike valley. Channel Featheredge

Hard substrate Example: Kaibab Formation

Figure 6. Diagrammatic representation of typical strike valley formed at foot of hard-over-soft scarp retreating (northeast) down structural slope. Also shown are the channel located near featheredge of slope-forming unit, basalt that in many places flowed down strike valleys, and debris derived from caprock. Looking northwest. Width of valley typically is a few km.

Figure 7. High-altitude (U-2) photograph looking west, down much of Grand Canyon. Shivwits Plateau is visible in the distance on the right. The short, steep, and immature tributaries to the Grand Canyon are evident, in contrast to the well-developed Cataract Creek in the middle distance on the left. The confluence of Kanab Creek with the Colorado River is in the middle right of the picture.

been functional for a substantial period of time because there is little evidence on the southwestern Colorado Plateau of wide-spread and long-lasting Tertiary ponding in most pre-rifting time. The conclusion is that the drainages must have been tributary to a master stream that presumably also reflected the effects of structure and stratigraphy (Fig. 4). The inference is that this was the ancestral Colorado River and that it flowed chiefly in northwest-trending strike valleys. The Colorado, however, did not flow through the western Grand Canyon region. Here, the presence of gravels and the Peach Springs Tuff shows instead that drainage near the edge of the Colorado Plateau was to the northeast.

Additional evidence that the river did not flow through the western Grand Canyon itself is provided by a knob on the south rim of the Grand Canyon near Separation Canyon (Fig. 2). The knob is composed of a mid-Miocene basalt flow that caps gravel and colluvium containing upper Paleozoic rocks (Young 1966; Lucchitta and Young 1986). In this area, such rocks crop out only on the Shivwits Plateau to the north. Today, the knob and the Shivwits Plateau are separated by the Grand Canyon, which could not have existed when the colluvium and basalts were emplaced (Young 1966).

Rifting

Mid-to-late Miocene basin-range rifting affected most of the western Cordillera and produced ranges separated by deep structural basins filled with thick sequences of deposits. The basins sank so rapidly that through-flowing drainage was not able to maintain itself, resulting in widespread interior drainage. In western Arizona, rifting began in mid-Miocene time and was essentially over by the end of that epoch.

The southwestern Colorado Plateau was much less affected by rifting than was the Basin and Range terrain to the west and south. Normal displacement occurred on many small faults and several larger ones, such as the Hurricane, but the intensity of faulting did not approach that in the Basin and Range. The most noteworthy effect of the rifting was the foundering of the formerly high areas south and southwest of the plateau. The foundering formed structural basins beyond the edge of the plateau and interrupted the old drainages that had deposited the rim gravels (Peirce and others 1979; Young 1966; Young and Brennan 1974; Lucchitta 1966, 1979). Ponding and deposition of lake beds occurred locally on the plateau, presumably as a result of minor warping. The most conspicuous such lake beds are the Bidahochi Formation of latest Miocene and Pliocene age and the Willow Springs Formation of Pliocene to Pleistocene(?) age. Such deposits, however, are minor in comparison with the ubiquitous interior-basin deposits of the nearby Basin and Range Province.

Figure 8. High-altitude (U-2) photograph looking about north from near mouth of the Grand Canyon (right foreground). Lake Mead is in foreground. North-trending valley in center of picture and west of the prominent Grand Wash Cliffs is the Grand Wash Trough. Iceberg Canyon is the narrow, north-trending arm of the lake in the lower left. Pierce Ferry is at wide area of the lake in the lower center of the picture.

Particularly interesting with respect to the history of the Colorado River is the Miocene Muddy Creek Formation, as exposed in the Grand Wash trough and the Pierce Ferry area (Figs. 2, 8) at the mouth of the Grand Canyon. No evidence of a river emptying into the Grand Wash trough is present in the lithologies and facies distribution of the Muddy Creek, which, instead, records interior drainage with derivation of clastic material from nearby highlands. Analysis of directions of transport and topographic closure leads to the same conclusion.

West-draining canyons are present in the Grand Wash Cliffs, a prominent, west-facing fault scarp formed during basin-range

326   GRAND CANYON GEOLOGY

deformation. The scarp marks the mouth of the Grand Canyon and defines the western edge of the Colorado Plateau (Figs. 2, 8). The west-draining canyons are short and steep. Where they debouch into the Grand Wash trough, the washes that carved the canyons have deposited fans of locally derived material. One such fan emerges from Pierce Canyon, whose mouth is only 1.5 miles (2.5 km) north of the Grand Canyon (Fig. 2, 8). The fan was deposited across the mouth of the present Grand Canyon. This could not have happened if the Grand Canyon and the Colorado River existed in their present location at the time.

The highest unit of the Muddy Creek Formation in the Grand Wash trough is the Hualapai Limestone, which interfingers downward with clastic rocks of the Muddy Creek. The Hualapai was deposited in a shallow lake, which extended about 25 miles (40 km) west from the Grand Wash Cliffs. As with other lithologies of the Muddy Creek, the Hualapai contains no evidence for a major river emptying into the lake in which the limestone precipitated. Instead, the lake waters were highly charged with calcium carbonate and other salts. Deposits younger than the Hualapai record through-flowing drainage.

The Hualapai was the youngest unit to be deposited in the area near the mouth of the Grand Canyon before the lower Colorado River was established. However, geologists have not yet dated it directly. The Muddy Creek Formation near Hoover Dam contains basalts that are five to six million years old (Anderson 1978; Damon and others 1978). A tuff about 1600 feet (500 m) below the Hualapai Limestone in the Pierce Ferry area has been dated by fission-track methods at eight million years (Bohannon 1984). Another tuff low within the limestone in an area about 25 miles (40 km) west of the mouth of the Grand Canyon also has yielded an age of about eight million years (Blair 1978). South of Hoover Dam (Fig. 1), the Bouse Formation (an estuarine deposit associated with the opening of the Gulf of California) records the presence of the Colorado River in its present lower course. The Bouse has been dated by the K-Ar method at 5.47 + 0.2 million years (Damon and others 1978). In the Pierce Ferry area, basalts intercalated with Colorado River gravels flowed down the valley of the Colorado when it was within 330 feet (100 m) of present grade (Lucchitta 1966, 1972). The flows have been dated by K-Ar methods at 3.8 million years (Damon and others 1978). This evidence indicates that the end of interior-basin deposition and the establishment of through-flowing drainage along the lower Colorado River in its Basin and Range course occurred between four and six million years. No lower Colorado River existed before that date.

Evidence of various kinds from the western Grand Canyon region leads to similar conclusions. According to Lucchitta (1975), the Shivwits Plateau is capped for the most part by upper Miocene lavas that overlie uppermost Paleozoic and lowermost Mesozoic rocks. A long and narrow finger of the plateau juts southward for about 20 miles (30 km) into the Grand Canyon (Fig. 7). This finger is surrounded on three sides by the Grand Canyon. Relief of nearly 1 mile (1.5 km) within a horizontal distance of a few miles between the top of the plateau and the Colorado River results in an extremely rugged topography of canyons, cliffs, and buttes. In contrast, the basalt capping the plateau, dated by the K-Ar method at 7.5 and 6.0 million years (Lucchitta and McKee 1975; Lucchitta 1975), overlies a remarkably flat and smooth surface with only a few meters of relief. In several places, cone-shaped vent areas for the basalt are truncated by the cliff-forming edge of the Shivwits Plateau, yet there are no remnants of Shivwits lava within the Grand Canyon.

Lavas of the Shivwits Plateau locally overlie the erosional featheredge of the Triassic Moenkopi Formation on the resistant Permian Kaibab Formation (Fig. 6). The featheredge formed northwest-trending strike valleys bounded on the northeast by a scarp capped by the Triassic Shinarump Member. This is a situation common on the Colorado Plateau. The Shinarump shed its pebbles into the valley in a way that preserved the pebbles both beneath and on top of the basalt (Lucchitta 1975). Today, the scarps are gone, and their place is taken by canyons tributary to the Grand Canyon (Figs. 6, 9).

On Grassy Mountain, which is on the Shivwits Plateau 15 miles (25 km) northeast of the Grand Canyon (Fig. 2), a six-million-year-old basalt overlies remnants of gravels that contain pebbles of igneous and metamorphic rocks. The gravels also contain metamorphosed volcanic rocks similar to rocks of Proterozoic age cropping out south of the present Colorado Plateau margin. The gravels rest on the Moenkopi Formation. The only reasonable source for the pebbles is to the south because in other directions, erosion has cut down only to Paleozoic and Mesozoic rocks. Neither the pebbles nor the arkosic

GRAND CANYON GEOLOGY

matrix of the gravels shows much sign of weathering; this suggests that the gravels are not reworked from older deposits. The lack of weathering within the gravels, together with the lack of a weathered zone between the gravels and the basalt, suggest that the two are of approximately the same age. The conclusion is that streams flowed northward across the present course of the western Grand Canyon as recently as six million years ago (Lucchitta 1975). These streams presumably were tributary to a larger stream that may well have been the ancestral Colorado River, but the course of this river would have been north or northwest of the western Grand Canyon. This interpretation is supported by other remnants of gravels that are composed of exotic lithologies and are found in several places in the Arizona Strip country north of the Grand Canyon.

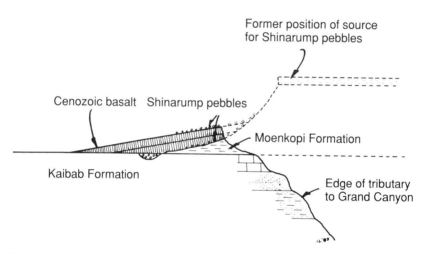

Figure 9. Diagram showing how scarp previously forming northeast side of Shivwits strike valleys occupied by basaltic lavas now is the site of a deep canyon tributary to the Grand Canyon. Looking approximately northwest

Collectively, the various features of the southwestern Colorado Plateau indicate that no rugged canyon topography and no Grand Canyon existed in the western Grand Canyon region as recently as six million years ago, the age of the young lavas in the southern Shivwits Plateau (Lucchitta 1975). The Colorado River must have become established in the western Grand Canyon after that date. It may, however, have flowed northwestward through the Strip country long before.

Post-rifting

The evidence summarized above indicates that the transition from interior drainage to the through-flowing drainage of the lower Colorado River probably occurred between five and six million years ago, or at the Miocene-Pliocene boundary. Miocene, Pliocene, and Quaternary deposits of the Colorado River are distributed widely along its lower course. Included are estuarine deposits (upper Miocene and Pliocene Bouse Formation); probable deltaic deposits (Miocene and Pliocene Imperial Formation); gravels of various ages and degrees of cementation, chiefly reflecting downcutting; and fine-grained deposits (Pleistocene Chemehuevi Formation) produced by temporary aggradation. Information on these deposits has been summarized by Lucchitta (1972), who interpreted the data in light of the history of the lower Colorado River.

The Bouse Formation is particularly informative. In the Yuma area (Fig. 1), the Bouse is distributed widely in the subsurface even well away from the Colorado. Upstream along the river, however, the formation is restricted largely to the river valley. Fossils in the Bouse indicate a brackish estuarine environment that became progressively less salty northward along the river valley (Smith 1970). The Bouse is an erosional unconformity with deposits that reflect interior drainage and that locally interfinger upward with gravels of the Colorado River (Metzger 1968; oral comm. 1969).

The evidence indicates that the lower Colorado River became established early in Bouse time (latest Miocene) along its present course and that its prograding deltaic deposits progressively filled the estuary in the southward direction. Eventually, the delta reached the Salton trough of California, where it is represented by the thick Imperial Formation. This formation contains a well-defined horizon above which Cretaceous coccoliths found elsewhere only in the Mancos Shale of the Colorado Plateau make their appearance; the horizon signals the capture of the old, upper Colorado River by the developing and headward-eroding lower Colorado.

A 3.8 million-year-old basalt flow associated with indurated gravels of the Colorado River occurs about 330 feet (100 m) above present river grade at Sandy Point (Fig. 2) in the upper Lake Mead area (Longwell 1936; Lucchitta 1966). This basalt is part of the extensive basalts of the Grand Wash (Figs. 2, 9) and flowed for several miles along the valley of the Colorado River, which must have been nearly as low then as it is now.

In the western Grand Canyon, basaltic flows occur at river level for a long distance. These flows, which have their source in the area of the Uinkaret Plateau (Fig. 2), are the intracanyon lavas of McKee and others (1968) and have been dated at about 1.2 million years. The lavas show that at the time of their emplacement, the western Grand Canyon was as deep as it is today.

The upper Cenozoic deposits along the lower course of the Colorado thus show that the river came into being five to six million years ago when the Gulf of California opened up. The river extended itself southward, or downstream, by filling its estuary with deltaic deposits. It also extended itself by headward erosion and the integration of former interior drainages. By the time the delta reached the Salton trough area, the head of the river had captured the old upper Colorado River. By 3.8 million years ago, the river was essentially at its present grade in the upper Lake Mead area, and by one million years ago (at the latest), it was at its present grade in the western Grand Canyon. According to these figures, the western Grand Canyon was excavated during the interval between six million years ago (younger Shivwits lavas) and one million years ago (intracanyon basalts). More likely, it began to be cut shortly before five million years ago and was nearly as deep shortly after four million years ago as it is today. The rapid downcutting almost certainly was aided by the nearly 0.6 mile (0.9 km) of uplift experienced by the western edge of the Colorado Plateau and adjacent Basin and Range terrain since the time when the Lower Colorado River was established in its present course (Lucchitta 1979).

Much of the course of the lower Colorado River in the western Grand Canyon and upper Lake Mead areas can be explained through structural or topographic features. Thus, the Virgin Canyon section (Fig. 2) probably is at the margin of an ancient fan spilling southward from the south Virgin Mountains; the Greggs Basin-Iceberg Canyon section (Figs. 2, 8) is along Wheeler fault; the Pierce Ferry section is through the lowest spot in the old Muddy Creek basin; the Hualapai Plateau section is along a strike valley developed at the foot of the upper Grand Wash Cliffs (southwest-facing rim of the Shivwits Plateau); and the section along the east side of the Shivwits Plateau is developed near the Hurricane fault.

SUMMARY

A striking conclusion that emerges from the study of the Colorado River is that even a canyon as intricate and immense as the Grand can be carved in a surprisingly short time—five million years, probably substantially less, in spite of the tough and remarkably undeformed rocks that make up its walls. Perhaps the key to understanding this phenomenon is the fact that in canyons the volume of material removed per unit of downcutting is small compared with that for more open valleys. In other words, the rate at which material is carved from a canyon is small in relation to the rate at which the floor of the canyon is lowered (Lucchitta 1966).

Another and perhaps even more striking conclusion is that one can draw a parallel between the development of physical systems, such as drainage networks, and the Darwinian concept of biological evolution based on survival of the fittest and natural selection. In river systems, the external agent that triggers change is tectonism rather than the random mutations of biology. Competition between drainages occurs through changes in gradient, whereby rivers whose gradient is increased are favored, and those whose gradient is reduced are handicapped.

The battles for survival are fought with headward erosion and stream piracy or capture. The ultimate result is a succession of drainage configurations that change with time in response to external forces—chiefly, the deformation of Earth's surface. Since any particular configuration is merely a still frame within the movie of evolution, ancestors may bear little resemblance to their descendants.

CHAPTER 16

HYDRAULICS AND GEOMORPHOLOGY OF THE COLORADO RIVER IN THE GRAND CANYON

Susan Werner Kieffer

INTRODUCTION

The Colorado River and its tributaries drain much of the southwestern United States, ultimately emptying into the Gulf of California in Mexico. Only the Mississippi River exceeds the Colorado in length within the United States. The potential use of the water of the Colorado River for irrigation, hydroelectric power, and domestic purposes was recognized more than a century ago. In 1905, severe floods on the Colorado River caused extensive damage in the Imperial Valley of California, and political pressure arose for construction of flood control/storage dams on the river. The development and regulation of the river expanded rapidly until, at the present time, the Colorado River is sometimes referred to as "the world's most regulated river" (National Research Council 1987, p. 18). Additional information about the history of the river's development is summarized in Kieffer and others (1989), and so further discussion in this chapter focuses on the portion of the Colorado River that lies mainly within Marble and Grand canyons—that is, between Glen Canyon Dam and Lake Mead.

It is convenient to divide the history of the Colorado River within the Grand Canyon into three (very unequal) time spans: (1) the time of unregulated flow, prior to the finishing of Glen Canyon Dam in 1963; (2) the time of reservoir filling from 1963 to 1983; and (3) the current operational period, when Lake Powell is filled to a capacity that optimizes requirements for water storage and power generation. The hydraulics and sediment transport capacity of the river have been very different in each of these times.

The downcutting of the Grand Canyon was accomplished by a river capable of carrying large sediment loads during the time of unregulated flow. Monitoring at the U.S. Geological Survey's gage station near Phantom Ranch from 1921 to 1962 showed that typical average daily discharges were near 17,000 cubic feet per second (cfs). The mean annual flood was 77,500 cfs, but larger floods were not uncommon. For example, a flood of 300,000 cfs occurred in 1884. Sediment loads were large during this time; the average daily sediment load past the Phantom Ranch gage exceeded 300 tons per day.

During the reservoir-filling period, the average daily discharge was reduced to 11,000 cfs. Sediment that originates upstream of Lake Powell has been trapped in the lake since 1963, and so the average sediment load has been reduced to about fifty tons per day. Much of the sediment now available downstream from Glen Canyon Dam is contributed by the two large tributary streams, the Paria (near Lees Ferry) and the Little Colorado River. Thus, the Colorado River (whose name in Spanish means "colored red" in reference to its former heavy sediment load) now often runs clear.

Between Lees Ferry (near the Arizona-Utah border) and Diamond Creek (about 220 miles downstream), the water surface of the Colorado River drops from 3116 feet in elevation (944 m) down to 1336 feet in elevation (405 m). The water surface does not maintain a constant gradient between these two elevations, but consists, instead, of a series of pools that are relatively flat and tranquil and rapids that are steep, fast, and turbulent. The bed of the river likewise is not uniform in gradient, but consists of relatively flat sections under the tranquil pools, debris dams (natural wiers) where tributaries enter, and scour holes. These characteristic configurations of the water surface and channel bottom have been created over thousands of years by hydraulic, hydrologic, and geomorphic interactions superimposed on the tectonic forces that drive the uplift of the Colorado Plateau.

The purpose of this chapter is to show how an understanding of hydraulic dynamics in individual stretches of the river near rapids can help us interpret the shape of the river channel and the geomorphic and hydrologic history of floods and erosion over the past thousands to perhaps hundreds of thousands of years.

A DESCRIPTION OF THE MAJOR FEATURES OF RAPIDS

The pool and rapid sequences of the Colorado River are very obvious in high-altitude photographs (Fig. 1a) because the rapids occur where the river is constricted by debris from side canyons, and these constrictions, as well as the white water of the rapids, are obvious in the photographs. Photographs taken at lower altitude (Fig. 1b) reveal that at the constrictions the water surface changes from a featureless pool into a white, wave-filled chute. At the bottom of the chute, the water squirts out of the constriction into the next quiet region like a jet from a fire hose. Adjacent to the jet are eddies and beaches, sites of some of the most critical ecologic zones and recreational beaches along the river.

These stretches of the river between one relatively tranquil pool (or main-stem segment) and the next are the famous rapids of the Colorado River. They reveal many fundamental fluid mechanical phenomena on a grand scale: waves that stand still while water flows through them; zones of smooth, tranquil flow where large boulders protruding through the water cause hardly a ripple; zones of turbulent, aerated flow where the large boulders cause major hydraulic features; and even zones where the water flows backwards.

Common features of rapids are illustrated in Figures 2 and 3. In describing the rapids, we use a mixture of terms from hydraulics and from the vocabulary of those who have explored and described the river.

Rapids in the Grand Canyon typically form where a debris fan from a tributary canyon constricts the river. Above each rapid, the river is wide, relatively deep, and tranquil. In this chapter, the term "pool" is reserved specifically for these tranquil sections of the river immediately upstream of a rapid (Fig. 2a). At most

Stone Creek

Deubendorff Rapids

Galloway Canyon

Bedrock Rapids

Spectre Rapids

128 Mile Rapids

a

GRAND CANYON GEOLOGY

Figure 1. (a). (on opposite page.) High-altitude photograph (U.S. Geological Survey Series GS-VFDC-C, #6-134, taken June 24, 1982) of the Middle Granite Gorge of the Grand Canyon. River flows from bottom of photo toward top. Note how each rapid is associated with a constriction in the river at the base of a side canyon (or two side canyons in the case of Deubendorff Rapids; see text). The alternation of quiet pools with high-velocity rapids, visible in four places in this photograph, has been called the "pool and rapid" sequence of the Colorado River (Leopold 1969). (b) A lower altitude aerial photograph centered on Deubendorff Rapids, showing the transition of the river from a quiet pool on the left into a foaming jet with standing waves in the constricted region, and the jet emerging like a firehose (labeled "tailwaves") into the next quiet region of the river. Photograph by U.S. Bureau of Reclamation, 1984

discharges, a pool is a hydraulic backwater (a concept discussed later in more detail). Conceptually, a pool can be thought of as a pond formed by the debris-fan dam. At discharges between about 7000 and 30,000 cfs, water velocity is low in the pools, typically less than 1 foot per second (0.3 m/s). Water in the pools is deep, typically more than 15 feet (5 m) at the low end of this discharge range and more than 30 feet (9 m) at the higher end of this discharge range (Kieffer 1987).

At the downstream end of a pool, water accelerates gradually in the constriction (Figs. 2b, c; 3a, b, c), reaching velocities more than an order of magnitude higher than the velocities in the pool, even at discharges as low as 7000 cfs (Fig. 3d). Velocities of this magnitude have been found at most of the major rapids where velocity measurements were performed (they were not performed at Cremation-Bright Angel and 24.5-Mile Rapids). The highest velocities measured were approximately 33 feet per second (10 m/s) at Hermit Rapids (Kieffer 1988, Map 1897 F). In the converging portion of the channel (where rapids are found), standing waves, or laterals, bound a tongue of smooth, accelerating water, upon which may stand smooth, undulating, nonbreaking waves called rollers. In the constriction and the diverging portion of the river channel, crisscrossing lateral waves typically intersect to form high-amplitude breaking waves or haystacks.

Below each rapid is a zone in which the depth is intermediate between that of the shallow constriction and the deep pool. In this zone, the so-called "runout" of the rapid, water velocity still is relatively fast, typically 10 to 15 feet per second (3-5 m/s), but it decelerates toward the ambient conditions in the next pool, the so-called "tailwater" conditions in hydraulics. Strong vertical motions occur in the water of this region. The shear that results can give rise to turbulent boils with as much as one foot (0.3 m) of superelevation, indicating a vertical velocity of at least 8 feet per second (2.4 m/s) (Leopold 1969).

Obstacles in the bed of the channel (such as rocks or bedrock protrusions) also cause a variety of wave patterns—including holes, curlers, rooster combs, and sculpted waves. A definition of these less commonly known terms follows (refer also to Fig. 2):

(1) Tongue: Smooth water between the first two strong lateral waves (right and left) at the top of a rapid.

(2) Roller: A wave that stands oblique, often perpendicular, to the current and breaks back onto the current; the term "nonbreaking roller" is used to indicate the smooth, rolling waves often found on the tongue.

(3) Lateral: A wave standing oblique to the current near the top of a rapid, usually emanating from shore.

(4) V-wave: The composite wave formed when opposing laterals intersect.

(5) Eddy fence: The shear zone between two currents with different velocity magnitudes or directions.

(6) Pourover: A zone where water "pours over" an obstacle, obtaining a large, downward component of velocity.

(7) Hole: A trough in a standing wave, usually deep.

(8) Haystack: A pyramidal wave (shaped like a haystack), usually breaking on top and sending spray in all directions.

(9) Rooster comb: A haystack elongated in the downstream direction.

(10) Runout: A zone of standing, generally nonbreaking (or weakly breaking) waves at the bottom of a rapid; more or less synonymous with "tailwaves."

(11) Rock garden: An area of rocks in the channel within or downstream of the diverging section of a rapid.

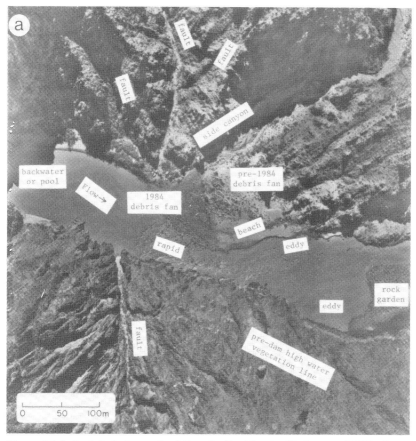

Figure 2 (a). Aerial photograph of Granite Rapids at a discharge of 5000 cfs, showing the geomorphic features common to many rapids. Terms defined in text. Photograph by U.S. Bureau of Reclamation, 1984. Regional faults from Dolan, Howard and Trimble (1978)

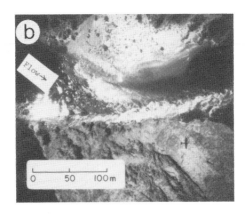

Figure 2 (b and c). Aerial photographs of Granite Rapids at 30,000 cfs showing typical wave structures in rapids. Photographs by U.S. National Park Service

To visualize some of these features and to understand the hydraulics of the rapids described and interpreted in the rest of this paper, a reader may rely on, or create, intuitive mental background pictures of the rapids by several methods. An excellent description of the river, its rapids, and the eddies and whirlpools below them can be found in Beer (1988), an account of two men who swam the whole length of the river. Alternately, many features of the rapids can be seen on video (Kieffer 1986). If

340   GRAND CANYON GEOLOGY

the reader has been to (or can hike to) Phantom Ranch via the Bright Angel or Kaibab trails, careful study of Bright Angel Rapids located at the bottom of these trails will be useful, as will study of the much smaller, but analagous, Bright Angel Creek. Other places where features of the flow that are discussed here can be seen are in small steep-gradient streams (particularly in the Pacific Northwest), in storm gutters along city streets when water is flowing, and in amusement parks that have water rides!

It also may be helpful for the reader who has not been around any rapids to spend a few minutes at the kitchen sink—using the faucet to simulate the river and a cookie sheet, some curved dishes or pie pans, and a few utensils to create minature rapids. Place a cookie sheet at an angle under the faucet, and turn on the water, watching the change in flow patterns as the discharge through the faucet is increased. Try confining the flow into a channel with a few straight-edged implements (yardsticks, knives, etc.), and then try making a converging-diverging flow by placing curved implements on the cookie sheet. As the flow changes character-istics, try placing the tip of a spoon in various flow regions to simulate an obstacle (like a rock) in a rapid, and note the different wave behaviors that can be obtained. This experiment has the potential to be messy, as well as instructive!

The interpretations of the hydraulics of the rapids in this chapter are based on a comparison of the rapids with slightly fancier and better-controlled versions of the kitchen-sink experi-ment—laboratory flumes. The comparison is only semiquantita-tive because the Colorado River channel is much more complex than laboratory flumes, and we have relatively little data about the details of the channel shapes. In the river channel, wave patterns are irregular on a microscale because the complex and detailed channel geometry disturbs regular wave patterns (for example, individual large boulders change the local energy of the flow and create locally complex hydraulic patterns). The wave patterns also are irregular in time because discharge changes with time.

The use of laboratory flume data involves scaling problems of several orders of magnitude since laboratory flumes typically are of meters in dimension, whereas the Colorado River dimensions are one or two orders of magnitude larger. This is descriptively conveyed by Larry Stevens (1985, p. 25): "Let's put it this way; at 32,000 cfs, 1000 tons of water are moving through the river channel every second. If an average elephant weighs about 5 tons, this means that the flow of the river is equal to 200 elephants skipping by every second. A hole in the river may take up about a third of the channel, so the hydraulic dynamics in that hole are about the same as 67 elephants jumping up and down on your boat." In spite of the difference in scale between the Colorado River and a laboratory flume, we can learn a great deal about the hydraulics of the river by applying basic open-channel hydraulic principles.

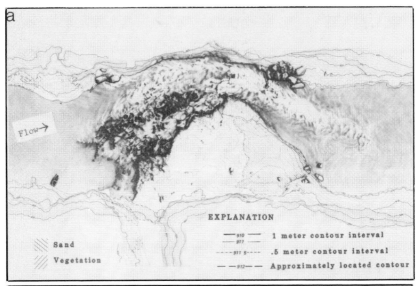

EXPLANATION

Sand

Vegetation

—910— —977—   1 meter contour interval

---977.5--- .5 meter contour interval

— —912— — Approximately located contour

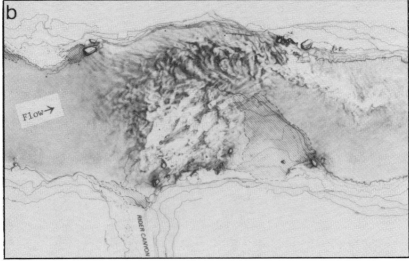

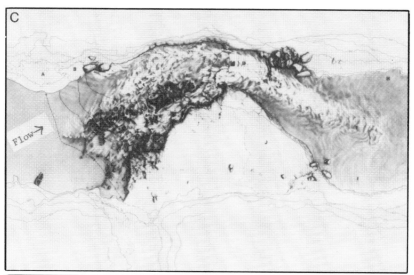

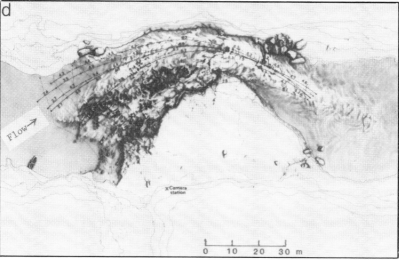

Figure 3 (a-d). Parts of preliminary hydraulic map I-1897-A, House Rock Rapids (Kieffer, in press). (a-d) Map views; Contour intervals indicated with solid lines are 1 meter; dashed lines on beach in (a) are 0.5-m contours. (a) Topographic contours and standing waves at 5000 cfs; (b) the same at 30,000 cfs; (c) water-surface profile at 5000 cfs; (d) water-surface velocities and streamlines of floats at 5000 cfs. In (a)-(d), flow direction is from left. In (d), numerals indicate velocities (in m/s) along the streamlines between the adjacent dots. Trajectories of the floats were determined from analysis of movies taken from the camera station indicated. The boat, shown only for scale, is a standard commercial motor raft that is 10 m (33') long. These maps are for schematic illustration of hydraulic features only and are not intended for navigation purposes.

In turn, increased knowledge of the properties of the river and its interaction with the channel margins can help us better manage the river corridor (as in the current U.S. Bureau of Reclamation Glen Canyon Environmental Studies project), understand potential hazards for recreational boating, and interpret the geomorphology of the river channel.

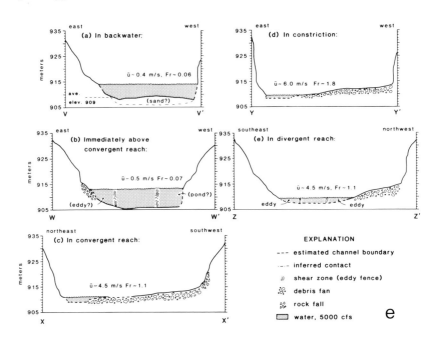

Figure 3 (e). Location of cross sections is approximately described above each cross section and view through cross section is downstream. Average water velocity is given by u, and Froude number by Fr.

LOCATION OF THE RAPIDS AND DESCRIPTION OF CHANNEL MORPHOLOGY

Many geologists have speculated on the origin of the pool-rapid-runout sequence (Leopold 1969; Dolan and others 1978; Howard and Dolan 1981). Important factors in determining the origin and location of the rapids are gradient and variations of gradient along the river, spacing of the rapids, and the relation of the rapid spacing to the location of structural controls. We can divide the channel of the river into stretches that have different geomorphic and hydraulic characteristics (Howard and Dolan 1981). The frequency and, to some extent, the magnitude of the rapids depends on their location in these stretches. Typical stretches are:

(a) a wide valley with a freely meandering channel (for example, miles 67-70, near Tanner Rapids);

(b) valleys of intermediate width with tributary fan deposits (in these valleys, the river usually has cut into soft sandstones or limestones; for example, the few miles just downstream of the Little Colorado River near mile 61.5);

(c) narrow valleys in fractured igneous and metamorphic rocks (for example, "Granite Narrows" through miles 77-112; Fig. 1);

(d) narrow valleys of roughly uniform width and few constrictions in massive Muav Limestone (for example, miles 140-165);

(e) nearly flat stretches where the channel bottom is sandy (for example, miles 1-10 and parts of Marble Canyon).

The rapids occur almost exclusively where floods in tributary canyons, controlled by local or regional faulting or jointing, have delivered large boulders into the river channel (Dolan and others 1978) (see the example of Fig. 2a, showing the faults and debris fan at Monument Creek where Granite Rapids is formed). Because the tributary canyons are much steeper than the Colorado River channel, floods in the tributaries can deliver boulders into the main channel that may be too large for even the large natural floods of the Colorado River to move (the discussions of Graf [1979, 1980] are relevant to this problem, though not based specifically on data from the Grand Canyon).

Debris fans from the tributary canyons can form on one or both banks of the Colorado River at the tributary junctions because the controlling faults cross the canyon (Howard and Dolan 1981). If meteorologic and drainage conditions are conducive, floods can occur in opposing tributaries, thus forming two debris fans. The relative size of the fans depends on the availablility of moveable material in the contributing drainages, on the frequency and magnitude of floods in each drainage, and on the tributary gradient (for a recent study, see Webb and others 1987). It is, however, more common to have one of the debris fans be significantly larger than the other (see Figs. 1 and 2). The river usually erodes through the weaker parts of the debris fan, but the erosion may extend into the bedrock wall at the distal end of the debris fan if this material is easily eroded. This process can result in the formation of a pronounced change in the course of the river, and many rapids occur on local curves of the river (Fig. 1b).

In spite of the variations in the nature of bedrock along the course of the river discussed above and the structural control exerted by faults and joints on the location of the rapids, the river channel at the rapids is remarkably uniform in shape where it cuts through the tributary debris fans (Fig. 4). The unconstricted channel is about 300 feet (90 m) wide at 10,000 cfs (the width varies with discharge but this effect is not significant at the level of overall generality considered in this paper). At each rapid, the channel narrows appreciably, typically to about one-half of the unconstricted width. Figure 4 shows the ratio of surface width of the river immediately upstream of a rapid to the width in the narrowest part at 54 major debris fans; this ratio is termed the "constriction" in this chapter. The remarkable feature of the histogram of Figure 4 is the uniformity of this shape parameter—sharply peaked with nearly half of the rapids at the value 0.5. Why is the channel so uniform in shape at so many rapids, and why is this value specifically 0.5? The discussion of hydraulic-geomorphic interactions presented in this chapter will suggest an answer to this question.

The values of constriction plotted in Figure 4 were measured on 1973 air photographs (Fig. 2a is taken from this 1973 series of photographs). Because Glen Canyon Dam was operating under daily fluctuating flow conditions when the 1973 air photograph series was taken, the discharge at different rapids along the photograph series varied from about 7000 to 30,000 cfs. In the analysis that follows in this chapter, I need to use an average width for the river at places above, in, and below the rapids. This average width must be one that, together with an average depth and an average flow velocity, satisifies the requirement for conservation of discharge. A problem is that the surface width of the water measured from the photographs and shown in Figure 4 is not identical to the average width of an idealized channel. For example, in the histogram of Figure 4, Crystal Rapids is the point at the furthest left, at a constriction of 0.33. However, idealization of the channel shape to a rectangular cross section suggests that the average channel constriction at Crystal was about 0.25 in 1973.

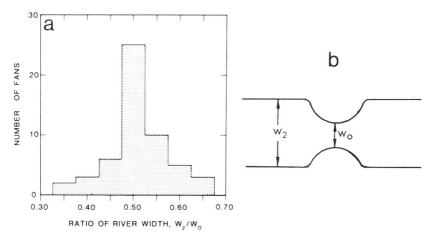

Figure 4. (a.) Histogram of constriction of the Colorado River as it passes through 59 of the largest debris fans in the 248 mile (400-km) stretch below Lees Ferry (Kieffer 1985). (b) Idealized sketch of converging-diverging geometry of a rapid, showing widths, W_o and W_2 used to define the "constriction" of the rapid in (a)

This average value, rather than the value based on surface widths, is used in the hydraulic discussions in this chapter.

At this time, we cannot make a histogram like Figure 4 based on actual channel cross sections or on average widths because we lack detailed surveys of the river channel. However, the author has prepared a series of maps of the ten largest rapids along the river (House Rock, 24.5-Mile, Hance, Cremation-Bright Angel, Horn Creek, Granite, Hermit, Crystal, Deubendorff, and Lava Falls; Kieffer 1988). These maps have a topographic base sufficiently accurate to allow the value of constriction and its dependence on river stage to be determined more accurately. They also have additional data that allow visualization of hydraulic features in the rapids. Figure 3 is an example of parts of the map of House Rock Rapids.

Each hydraulic map contains the following: (a) topographic contours of the channel (Fig. 3a-d); (b) hydraulic information at two or more discharges (compare Figs. 3b and c); (c) water-surface elevations at different discharges, that is, rating curves and water surface profiles, shown implicitly by comparison of the shorelines in Figs. 3b and c, and explicitly in Fig. 3b; (d) velocity and streamline data at one or two discharges (Fig. 3d); (e) channel cross sections, showing both the shape of the channel and the hydraulic conditions of the water (Fig. 3e); and (f) a detailed discussion of the characteristics of the rapid and interpretation of the data (not shown in Fig. 3).

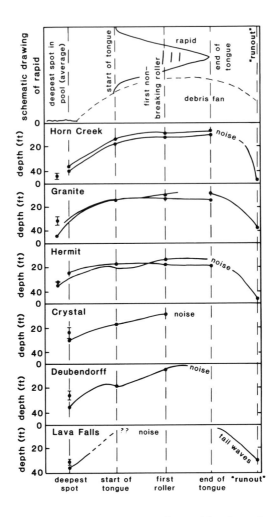

Figure 5. Summary of fathometer tracings obtained by the author, Julie Graf (U.S. Geological Survey, personal comm., 1986), and Owen Baynham (river guide, personal comm., 1986). No horizontal dimensions are implied. The top drawing is a schematic map view of a typical rapid showing the maximum depth measured in the upstream pool, the start of the tongue, the location of the first nonbreaking rollers, the end of the tongue, and the runout of the rapid. Where known, discharges are indicated. Similar profiles occur in every rapid for which fathometer data were obtained.

GRAND CANYON GEOLOGY

Careful study of the channel shape in three dimensions reveals that at rapids the channel is constricted both laterally and vertically. In map view (for example, as in the air photographs of Figs. 2a and 2b, or on the topographic maps of the channel in Figs. 3a and 3b), the lateral constriction can be seen easily if the discharge is low enough that the river does not cover the debris fans. In vertical cross sections that are perpendicular to the flow direction (such as those of Fig. 3e) or those that are parallel to the flow directions (such as the fathometer tracings of Fig. 5), we can see that the channel of the Colorado River is constricted also by a vertical bulge in the bed, caused by the debris fan.

Detailed studies of the rapids show that the constrictions typically begin underwater well upstream of the location of the first waves in the rapid. The water gets shallower and faster as the constriction gets tighter, and the most complex and largest waves in a rapid occur in the shallowest and narrowest part of the channel. The channel deepens again under the runout or tailwaves. Although the water surface drops several meters in most rapids, the elevation of the channel bottom is often the same upstream and downstream of the rapid to within experimental accuracy (~1 m). Thus, the rapids are not formed by sudden drops in channel bed elevation, but rather are a result of vertical and lateral constriction.

WHY ARE THERE WAVES IN RAPIDS? TWO IMPORTANT HYDRAULIC PARAMETERS AND THEIR INTERPRETATION

In the previous section, I defined the nomenclature used to describe the rapids. In this and the next sections, I use hydraulic theory of open-channel flow to relate the descriptive terminology to hydraulic conditions in the rapids.

We can calculate two important hydraulic parameters in the pools, rapids, and runouts below the rapids from velocity and depth data presented above: the Reynolds number ($Re = u\,D/\mu$, where u is the average flow velocity, D is the average depth, and μ is the viscosity), and the Froude number ($Fr = u/(gD)^{1/2}$, where g is the acceleration of gravity). These dimensionless numbers indicate the state of the flow. The Reynolds number indicates whether the flow is laminar ($Re < 10^4$) or turbulent ($Re > 10^4$), and the Froude number indicates whether the flow is subcritical (Fr <1), critical (Fr = 1), or supercritical (Fr > 1).

Under most flow conditions, the Reynolds number is greater than 10^5, indicating turbulent flow. For example, consider the Reynolds number of a backwater—where the flow looks tranquil and nonturbulent. If the discharge is greater than a few thousand cfs, the water velocity typically is greater than 0.3 feet per second (10 cm/s). The depth typically is greater than 3 feet (100 cm). Water viscosity is on the order of 0.01 cm^2/s. Even for these apparently tranquil backwater conditions, the Reynolds number is greater than 10^5, and the flow thus is turbulent. Appearances are deceptive! In other parts of the river, u increases by up to an order of magnitude, and D can decrease by the same amount, but we generally find Reynolds numbers well in excess of 10^4. Such high Reynolds numbers imply that turbulent conditions exist nearly everywhere in the river. The turbulence has important implications for the mixing of sediment and nutrients. However, since the river is turbulent nearly everywhere, variations in the Reynolds number cannot explain the dramatic differences we find in the water structure of the pools, rapids, and runouts.

The Froude number is the dimensionless number that we need to look at to investigate differences in stability of waves and hydraulic regimes. It characterizes two states of flow that are dramatically different in energy balance and wave stability. The numerator of the Froude number depends on velocity—that is, on momentum or the square root of the kinetic energy. The denominator depends on potential energy—that is, on depth. The Froude number, therefore, is a measure of the relative importance of the kinetic and potential energies. The denominator also defines a characteristic velocity, called the critical velocity [c=(gd)$^{1/2}$], that depends only on water depth. The stability of standing waves depends on the ratio of the fluid velocity (numerator) to this characteristic velocity (denominator).

Dramatic changes in flow regime occur as the Froude number changes from less than one (subcritical) to greater than one (supercritical). These changes are similar to those that occur in transitions from subsonic to supersonic flow in gas dynamics, and this analogy may be useful for the reader in thinking about the standing waves in the rapids.

Changes in Froude number near a rapid are dramatic. In a typical backwater, u ~ 1 m/s and D ~ 10 m, so Fr ~ 0.1 or even less (a subcritical condition). In a rapid, on the other hand, u > 5 m/s, D < 3 m, so Fr ~ 1 or Fr > 1 (critical and supercritical conditions). Measurements show that Froude numbers exceeding 2 can occur in the rapids. Figure 3e shows Froude numbers for different parts of House Rock Rapids. Measurement of the Froude numbers in different parts of the rapids therefore suggests that the dramatic change in flow regime from backwater to rapid is caused by differences in the balance of kinetic and potential energy. These differences change the stability of waves in the channel. The general principles that apply to analysis of shallow-water flow in these different flow regimes comprise the classic theory of open-channel hydraulics (Bakhmeteff 1932; Ippen 1951; Ippen and Dawson 1951; Rouse and others 1951; or Chow 1959).

At this point, the reader can develop an intuitive feeling for the significance of Froude numbers for different flow regimes and about the concept of standing waves by returning to the kitchen sink and putting a flat plate on the bottom of the sink under the faucet (this experiment is described in Thompson 1972, p. 525). As the faucet is turned on and the downgoing jet of water strikes the plate, water spreads laterally toward the edges. Depending on the rate of discharge from the faucet and the position of the plate, an inner ring of water that is moving rapidly outward in a radial direction surrounds the impact point where the jet hits the plate. If the reader has produced the correct experimental conditions, a ridge of water surrounds this inner ring a few inches radially outward from the jet. Beyond this ridge, the water is deeper and moves more slowly toward the edge of the plate. The ridge or ring where the water depth and velocity change is a standing wave that separates an inner region of supercritical flow from an outer region of subcritical flow. This standing wave, which is called a "hydraulic jump," is circularly symmetric about the descending jet of water. In a linear channel, however, hydraulic jumps commonly are linear, though they may stand perpendicular or oblique to the flow direction.

If you insert a probe, such as a finger, spoon, or small pebble, into the two different regions of flow on the plate, very different wave phenomena occur. In the inner, supercritical region, standing waves will be formed as "wakes" to the object, whereas, in the outer, subcritical region any waves formed by the object migrate upstream or downstream through the fluid. Thus, they are traveling, not standing, waves. Wave behavior in the critical region is highly unstable (by analogy, it is well known that wave instability makes maneuvering aircraft in the transonic regime difficult, but that maneuvering becomes more stable under supersonic conditions). In all of these regimes, eddies caused by shear in the fluid can form downstream of an object immersed in the flow. These eddies must be distinguished from standing waves.

Hydraulic jumps are the basic standing waves in the Colorado River, and they occur in a variety of geometries. Large, oblique hydraulic jumps emanate from shore and are oriented downstream at an angle to the current (see Fig. 1b, or 2b and c). Normal hydraulic jumps stand perpendicular to the current. Finally, miscellaneous hydraulic jumps of various geometries stand around rocks and obstacles on the bed of the channel in regions of supercritical flow in the rapids.

The same change of conditions that gave rise to standing waves in the inner ring under the kitchen faucet and yet produced no standing waves in the outer, deeper water occur in the rapids of the Colorado River. The contrast in stability of standing waves between supercritical and subcritical flow explains why there are strong waves in rapids where the flow is supercritical, but there are no standing waves in backwaters where the flow is subcritical. Just as you can manipulate the strength, stability, and position of the circular hydraulic jump under the faucet by increasing or decreasing the flow rate through the faucet or by changing the position or angle of the plate, so the behavior of waves in the rapids depends on the flow rate of the Colorado River, the shape of the channel, and the gradient of the bed. Using these concepts, I now show how the different parts of a rapid and the waves can be semiquantitatively explained in terms of flume hydraulics.

THE PIECES AND PARTS OF A RAPID: BACKWATERS (POOLS)-RAPIDS-RUNOUTS

Backwaters (Pools)

Pools and backwaters form upstream from a rapid if changes in channel shape produce local conditions in which the given discharge cannot be accommodated in the channel cross section without a transition from subcritical to supercritical flow. To clarify this, consider a specific example of flow at 30,000 cfs. If the main channel is 100 feet wide (30 m) and 30 feet (9 m) deep, the flow per unit area is 10 ft/sec (3 m/s), that is, the flow velocity required to accommodate the discharge is 10 ft/sec. The flow is subcritical because the critical velocity is 17.1 ft/sec (5.2 m/s) and the Froude number is 0.6. If the channel narrows to 30 feet (9 m) in width and maintains the same depth, the flow per unit width or flow velocity would have to increase to 33.3 ft/sec (10 m/s) to accommodate the discharge. The flow would be supercritical with a Froude number of 1.9. Conservation of energy would not allow the fluid to accelerate in this simple way because of the change in flow regime. Instead, water would pond upstream of the rapid in a backwater to increase the depth and, therefore, potential energy of the flow, and the river would adjust itself so that the Froude number would just equal unity in the constriction.

In effect, the cross-sectional area of the whole backwater-rapid system is increased by ponding in the backwater or "pool" above a rapid (see the cross sections in Fig. 3e). The potential energy of the deepened pool is available for conversion to velocity in the constriction (compare depth and velocity in cross sections (a) and (b) with those in (c) and (d) in Fig. 3e). Within the backwater itself, the increased depth caused by the constriction reduces the velocity compared to that in an unconstricted channel of the same diameter (that is, compared to a normal main-channel flow that enters the backwater). As passengers on a raft float from the main-channel current into a backwater above the rapid, they often lose the sense of "floating" and may feel the need to row across the tranquil pond—especially if there is any upstream wind to halt all progress! For example, the backwater above Crystal Rapids, which extends approximately a mile back to Boucher Rapids, is known affectionately by river runners as "Lake Crystal." Velocities of only a few tenths of a foot per second and Froude numbers as low as 0.01 can occur in the backwaters, indicating conditions dominated by potential energy.

The Converging Section of Rapids:
Tongues, Nonbreaking Rollers, and Oblique Lateral Waves

In this section, I show how features in laboratory flumes of relatively simple geometry and well-controlled discharges (Figs. 6, 7, 8) relate to the more complex and variable patterns of waves in the rapids of the Colorado River (Figs. 9 and 10).

A simplified illustration of hydraulic features in flumes with converging-diverging geometries is given in Figure 6. The top

(a) Subcritical conditions:

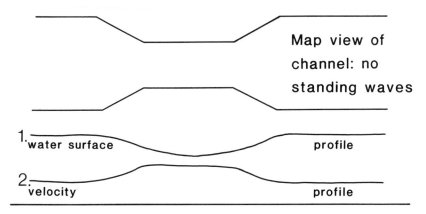

Map view of channel: no standing waves

1. water surface profile

2. velocity profile

(b) Supercritical conditions:

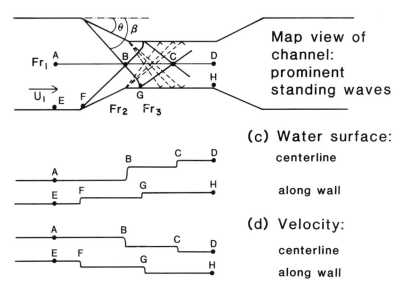

Map view of channel: prominent standing waves

(c) Water surface: centerline along wall

(d) Velocity: centerline along wall

Figure 6. Comparison of the flow fields in subcritical and supercritical flow (compiled from Chow 1959 and other basic hydraulics texts). (a) Schematic map view of subcritical flow through a constriction. Schematic profiles (1) and (2) show how the water surface elevation and velocity change along the channel. (b) Schematic map view of supercritical flow through the same constriction. (c) Changes in water surface elevation and (d) local velocity through the channel, respectively. The profiles of these quantities are along the paths A-B-C-D, and E-F-G-H.

GRAND CANYON GEOLOGY

(a)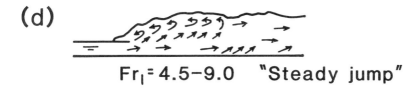

$Fr_I = 1.0-1.7$ "Undular jump"

(b)

$Fr_I = 1.7-2.5$ "Weak jump"

oscillating jet

(c) roller

$Fr_I = 2.5-4.5$ "Oscillating jump"

(d)

$Fr_I = 4.5-9.0$ "Steady jump"

(e)

$Fr_I > 9.0$ "Strong jump"

idealized
water surface

(f)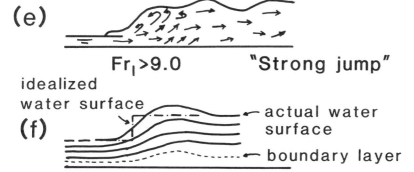

actual water
surface

boundary layer

Figure 7 (a - e). Idealized cross sections of hydraulic jumps with the entering flow at different Froude numbers as shown. From Chow (1959, p. 395). (f) Schematic cross section of idealized and actual hydraulic jumps, after Ippen (1951, p. 339)

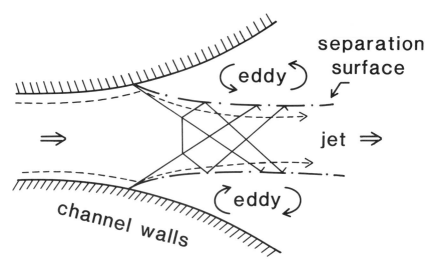

Figure 8. Illustration of the structure of a supercritical jet emerging from a constriction. From Chow (1959, p. 471; originally from Homma and Shima 1952)

part of Figure 6a is a map view of subcritical flow in an idealized converging-diverging laboratory flume, and it is singularly uninteresting because there are no standing waves. The two lines below the map view of the flume are (top) water-surface elevation and (bottom) velocity profile. The increase in velocity and decrease of water-surface elevation in the constriction in subcritical flow conditions is exactly analogous to the well-known venturi effect that gas flow shows in converging-diverging nozzles. Figure 6b is a map view of the same flume with supercritical flow conditions. This case is much more interesting than the subcritical case because patterns of oblique, criss-crossing waves occur in the converging and constricted sections. Such waves also can occur in the diverging section but for simplicity are not shown here. The wave patterns (specified by the angle ß) depend on the channel shape (specified by the angle φ). The height of the waves can vary across the channel—particularly where waves intersect each other. At the points of intersection, the wave heights can either be added to or subtracted from each other, depending on the type of wave.

Smooth water from the backwater extends farthest downstream into a rapid along the tongue in the converging part of the rapid (compare the tongue of Granite Rapids in Fig. 2 with the centerline A-B in Fig. 6b). A characteristic of most tongues is the presence of smooth, nonbreaking rollers (Fig. 1c). The length of the tongue and the angle of the oblique waves change with discharge and channel shape.

The converging part of a rapid is a region in which the flow changes in a complex way from subcritical conditions in the backwater to supercritical conditions in the constriction downstream; this region is often referred to as the "top" or "head" of the rapid. The measurements of water surface velocities and depths in the rapids suggest that the Froude number of the flow reaches a value of unity near the beginning of the tongue. For this reason, I interpret the nonbreaking rollers seen on most tongues as undular hydraulic jumps typically found when the Froude number is in the range of 1 to 1.7 (Fig. 7a).

In the supercritical region of the rapid (to the sides of and downstream of the tongue), complex waves appear. The stronger breaking waves (the oblique lateral waves and the haystacks) suggest higher Froude numbers than do the undular waves on the tongue. Values greater than 2 have been measured (Fig. 3e). The measured values of Froude number are probably lower limits for several reasons. The Froude number depends on water velocity, and, properly defined, it must be an average velocity for the whole water depth. When floats are used to measure velocities, several effects cause the measured velocities to be lower than the average velocity—drag between the water surface and air, drag between the float and air, and "surfing" and bouncing of the floats on the waves. It seems probable that the floats we used to measure surface velocities traveled appreciably slower (10-20%?) than the mean water velocity. Nevertheless, the measurements are internally consistent, and they show that the velocities and Froude numbers increase continuously from the backwater into the constriction and, sometimes, on into the diverging region of the rapid (Fig. 3d).

The water's energy is dissipated in the converging and constricted part of a rapid by wave action, bottom friction, and air entrainment. The noise of the rapids is part of the mechanism by which potential energy that was stored in the water in the backwater (and that was converted to kinetic energy in the converging section of the rapid) is dissipated as the water begins to decelerate back toward tailwater conditions.

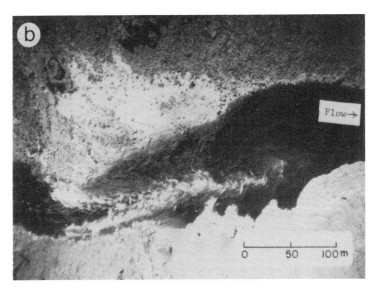

Figure 9. 24.5-Mile Rapids at discharges of (a) 5000 cfs and (b) 30,000 cfs. Note the dramatic change in the orientation of the tailwaves and the sizes of the eddies on each side of the tailwaves with discharge. Photograph (a) by U.S. Bureau of Reclamation, 1984; (b) by National Park Service, 1986

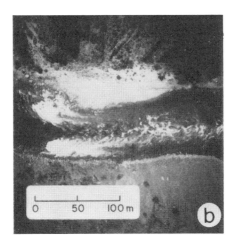

Figure 10 (a). Hermit Rapids, showing the tongue and lateral waves at 5000 cfs and (b) view of part of the same rapid at the same scale showing these features at 30,000 cfs. Note differences in tongue lengths, angle of lateral waves to shore, and nonbreaking rollers on the tongue. Photograph (a) by U.S. Bureau of Reclamation, 1984; photograph (b) by National Park Service, 1986

The Diverging Section:
Haystacks, Tailwaves, Eddies, and Beaches

Some very nonintuitive phenomena occur in the diverging part of a constricted rapid because the flow is supercritical. Our intuition, typically based on subcritical flow, would lead us to believe that the water would decelerate as soon as the river channel widens (Fig. 6a). Flow velocity measurements show, however, that the velocity can increase not only into the constriction, but well into the diverging part of the channel (this is not dramatic at House Rock Rapids shown in Fig. 3, but can be seen in the data from other rapids in the maps of Kieffer 1988). The flow velocity and the Froude number both increase until a hydraulic jump is encountered—the hydraulic purpose of the jump is to return the flow toward ambient main-channel conditions appropriate to the downstream reach. This process is nonlinear; the flow does not adjust smoothly to the convergences and divergences of the channel. Instead, changes occur discontinuously through standing waves (hydraulic jumps). When sufficient energy has been dissipated by the mechanisms mentioned above, the water returns to subcritical conditions and must adjust its depth and velocity to the tailwater conditions downstream.

Another nonintuitive aspect of the flow in the rapids is the influence—or, rather, lack of influence—of the downstream tailwater conditions on the upstream flow conditions. In a subcritical constriction, the velocity and depth of water in the constriction and even upstream of the constriction are influenced by the depth in the downstream tailwater—that is, the flow adjusts continuously and smoothly to changes in channel geometry. In contrast, in supercritical flow, the tailwater conditions can only influence flow conditions in the lower part of the divergence below the hydraulic jump where flow returns to subcritical conditions. Flow in the converging and constricted parts of the rapid is uninfluenced by conditions downstream of the major waves.

Channel expansion typically is very abrupt in the diverging section (Fig. 2). When changes in channel shape are abrupt, flow streamlines generally cannot follow channel boundary curvatures (a condition known as rapidly varying flow in hydraulics). Because the streamlines cannot follow the channel boundary, the narrow high-speed jet formed in the constriction squirts into the divergent section with nearly the same diameter that it acquired in the constriction. The jet in the divergent reach is marked by a train of gently breaking to nonbreaking tailwaves (Fig. 2c). Measurements show that tailwaves generally are regions where the Froude number is very close to 1 (see Fig. 3d). Thus, we can interpret the tailwaves as nonbreaking, undular hydraulic jumps, and we note that they have a structure similar to, but not identical with, the nonbreaking rollers on the tongue in the converging part of the rapid where the Froude number was also estimated to be 1. A major difference between the tongue and the tailwaves is the width of the main current: in the converging part of the rapid, nearly the whole discharge is gradually funneled into the tongue, and the flow accelerates until F~1 conditions are obtained. In contrast, in the diverging part of the rapid, the flow does not expand back to the full width of the channel.

The region between the jet and the channel boundary fills with recirculating flow of relatively low velocity—an eddy (Fig. 8). Eddies always are found on at least one side of a strongly diverging river channel below a rapid, and they often are found on both sides (Figs. 1b, 2c). These recirculating zones, in turn, allow sand deposition. For this reason, many important ecologic sites and pleasant camping sites are downstream of the rapids. Details of the expansion of the constricted jet back to ambient conditions in the tailwater are important in determining sediment transport to these beach sites (Fig. 2c).

The shape of the main jet of water emerging from the constriction can be defined quite well because separation surfaces between the jet and the recirculating zones are manifested as strong shear zones (eddy fences) and "boils" (small whirlpools). The shape of the jet in the tailwater cannot be predicted accurately from theory. However, laboratory data suggest that the distance downstream that the jet will maintain constant diameter is proportional to the diameter of the constriction in which it was formed. A jet typically will maintain constant diameter for a distance downstream of several constriction diameters. In detail, this relation depends also on the Froude number of the jet in the constriction and divergence. For example, with a Froude number of two, an ideal laboratory jet would maintain constant diameter for roughly three constriction diameters downstream. As an example, at House Rock Rapids, the constriction diameter is about 30 m when the discharge is 5000 cfs (Fig. 3a), and the jet maintains a strong identity (as evidenced by tailwaves) for at least 100 m downstream. Within this distance, the jet velocity appears to remain constant at 10 to 15 ft/s (3-5 m/s), and the Froude number appears to be near unity. The length and orientation of the jet and tailwaves can change dramatically with discharge (Fig. 9), a fact that has important implications for the formation and stability of beaches under different discharge conditions.

Much of the energy that must be dissipated in the rapid (the excess backwater head and the vertical elevation drop) is expended in the waves and in bottom friction. Nevertheless, the fact that the flow often has a high velocity at the bottom of the rapid indicates that not all of the water's excess energy has been dissipated. Additional dissipation occurs through mixing with the relatively stagnant water of the eddies that bound the jet in the tailwater. The motion of the jet induces circulation in the eddies, and the two flows (jet and eddy) mix in a zone that expands around the separation line (Landau and Lifschitz 1959, p. 131).

In the photographs and airbrush illustrations of jets in this chapter (Figs. 1, 2, 3, 9, 10), you can see the narrowing of the tailwaves with distance downstream. The converging lines that bound the tailwaves can be taken as evidence of the extent to which the mixing zone has extended into the main current (the jet). To a first approximation, one can say that the jet has decelerated to and been mixed back into main channel (or tailwater) conditions by the end of the tailwaves. The mixing not only decelerates the jet, but it allows suspended sediment carried in the high-velocity main channel current to be transported laterally toward the channel boundaries, into the recirculating eddy, and, ultimately, onto the beaches. Flow in the recirculating zone typically is very slow, about 1 foot per second (0.3 m/s). Thus, sediment , if available, can be deposited within these zones (the rate of sedimentation will depend on particle size and density, fluid velocity and density, and other factors). Detailed studies of this zone are in Schmidt and Graf (1987).

Below the eddy-beach system, another pool and rapid sequence begins, and the hydraulics described here are repeated again, each time like a theme and variations, along the length of the Grand Canyon. Each rapid is unique, yet all are similar.

With this background, we can ask how the characteristic configuration of backwaters, rapids, and fast-water runouts has been created and evolved over thousands of years of hydraulic, hydrologic, and geomorphic interactions superimposed on the tectonic forces that drive the uplift of the Colorado Plateau and the downcutting of the Grand Canyon. A glimpse into these processes was offered by a series of unique events in 1983.

THE 1983 FLOODS: A UNIQUE WINDOW TO HYDRAULIC-GEOMORPHIC INTERACTIONS

In 1983, unusually high releases from Glen Canyon Dam, which controls the flow through the Grand Canyon, provided the opportunity to observe how channel geometry at a rapid changes as the discharge history of the river changes. Most of the interesting events were at Crystal Rapids (Fig. 11). This rapid was relatively insignificant before 1966 (for example, it was not mentioned in Powell's 1875 report). Figure 11a shows the configuration of the rapid in 1963, the year that Glen Canyon Dam was closed. Three years later, a large storm over the north rim of the Grand Canyon caused flash flooding in a number of tributaries (Cooley and others 1977). In particular, debris poured down Crystal Creek and Bright Angel Creek, which is 10 miles (16 km) upstream from Crystal Creek.

The debris flow down Crystal Creek made Crystal Rapids one of the most dangerous rapids on the river for recreational boating. Large boulders carried in the debris flow tightly constricted the river channel at Crystal Rapids (Fig. 11b). Discharges at Glen Canyon Dam between 1966 and 1983 varied between 3000 and 35,000 cfs. These discharges were sufficient to cut a narrow channel through the distal (south) end of the debris flow and to cause occasional shifts of boulder positions within the channel. Overall, however, the channel at Crystal remained tightly constricted, and it retained the general shape documented by the 1967 photograph shown in Figure 11b. The shape parameter or "constriction" defined in Figure 3 (but modified as discussed in the text) was about 0.25 in 1973.

During the spring of 1983, rapid snowmelt at the headwaters of the Colorado River forced operators of Glen Canyon Dam to increase discharges to 92,000 cfs to keep Lake Powell from overtopping the spillways of the dam. This discharge was nearly three times larger than any discharge through Crystal Rapids since the 1966 debris flow, and it was comparable to the annual Colorado River floods prior to Glen Canyon Dam. Thus, a relatively young debris fan was subjected to its first "flood." The changes in Crystal Rapids (compare Fig. 11c with 11d) during these high flows

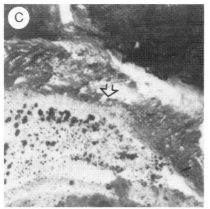

Figure 11. Crystal Rapids at three different times. (a) March 1963, discharge approximately 5000 to 6000 cfs. Photograph by A. E. Turner, Bureau of Reclamation. (b) Same view, March 1967, approximately three months after a debris flow in Crystal Creek. Discharge is 16,000 cfs. Photograph by Mel Davis, Bureau of Reclamation. Note that even though the stage at 16,000 cfs is higher than at 6000 cfs, the debris fan is larger in (b). (c and d) Pair of aerial photos, showing the configuration of the debris fan and the rapid after the 1983 flood. Discharge approximately 5000 cfs in both photographs. The left photograph is from (a). About 10-15 m of lateral erosion has taken place along the shore adjacent to the rock indicated by the arrow. Right photograph by Bureau of Reclamation, October 1984

provided an opportunity to observe some of the dynamic processes that contour the river channel over geologic time.

During the 1983 flood, the channel at Crystal Rapids was widened, and the shape parameter increased significantly from 0.25 to 0.4. This corresponds to an increase of 35 to 50 feet (10-15 m) in width, and this increase dramatically altered the hydraulic characteristics of the rapid. Nevertheless, the rapid is still quite different from the other rapids that have constrictions of 0.5 (remember from Fig. 3 that most rapids have constrictions of 0.5). Local gradients within Crystal Rapids are steeper than those generally found at other rapids, waves are larger, and the dependence of wave structure on discharge is more variable. By watching the evolution of this rapid over two decades and comparing the evolution of the rapid toward the configuration of the more mature rapids, I have worked out the following ideas on the hydraulic-geomorphic interactions between the Colorado River and the debris dams that episodically block its course.

EROSION OF THE DEBRIS FANS BY THE RIVER

The boulder deposits that constrict the river to form rapids are emplaced by debris flows from steep tributary canyons. The steep gradient of these canyons allows these streams to carry large boulders into the main channel (which has a much smaller gradient than the tributaries). These boulders cannot be moved by typical main-stem flows in unconstricted reaches of the river. Although the initial size distribution of material in the debris flows is quite broad (Webb and others 1987), the smaller material, up to large cobble size, is washed away by the river at relatively low rates of discharge. Excellent descriptions of sediment transport through the canyon can be found in Howard and Dolan (1976, 1981), and the U.S. Bureau of Reclamation Glen Canyon

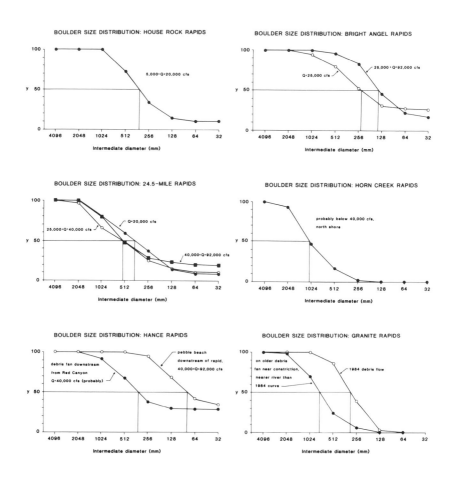

GRAND CANYON GEOLOGY

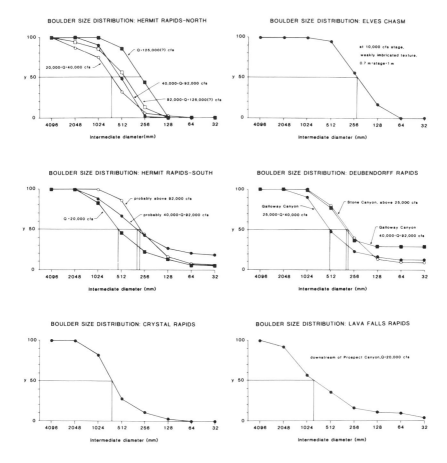

Figure 12. (Part I opposite page, Part II above) Particle size distributions measured at the places indicated at rapids (Kieffer 1987). The twelve graphs are arranged in downstream order. The ordinate, y, is the percent of particles smaller than the given (intermediate) diameter. The horizontal and vertical lines in each graph are to guide the reader's eye to the median diameter of the particles at the rapid.

Environmental Studies (1987) contains recent detailed studies of sand and silt transport in the main channel. Boulders 3 to 10 feet (1 - 3 m) in diameter are common on the debris fans (Fig. 12). These boulders resist erosion (either by chemical or mechanical abrasion processes or by movement at low discharges), and so they stabilize the debris fans with the geometry shown in Figure 3. Because at least half of the channel generally is cleared of all but the very largest boulders at mature rapids, it is obvious that the boulders can be moved under some conditions.

Although hydrologists have documented the transport of sand and silt-sized sediment past the Lees Ferry and Grand Canyon (Bright Angel) gage stations, we know very little about the mobility of large particles in the vicinity of rapids. Specifically, quantitative modeling of the movement of boulders in the Grand Canyon has not been done. Laboratory experiments on much smaller particles and theoretical studies have shown that the size and amount of material transported are related to water velocity, depth, and, therefore, rate of discharge. Graf (1979, 1980) analyzed the stability of boulders in the Green River in Utah and concluded that the largest boulders were stable, even against motion during the largest floods. Because the Colorado has many features in common with the Green River, it generally has been assumed that large boulders, once emplaced, also are relatively stable in the rapids in the Grand Canyon.

The ability of the river to clear out debris emplaced in the channel (its "competence") is proportional to variations in velocity and water depth. The discussion in the first half of this chapter demonstrated that changes in these parameters of more than an order of magnitude occur within a rapid. There are, therefore, substantial differences in transport capacity from one section of a rapid to another. The photographic documentation of changes in Crystal Rapids shown in Figure 11 and the statistical analysis of rapid shapes shown in Figure 4 show that the water can move large boulders in the channel—at least until the river has cleared itself to about one-half of the characteristic unconstricted width. This capacity of the river to clear out large debris from at least half of the channel width produces the characteristic "nozzle-like" geometry seen at the rapids. The processes that accomplish this erosion involve a complex feedback between the hydraulics of the river (discharge as a function of time; depth and velocity as a function of position in the river) and the nature of the material in the bed (grain size and position in a bed of variable cross section).

One criterion widely used for predicting the transport of smaller material in a river is the Hjülstrom relation (Hjülstrom 1935; Strand 1986), which relates water velocity to the size of the largest particles that can be transported (Fig. 13). This Hjülstrom relation, extrapolated to large boulder sizes in Figure 12, suggests that a velocity of 20 feet per second (6 m/s) would be capable of moving a boulder 1 to 2 feet (0.5-m) in diameter and that it would be possible to move material up to 3 to 7 feet (1-2 m) in diameter.

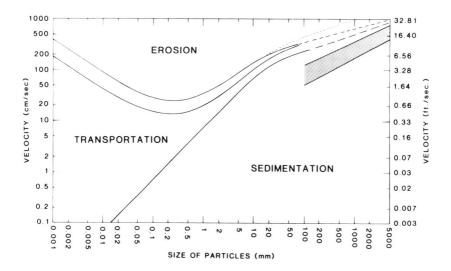

Figure 13. Summary of the relations between stream velocity and size of moveable boulders. The lines on top represent the Hjülstrom criterion (Hjülstrom 1935; as reported in Strand 1986), extrapolated beyond particle diameters of 50 mm. The Hjülstrom criterion is for particles in a uniform bed. Bed roughness and particle shape appear to cause particles to move at lower velocities. The lower stippled area to the right represents criteria developed from field work by Helley (1969).

These values depend on how the Hjülstrom curves are extrapolated.

As discussed in the previous section, water in the highest-velocity part of the rapids can have velocities exceeding 30 feet per second (10 m/s) even at the lower end of the range of discharges from Glen Canyon Dam generators (about 3,000 cfs). By the Hjülstrom criterion, flows at these discharges should be capable, therefore, of clearing the channel of boulders in the 3 to 7 feet (1-2-m) size range.

A second criterion for boulder transport is the concept of unit stream power, originally introduced by Bagnold (1966, 1980) and recently applied to paleohydrogeologic problems by O'Conner and others (1986). Unit stream power is the stream power (rate of expenditure) per unit area, Ω. Bagnold (1966) defined it as

$$\Omega = x \, Q \, S_f \, / w = \beta \, u$$

where x is the specific weight of the fluid (assumed to be 9800 N/m^3 for clear water), Q is the discharge (specifically, that component of the discharge carried by the main current), S_f is the friction slope, w is the channel width, ß is the total channel shear, and u is the mean channel velocity. A more convenient form of this equation is

$$\Omega = x \, n^2 u^3 / R^{1/3}$$

where n is the Manning coefficient of roughness (assumed for these calculations to be 0.035) and R is the hydraulic radius of the channel, taken here to be the average depth.

For example, the reader can use Figure 3 to calculate the unit stream power in House Rock Rapids at 5000 cfs discharge. Although the maximum water surface velocity reaches 25 feet per second (7.5 m/s), we will assume an average velocity of 21 feet per second (6.5 m/s) for a conservative estimate. The average depth is about 3 feet (1 m) in the narrowest part of the rapid. The unit stream power, therefore, is on the order of 3300 Newtons per meter per second. Unit stream power and sediment transport relations (from Williams 1983; summarized in O'Conner and others 1986) suggest that a river with this unit stream power could transport boulders up to about 7 feet (2 m) in diameter. This conclusion agrees with the inferences from an extrapolation of the Hjülstrom diagram. Both criteria suggest that the Colorado River within the Grand Canyon is capable of moving boulders comparable in size to those moved during the largest floods known from paleohydraulic reconstruction techniques (for example, the Missoula flood, Baker 1973, 1984; and Katherine Gorge floods in Australia, Baker and Pickup 1987).

These two criteria also lead to the conclusion that the river can move large boulders even at low discharges in which the flow occupies only a relatively small part of the total river channel (note how little of the river channel is occupied by water at the 5000 cfs discharge conditions shown in the cross sections of Fig. 3). Another way to state this is to say that the relatively shallow, high-velocity flow characteristic of high-gradient rivers can move rather large boulders. Thus, that part of the channel of Crystal Rapids that was exposed to discharges up to 35,000 cfs was depleted of small to relatively large material prior to the 1983 flood. During larger floods, the deepest part of the main channel is cleared of even larger boulders. However, even the parts of the channel exposed to relatively shallow overflow during the floods have boulders removed, as shown by the size distributions of debris fan material that has been exposed to main-stem floods (Fig. 12). Preliminary measurements by the author of boulder size distributions above the river shoreline that correspond to the 92,000 cfs discharge of 1983 showed depletion of small boulders, suggesting larger floods in the past. These observations suggest that a detailed study of boulder size distributions at different places on the debris fans and talus slopes along the Colorado could be used to infer flood histories.

There are, however, serious complications in interpreting boulder size distributions in terms of simple erosion by the main stem because each debris fan has a complicated erosional and depositional record. If a debris fan had a wide variety of particle sizes when it was emplaced, and if it was not graded in size laterally, then erosion of this fan by a large flood would be expected to remove larger particles low on the fan (where the flow is the deepest and fastest) and to remove progressively smaller particles higher on the fan. The size distributions measured on the north and south banks of Galloway Canyon at Deubendorff Rapids (Fig. 12) are consistent with such a simple emplacement and erosion model. However, material removed from the upstream parts of the fan may be deposited on the downstream sides in the recirculating zones discussed in the first part of this chapter; thus, the size distribution depends not only on the vertical elevation on the fan, but on the relative upstream-downstream location (for example, Hance Rapids, Fig. 12).

Boulder size distributions on other debris fans are rarely this simple to interpret. In addition to the erosional and depositional processes mentioned above, complexities arise because (a) initial particle size distributions in the debris flow are not known; (b) winnowing or piping of fine particles may be important; and (c) the initial history of the damming and breaching of the fan generally is unknown. For example, photographs of the Crystal Creek debris fan taken in 1967 (Fig. 11b) show that it had a surface veneer of boulders nearly all of the way up to the mouth of Crystal Creek—far above any stage reached by Glen Canyon Dam discharges between 1966 and 1967. Because of this absence of known floods prior to the time the 1967 photograph was taken, I speculate that the 1966 debris dam caused ponding of water to this level and that the overflow of this original pond removed a substantial number of boulders all the way to the top of the debris fan. I discuss in detail the implications of each size distribution shown in Figure 12 in Kieffer (1987).

These many erosional and depositional processes are important in the details of the channel shape and particle size distribution where the river cuts through a debris fan to form a rapids. The geometry of the channel is the result of a balance between the local increase in the river's erosive power within the constricted zone and its decreased transport capacity in slower parts of the channel. Water accelerates from velocities on the order of a fraction of a foot per second (a few tenths of a meter per second) in the backwater above a rapid to 15 to 25 feet per second (4 to 6 m/s) on the tongue of the rapid, to values greater than 30 feet per second (10 m/s) in the constriction and part of the diverging section. Velocities of about 15 feet per second (4 m/s) then are maintained through the tailwaves. Thus, in the constriction and the diverging section of the rapid immediately below the constriction, velocities are more than adequate to clear the channel of boulders.

The position of the constriction in a rapid can change as the rate of discharge changes because of variations in topography. The position and shape of the constriction also changes with time as the river channel evolves in response to floods. An excellent example of how the position of a constriction depends on discharge can be seen at Horn Creek Rapids. At low discharges (less than 30,000 cfs), the constriction is near the top of Horn Creek Rapids. At 92,000 cfs, this constriction is nearly drowned out by the backwater from a constriction about 1000 feet (several hundred meters) downstream (Kieffer, I-map 1897-E, 1988).

The position and shape of a constriction in a rapid also changes as a result of tributary flash floods, as at Crystal and Bright Angel Rapids in 1966. In 1983, the position of constriction at Crystal Rapids again changed, moving upstream in response to high discharges. These changes are examples of why observations of rapids over a large range of discharges and, by implication, a long period of time are necessary for compiling data on channel processes and their rates.

Because there is an increase in erosive power in the constriction and in the diverging section of a rapid, boulders move downstream until the channel widens and deepens sufficiently for the water to decelerate (typically by a transition from supercritical to subcritical flow). Large boulders are moved out of the constrictions of rapids, transported through all or part of the divergent sections, and then deposited hundreds of feet (in some instances, up to 1 km) downstream to form the "rock gardens" or cobble bars found below rapids (Fig. 2a). A rapid, therefore, evolves into two parts: the original debris deposit (reworked, at least on the surface, by overflow) and the rock garden (or cobble bar), usually found downstream of the initial deposits.

Crystal Rapids, 1966-1987

In 1983, we were able to document changes in the hydraulic behavior and channel shape at Crystal Rapids; the following summary is taken from Kieffer (1985). The rapids in the Grand Canyon were exposed to discharge levels three times that which had occurred since 1963, and this was a particularly significant event for Crystal Rapids because of the large debris fan emplaced in 1966. In addition to their geomorphic significance, the hydraulic events during 1983 had a significant effect on commercial and private rafting in the Grand Canyon. About 10,000 people each year navigate the 250-mile (400-km) stretch through the Grand Canyon. The debris flow in 1966 and the flood of 1983 both emplaced boulders and caused waves and eddies in Crystal Rapids that have made this area difficult to navigate.

Although the geometry of the river channel prior to the studies of Kieffer (1988) is largely unknown, studies of air photographs taken in 1973 and calculations of plausible cross-section shapes (Fig. 14) suggest that the constriction of the channel was about 0.25—that is, the surface width of the rapid in its narrowest part was only 1/4 of the width of the main channel upstream of the rapid. The water-surface elevation dropped about eight feet (2.5 m) between the head of the rapid and a large obstacle known as "The Crystal Hole" that was several hundred meters downstream from the entrance to the rapid (locales are shown on Fig. 14). The water surface dropped about another 8 feet (2.5 m) through the rock garden below the rapid (Leopold, personal comm., 1984). Because the Crystal Creek flood of 1966 appears to have been large enough to have strewn boulders across the entire river channel, it appears that either the initial breaching event (when the river cut through this debris dam) or power plant releases up to 35,000 cfs from Glen Canyon Dam between 1966 and 1973 were sufficient to move rocks out of the distal end of the debris flow into the rock

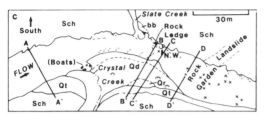

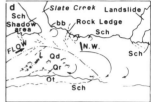

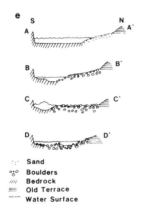

EXPLANATION

Qd - 1966 debris fan

Qr - Older debris fan

Qt - Older terrace

Sch - Schist

X - Rock or rock-caused
    wave (I)

--- Oblique wave (II)

— Normal wave (III)

⌒ Eddy

A —— A',etc., Cross-sections

—L Stranded log in (b)

bb - Boulder-bar

⌒⌒ Tamarisks

Sand
Boulders
Bedrock
Old Terrace
Water Surface

Figure 14. (a). Crystal Rapids on June 16, 1973 (U.S. Geological Survey Water Resources Division air photograph.) (b) View of part of the rapid at approximately the same scale during the high flow of 1983. (c) and (d) Keys to features on (a) and (b).and (c) (e) Schematic cross sections along lines A-A', B-B', C-C', and D-D'. Relative widths correct; vertical scale exaggerated. Rise of debris fan from the river level in (a) to the old alluvial terrace indicated by Qt is about 5.5 m. The stage at 92,000 cfs is just at the base of this terrace. Assumed boundaries for deep channel are shown by light dashed line in (a) and (c). P-P' was the preferred navigation route through the rapids prior to 1983. The Old Crystal Hole, formed by a normal wave (hydraulic jump) is indicated by N.W. This wave is not easily visible in (a) because of photo scale. The white line in (b) indicates the path of kayaks whose velocities were measured at approximately 9 m/s.

GRAND CANYON GEOLOGY

garden. The result was that by the time the 1973 air photographs on which the data of Figure 3 were based, the river had carved a channel with the value of constriction equal to 0.25.

A significant clue to the channel hydraulics in 1983 was the comparison of the waves in the mature rapids with the behavior of the wave that occupied the region of the Old Crystal Hole when discharges in late June and early July 1983 increased above earlier maximum discharge levels. Waves in the more mature, less constricted rapids than Crystal disappeared at discharges on the order of 90,000 cfs, unless the topography covered by the water at higher stages formed a different constriction in three dimensions. This phenomenon of the disappearance of local waves within rapids is the so-called "drowning out" of a rapid at high water. Instead of drowning out, the wave associated with the Old Crystal Hole in Crystal Rapids increased in height as discharges increased. At discharges in the range 50,000 -70,000 cfs, this wave formed a formidable barrier across the river: it was more than 15 to 20 feet in height   (5-6 m) and spanned nearly the entire navigable channel (Fig. 15).

The size, location, and even the sound of this wave changed with discharge. In the years prior to the 1983 flood, the trough-to-crest height had been about 10 feet (3 m) at 20,000 cfs and about 3 feet (1 m) higher at 30,000 cfs. Between 1966 and 1983, the wave was associated with a large rock in this location rather than with its critical position in the neck of the constriction. As discharges rose to around 50,000 to 60,000 cfs in June 1983, boatmen and passengers reported that the wave surged to a height between 15 and 30 feet (5 and 9 m); it was verified photographically at about 15 to 20 feet (5-6 m) (Fig. 15). Perhaps most interestingly, as discharges reached 92,000 cfs in early July, river observers noted that the wave height decreased to 10 to 15 feet (3-4.5 m). At discharges over 50,000 cfs, the wave appeared to be located about 100 feet (30 m) downstream from its pre-1983 position at 30,000 cfs (compare the position of the wave, labelled N.W. in Figs. 14c and d). Observers reported that at 50,000 to 60,000 cfs the wave emitted a low roar like a jet engine, but it did not generate the same loud roar at 92,000 cfs, though loud booms were clearly audible every few seconds.

After the 92,000 cfs discharges of 1983, surface wave patterns within the rapid altered dramatically, and the local gradient within the rapid changed (Fig. 1c). This new hydraulic situation

Figure 15. Photograph of the wave, interpreted as a normal hydraulic jump, that formed across much of the main channel in late June 1983, when discharges were raised to about 60,000 cfs. Photograph taken June 25, 1983; copyrighted by Richard Kocim, reprinted with permission. Pontoons on the raft are each 1 m in diameter; midsection is about 3 m in diameter. More than 30 passengers were on board; one head is visible on the lower side of the raft. From the scale of the raft, the trough-to-crest height of this wave can be estimated at more than 5 - 6 m.

has persisted. The drop in elevation of 5 to 10 feet (2-3-m) that was spread between the top of the rapid and the old Crystal Hole is concentrated now in a narrow zone of only a few tens of meters near the top of the rapid. As a result, the oblique waves on the right side of the tongue at the entrance to the rapid have increased dramatically in height. The change in bed slope and water-surface gradient at the head of the rapid suggests that about 30 feet (100 m) of headward erosion occurred during the high discharges.

In addition, the channel widened by 30-50 feet (10-15 m) at its narrowest point during this flood (compare Figs. 11c and 11d). This means that the constriction value changed to 0.40—approaching the 0.5 value typical of the older, more mature debris fans. Widening occurred in and downstream of the constriction in the zone of supercritical flow. Widening did not occur solely during peak flows, but it appears to have begun as soon as the discharges exceeded the controlled flows of the prior two decades. This conclusion is unsupported by direct measurement. However, a very similar series of events—including a debris flow in 1966—occurred at Bright Angel Creek, 10 miles (16 km) upstream from Crystal Rapids and in a similar geologic setting. There, as the discharge rose through the range of 60,000-70,000 cfs, the bed at Bright Angel gage station was scoured by about 8 feet (2.4 m). Presumably, a scour of similar magnitude occurred at the same discharges at Crystal because of the similarity in channel morphology and material at the two locations.

Water velocity varies throughout the length of a rapid because of geometry and gradient changes. Unfortunately, no systematic measurements of water velocity could be made at Crystal during the flood; however, on June 27, when the discharge peaked at 92,000 cfs, kayaks were filmed going through the rapid (approximately along the white line shown in Fig. 14a). Analysis of the films showed maximum velocities of 28 to 32 feet per second (8.5-9.8 m/s). Although there can be no rigorous correlation of these velocities with average water velocity, the kayaks appeared to be moving with the current. If the average water velocity was even close to these values, the river was capable of moving boulders several meters in diameter (Fig. 13). Large, moving boulders presumably were the source of the loud, booming noises heard by the author.

The velocities measured, the depths inferred from measurements of stage, and the behavior of the large wave that developed in Crystal Rapids all indicated conditions of supercritical flow during the flood. The flow was forced into supercritical conditions by the geometry of the converging-diverging channel. The large wave that stood across the channel had characteristics generally associated with a "normal hydraulic jump". Shallow-water flow theory allowed me to analyze the relation between discharge and backwater energy—conventionally expressed as depth (Fig. 16a), wave height (Fig. 16b), velocity in the constriction (Fig. 16d), velocity in the diverging supercritical section of the rapid and in the diverging subcritical section below the hydraulic jump (Fig. 16c), and the change in water velocity through the hydraulic jump (Fig. 16e).

Figure 16b shows that the measured wave height increased with discharge until the discharge reached about 60,000 cfs; the

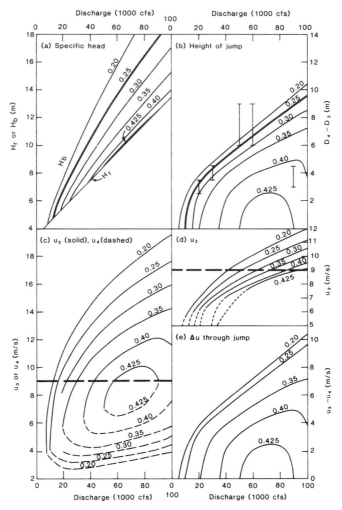

Figure 16. Model calculations for flow through Crystal Rapids (from Kieffer 1985). (a) Specific head (Hr) measured at Bright Angel Creek vs. discharge, with backwater heads (Hb) calculated for Crystal Rapids for the constrictions, w2/w0, (0.20, 0.25,...) indicated. The specific head can be thought of as the normal depth of the river, and the backwater head as the depth caused by the constriction. (b) Calculated height of the hydraulic jump for constrictions indicated. Bars denote observed values. (c) Calculated values of flow in the supercritical region of the diverging section of the rapid (u3), and in the subcritical region of the diverging section (u4). These two sections are separated by the hydraulic jump. Dashed line at 9 m/s indicates the velocity at which larger boulders at Crystal Rapids were assumed to be transportable by the current. (d) Calculated values of velocity in region 2 (the constriction) for constrictions indicated. Flow subcritical where dashed. (e) Velocity change through hydraulic jump that separates regions 3 and 4

378   GRAND CANYON GEOLOGY

calculations indicate that this behavior would be expected for a channel with a pre-flood constriction of 0.25. At higher discharges, the model calculations suggest that the wave height should have continued to increase, but, instead, it decreased. The curves in Figure 16b suggest that changes in the shape parameter caused the decrease in wave height. In particular, it appears that by the time discharges reached about 60,000 cfs, the channel had begun widening and that by the time of peak discharges, the constriction had changed to about 0.40.

Although the parts of Figure 16 look complicated, a careful study of these figures reveals the complex interactions that were occurring. The meteorological situation forced the engineers to increase the discharges constantly, causing water velocities to increase. Channel widening occurred as a result of erosion at the high velocities. The channel widening alone would have allowed a decrease in velocity, but the increasing water level of Lake Powell necessitated further increases in discharge.

During this flood event, the Colorado River continued to contour the channel at Crystal Rapids into a shape in which the velocities in the most highly constricted portion of the channel were equal to the threshold velocity for the transport of the major boulders. Material removed from the constriction was transported several hundred meters downstream into the area of the rock garden. This section of the river was modified substantially by the 1983 flood.

For nearly a year following the peak flows in June and July of 1983, high discharges prevented direct observations of the effects of the flood. By the time that Crystal could be examined again at low water (October 1984), the record of the 1983 events was partially overprinted by sustained discharges at 60,000 and 25,000 cfs. Nevertheless, field evidence indicated that the erosion postulated on the basis of the open-channel hydraulics theory did indeed occur (Figs. 11c and 11d).

Observers found that the eroded section of the channel at Crystal Rapids had a fresh cutbank in the boulders. Similar cutbanks have been observed after the 1984 debris flows down Monument Creek at Granite Rapids and at Elves Chasm in 1984. These cutbanks are evidence of the action of the river contouring its own channel, and if they can be related to specific discharges, they provide valuable clues about the geomorphic evolution of the debris fans.

A MODEL FOR THE GEOMORPHIC-HYDRAULIC
EVOLUTION OF THE RIVER CHANNEL
AT DEBRIS FANS

The two decades of observations at Crystal Rapids are but a glimpse into the history of much larger debris flows (e.g., from Prospect Canyon at Lava Falls) and much larger flood events that occurred throughout the history of Grand Canyon downcutting. Figure 17 shows a generalization and extrapolation of the ideas developed at Crystal Rapids. This figure should be interpreted to represent but one cycle of recurring episodes of deposition and modification.

In this model, I have arbitrarily chosen the beginning of the sequence as a time when the main channel is relatively uncon-stricted—that is, major floods are assumed to have occurred on the Colorado River since the last time this tributary canyon had a major debris flow (see Fig. 17a). Between Figures 17a and 17b, it is assumed that unusual weather, climate, or an accumulation of debris within the catchment basin caused a large debris flow to emerge suddenly from the tributary and to dam the Colorado River channel. The flow in the river is disrupted and ponded by the emplacement of a debris dam (Fig. 17b). A "lake" forms behind the debris dam, and at some time, depending on the height of the dam and the discharge in the main stem, a waterfall forms as this dam is overtopped. Schuster and Costa (1986) suggest that the dam would be breached (typically at the distal end of the debris fan) nearly instantaneously (hours to days)—and perhaps catastrophically. The evolution of a rapid from a lake and water-fall then would begin.

Unless there is a major breach in the debris dam with the first breakthrough of the ponded water (for example, more than fifty percent of the material is removed), the constriction of the main stem initially is severe (Fig. 17b). Floods of differing size and frequency then erode the channel to progressively greater widths (as shown in Fig. 17c,d, and e). Small floods (Fig. 17c) enlarge the channel slightly (again, a nonquantitative term that depends on the prior history of the fan during breaching and on the material composing the fan). At first, the depletion of fine material through the debris fan may undermine the positions of large boulders. This undermining can cause rather dramatic changes in the scale of individual boulders and waves within the rapid. Indeed, it may

GRAND CANYON GEOLOGY

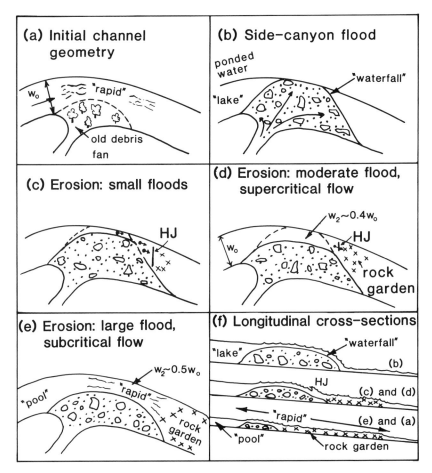

Figure 17. Schematic illustration of the emplacement and modification of debris fans, the formation and evolution of rapids, and the formation of rock gardens. See text for further explanation.

be the reason that changes in wave behavior in Crystal Rapids seemed rather frequent in the late 1960s and early 1970s. As discussed above, the movement of a single, large boulder can cause dramatic changes in local wave structure in a rapid within regions of supercritical flow.

Moderate floods (Fig. 17d) enlarge and widen the channel further. The channel may become wide enough that at low discharges the flow is weakly supercritical or even subcritical. For example, the 1983 discharges of 92,000 cfs at Crystal Rapids (considered high by standards of Glen Canyon Dam discharges) widened the constriction at Crystal sufficiently that the strength of the Old Crystal Hole was diminished during subsequent lower flows at which it previously had been a substantial obstacle. On the other hand, the lateral waves became stronger because of headward erosion, which created additional potential energy at the top of the rapid. These waves, rather than the Old Crystal Hole, now form the main rafting hazard, though the hole is still a strong wave at discharges above about 30,000 cfs. Lateral widening, vertical scouring, and headwall erosion of the underwater debris dam occur simultaneously (Fig. 17f). Large floods further widen the channel at the debris fans and erode upstream through the fans (Fig. 17e). This cycle (a-e) can be repeated over and over through geologic time as floods in tributaries and on the main stream occur.

In summary, the local geometry of the river channel is subject to change if a substantially larger discharge is put through a rapid. This discharge can occur from flooding caused by meteorological events, from flooding caused by local emplacement and the breaching of natural dams in the river (debris of lava flows), or, now that Glen Canyon Dam is in place, from operational procedures at the dam. When the local gradient in the channel changes from such flood events, new waves can arise and old waves can become small or disappear.

The Rapids: Past, Present, and Future

The shape of the river channel at a debris fan at any instant of geologic time reflects contouring by different flood events— including any flood event that may have accompanied the emplacement and breaching of the debris dam. Even discharges as low as several thousand cfs appear to have sufficient velocity to clear the channel of large boulders, though such discharges are typically only a few percent of the total channel cross-sectional area. Fine-grained, transient sediment may partially mask the larger scale erosion (e.g., Howard and Dolan 1981, Fig. 7).

We now return to the questions raised earlier in this chapter: "Why is the shape of the channel, i.e., nozzle, eroded through the debris fans so uniform, and why is the value of constriction specifically 0.5?" The answer lies in Figure 16d (and in extrapolations of this figure to higher discharges found as Fig. 13 in Kieffer's 1985 publication).

The fact that the Colorado River is less constricted at most of the tributary debris fans than it is at Crystal Rapids suggests that discharges higher than 92,000 cfs have occurred in their history. We know this to be true—a flood of 220,000 cfs occurred in 1921, and a flood estimated at 300,000 cfs occurred in 1884. We can reasonably assume that even larger floods have occurred since many of these debris fans formed, a time that may exceed 10^4 years (Hereford 1984).

Because of the higher velocities associated with higher discharges, larger floods will make the channel at a debris fan wider. Extrapolation of the calculations done for Crystal Rapids, based on a threshold transport velocity of 30 feet per second (9 m/s), suggest that floods as large as 400,000 cfs are required to open the channel up to the constriction of 0.5. The accuracy of this estimate cannot be stated because we know too little about the threshold velocity for erosion (or other similar criterion). There are other variables to consider as well: (1) the reliability of our extrapolations using standard power-law functions of the dependence of depth, velocity, and head-on discharge; (2) our lack of knowledge on the rate at which vertical cutting and headward erosion occurred (we are assuming that the channel comes to an equilibrium shape during each flood); and (3) our inability to consider the true geometry of the river channel. As further data become available, we will be able to construct more accurate models.

Despite these uncertainties, we know that discharges an order of magnitude greater than discharges from the power plant at Glen Canyon Dam (and approximately a factor of five greater than the 1983 flood levels) contoured much of the river channel at the debris fans and gave the rapids their characteristic configuration. Without floods of this magnitude in the future, the character of the rapids will change as tributaries flood. The change will be toward more highly supercritical conditions as the constrictions become narrower, both laterally and vertically.

CHAPTER 17

LATE CENOZOIC LAVA DAMS IN THE WESTERN GRAND CANYON

W. K. Hamblin

INTRODUCTION

We have no difficulty as we float along, and I am able to observe the wonderful phenomena connected with this flood of lava. The canyon was doubtless filled to a height of 1,200 to 1,500 feet, perhaps by more than one flood. This would dam the water back; and in cutting through this great lava bed, a new channel has been formed, sometimes on one side, sometimes on the other.... What a conflict of water and fire there must have been here! Just imagine a river of molten rock running down a river of melted snow. What a seething and boiling of waters; what clouds of steam rolled into the heavens!

J.W. Powell, Aug. 25, 1869

From the time Powell first viewed the remnants of basalt adhering to the walls of the inner gorge in the western Grand Canyon over one hundred years ago, relatively few people have had the opportunity to see and study this isolated area. But more and more visitors are discovering the viewpoint at Toroweap, where they are privileged to see one of the most spectacular displays of volcanism in North America (Fig. 1). Thousands more see remnants of the ancient lava dams from float trips down the Colorado River or glimpse the scene from chartered air flights.

Figure 1. Volcanic features in the western Grand Canyon. View looking northeast at the cascades and remnants of lava dams near the mouth of Toroweap Canyon. Vulcan's Throne is perched on the rim of the inner gorge. Recent extrusions of basalt flowed across the Esplanade platform and cascaded into the inner gorge where they cap large remnants of major complex lava dams. Smaller remnants of other dams can be seen high on the north wall of the canyon.

The volcanic features of this area are much more complex than one might first imagine. What Powell observed during his epic trip down the Colorado River was only a small fraction of the region's volcanic phenomena. Over 150 lava flows have poured into the canyon during the last 1.5 million years. They have left an incredible record of volcanic events and their influence on the Grand Canyon. Some flows were extruded on the Uinkaret Plateau and cascaded over the north rim of the canyon into Toroweap Valley and Whitmore Wash. Others were extruded within the canyon itself and spread out over the Esplanade Platform before forming spectacular frozen lava falls that plunged over the rim of the inner gorge into the Colorado River 3000 feet (900 m) below. In several places, volcanoes are perched precariously on the very rim of the canyon, and remnants of others cling to the steep walls of the inner gorge. In addition, the dikes, sills, and volcanic necks exposed in the canyon are all associated with the complex sequence of recent volcanic events.

The spectacular lava falls that spill over the Esplanade into the inner gorge cap remnants of an older sequence of flows that formed huge lava dams, some of which were over 2000 feet (600 m+) high. One was more than 84 miles (135 km) long. The lava dams backed up the water of the Colorado River to form temporary lakes. Several of these lakes extended upstream through the Grand Canyon into Utah, slightly beyond the present extent of Lake Powell. As the lake behind the barrier overflowed, a new gorge was eroded through the lava dam, leaving only remnants of the basalt clinging to the walls of the canyon. Later eruptions formed new dams, which subsequently were breached and then destroyed by the overflow of the Colorado River.

At least twelve major lava dams were formed in the Grand Canyon during the last million years. These dams, together with remnants of the sediment deposited in the lakes behind the dams, provide a fascinating record of this unusual and most recent series of events in the history of the Grand Canyon.

METHODS

The volcanic phenomena in the western Grand Canyon region are exceptionally well exposed, but they present some unusual problems to the field geologist because of the extremely rugged and largely inaccessible nature of this part of the canyon. Most of the basalt remnants exist as thin slivers clinging to the vertical cliffs of the inner gorge. This means that they are inaccessible for the most part and cannot be reached by trails from the canyon rim.

To study the exposures of lava throughout their 84-mile extent along the river, we had to float equipment and supplies 180 miles (288 km) downstream from Lees Ferry at the beginning of each field season and establish a series of temporary base camps along the river. This allowed us to enter and leave the canyon and work slowly downstream so that our river operations could last the entire season.

However, working from the river in this manner presented its own problems. Although the basalts are almost one-hundred percent exposed and stand out in stark contrast to the tan and reddish Paleozoic strata, many of the exposures are largely inaccessible because they exist as vertical cliffs hundreds to thousands of feet above the river. In an attempt to solve this problem, we photographed the entire canyon wall from view points on the opposite side of the river using a hand-held aerial camera. Enlarged prints were made and fitted together to produce a photo mosaic of the entire canyon wall. This then served as a base map (cross section) on which we could plot all of our measurements and geologic observations. Elevations of the contacts between flow units were measured from the river using a theodolite.

We plotted our original data on the enlarged mosaics of the canyon wall and on the enlarged vertical aerial photographs. Later, we transferred our mapping to seven and a half minute topographic maps as the maps became available.

We also established several base camps on both the north and south rims of the Esplanade to study the more complicated areas between Toroweap and Whitmore Wash. We obtained additional data by using light aircraft and a helicopter.

Although some tantalizing questions about the details of certain relationships among the lava flows and the significance of various flow units remain, the study we made has established a significantly large data base. And from this base, reasonably safe interpretations can be made about the late Cenozoic history in the western Grand Canyon.

Methods of Determining Relative Ages

The relative age of most of the flow remnants in the canyon is expressed clearly either by superposition or by juxtaposition. The process by which juxtaposition of the flow remnant is produced is shown in a series of diagrams in Figure 2. The idea is simple. The first intracanyon flow entered the canyon from cascades or from centers of extrusion in the canyon itself. The lava partially filled the canyon, causing a lake to form upstream. Eventually, the backwater overflowed the barrier and eroded most of the basalt, leaving only thin vestiges of the lava flow adhering to the canyon walls. A subsequent flow entering the canyon would be juxtaposed against the older remnants. With renewed erosion, thin remnants of the two flows would remain stacked side-by-side. The older flow would be next to the canyon wall and the younger flows juxtaposed in sequence against the older.

The remnants of lava flows in the Grand Canyon may be recognized and correlated throughout the canyon on the basis of several criteria. Although the petrography of some flows are similar, others are unique. Therefore, some flows can be recognized without difficulty on the basis of petrographic characteristics alone—even in a small outcrop. Diabase flows, for example, are readily distinguished from the dense, black, aphanitic basalt that characterizes other flow units.

In addition to petrographic characteristics, the internal structure of a number of flows is a distinguishing characteristic. One flow can be recognized by its abnormally thick joint columns in the basal colonnade. These range from seven to eighteen feet (2 to 5 m) in diameter. Others have unique jointing in the entablature, not only in size but also in the geometry of the columns. Stratigraphic sequences also are useful in correlation as well since river gravels of a specific nature (size, composition, and sorting) may be stratified within a specific sequence of flows. In addition, the elevation and gradients of the top of all flow units were measured carefully with a theodolite. This provides an important means of correlating flow units on the basis of geomorphic relations.

Using these methods, we were able to map remnants of twelve major lava dams that were formed and subsequently destroyed in the Grand Canyon during the last one to two million years.

CHARACTERISTICS OF LAVA DAMS

It is apparent from the sequence of basalt preserved in the inner gorge of the Grand Canyon that four different types of dams were constructed during the period of Late Cenozoic volcanic activity in the area. The nature of these structures is shown in Figure 3. Lava dams were simply intracanyon lava flows that

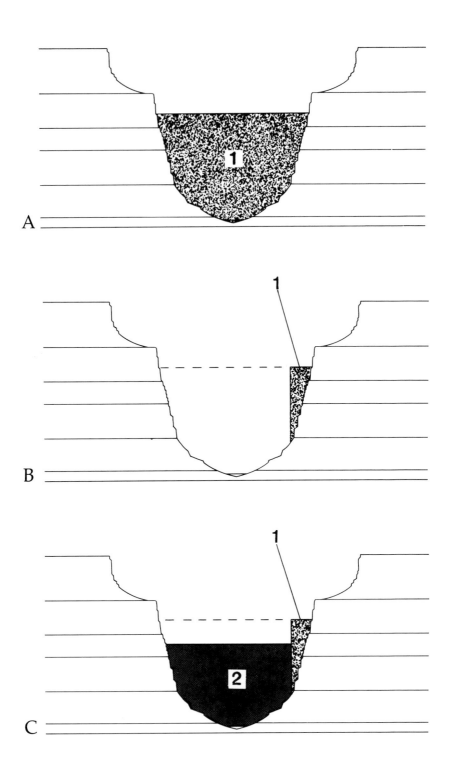

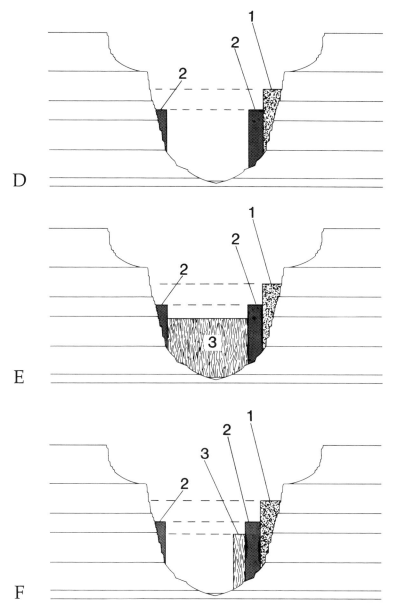

Figure 2. Diagrams showing the development of juxtaposed flows in the Grand Canyon. (A) Flow partly fills the canyon. (B) Erosion leaves small remnants of flow 1 adhering to the canyon wall. (C) Flow 2 refills the canyon. (D) Erosion removes most of flow 2, leaving remnants juxtaposed against flow 1 and against the canyon wall. (E) Flow 3 fills the canyon. (F) Erosion of flow 3 leaves remants of flows 1, 2, and 3 stacked side-by-side according to relative age.

attained a thickness sufficient to form a significant barrier to the flow of the Colorado River because of the restriction of narrow canyon walls. These lava dams were asymmetrical structures, steep on the upstream margin and gently inclined downstream. The length of the dams ranged from 20 to more than 86 miles (32

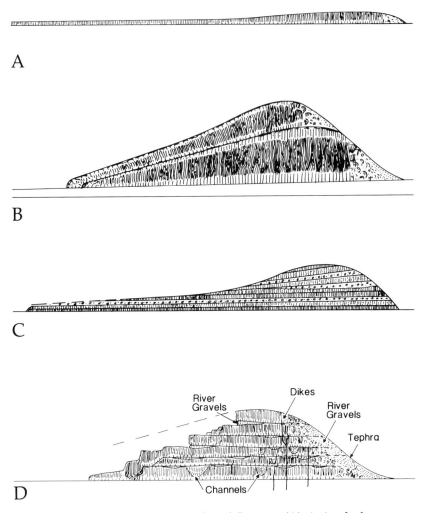

A

B

C

D

Figure 3. Types of lava dams in the Grand Canyon. (A) A simple dam formed by a single flow 150 feet to 600 feet thick. (B) High dam formed by massive flows over 800 feet thick. (C) Compound dam composed of numerous flows 10 to 30 feet thick. (D) Complex dam built from multiple flow units 50 to 200 feet thick

GRAND CANYON GEOLOGY

to 137 km), and the height ranged from 150 to more than 2000 feet (45 to 600 m). The rate, volume, and viscosity of the lava extruded are primarily responsible for differences in the nature of the dams.

Thin Single-Flow Dams

The simplest type of barrier in the Grand Canyon was a dam constructed of a single lava flow, 150 to 600 feet thick (45 to 180 m). This type of barrier was exceptionally long, extending downstream for several tens of miles (in one instance, more than 86 miles or 137 km). The internal structure of this type of flow is characterized by the classic columnar jointing that consists of a basal colonnade, an entablature, an upper clinker surface, and / or an upper colonnade. Most of these dams show little or no evidence of erosion on their upper surfaces.

The type of eruption producing such a barrier probably consisted of a rapid extrusion of a moderate amount of fluid lava. Such a dam could be constructed by lava extruded within the canyon or from lava that was extruded on the canyon rim and subsequently cascaded into the river. The volume of lava forming these barriers ranged from 0.03 to 0.5 cubic miles. This is comparable to the average volume of flows in the volcanic fields on the Uinkaret Plateau. Judging from historic eruptions in Hawaii, Iceland, and Mexico, these dams were constructed within a period of several weeks.

Massive dams

A distinctly different type of dam was formed in the canyon by abnormally thick, massive basalt flows. These flows were over 800 feet (240 m) in depth. Because these are the oldest dams to form in the canyon, only a few remnants are visible, but the elevation of the upper surfaces of the remnants indicate that these barriers probably were not more than twenty miles (32 km) long. It also appears that the lava was extruded faster than the backwater of the Colorado River could overflow. The best estimates indicate that the volume of these dams was 0.5 to 1.2 cubic miles. This is only twice the volume of the average single-flow dam.

Compound Dams

Compound dams were constructed of a sequence of numerous flow units ten to thirty feet thick (3 to 9 m), all of which were deposited in rapid succession. These dams ranged from 300 to 900 feet (90 to 270 m) in height. The overall geometry of the compound dams was much like the simple single-flow dams, consisting of a steeply inclined front (upstream slope) and a long, gently inclined downstream slope. The volume of these dams ranged from 0.06 to 0.23 cubic miles.

Complex Dams

Complex dams were built from a series of multiple flow units 50 to 200 feet (15 to 60 m) thick. The general characteristics of each individual flow unit are similar to those of the single-flow dams, but the upper colonnade or clinker surface on flows within the complex dam often is eroded. Locally, deep channels are cut in the flow. These channels are filled with younger flows, ash, and sediment. Lenses of river gravel, sand, and, in some cases, ash are found separating the major units of basalt. The upstream slope of the complex dams consists of flow units that are inclined in a steep angle upstream. These pass rapidly into rubble, agglomerate, and tephra.

The erosion of the flows, together with interstratified river gravels, clearly indicate that the lake which formed behind the dam overflowed during the period of extrusion. Thus, most flows were eroded to some extent between periods of eruption.

The complex dams were quite high, ranging from 600 to more than 1400 feet (180 to 420 m) above the present gradient of the Colorado River. The steep gradient on the preserved remnant suggests that the complex dams were no more than twelve miles long. In many cases, it may have been much less because the lower end of each flow would have been eroded by upstream migration of waterfalls and rapids before the subsequent flows were extruded.

CHARACTERISTICS OF LAKE SEDIMENTS

The lakes that formed in the Grand Canyon behind the lava dams were unusual from the standpoint of their geomorphic setting, origin, and history. They had little in common with natural lakes formed in glacial terrains or in low-lying coastal regions. These were lakes formed in a deep canyon and identical in nearly every respect to the present manmade reservoirs such as Lake

Mead and Lake Powell. The nature of the sediment filling these two reservoirs provides the best insight into the sediment deposited in the reservoirs behind the lava dams.

When the Colorado River was blocked by a lava dam, coarse sand and gravel were deposited as deltaic sediments at the point where the Colorado River entered the lake. Fine-grained sediments were carried by turbidity currents far into the deeper part of the reservoir, where they were deposited as graded beds of silt and mud. In the absence of large tributaries, the main filling of the lake was accomplished primarily by deposition from the Colorado River.

The lakes formed in the Grand Canyon were surrounded by steep canyon walls. Large beaches rarely developed along the shore, and without significant input from the tributaries, the major process operating along the shores of the lakes was mass movement. Slope wash, rock falls, and the general downslope migration of colluvium (which presently operates along the canyon walls) would have continued during the lifetime of the lakes. A thin mantle of coarse, subaqueous slope debris was deposited close to the canyon walls contemporaneous with the deposition of lake silts. Each tributary continued to transport and deposit this material into the lakes at the heads of tributary bays.

Beyond the zone affected by slope processes, the major types of sediment deposited were mud and silt. Gravel was a significant facies in some major tributary channels as a result of flash floods, and some of the larger tributaries near the upstream reaches of the lake probably constructed small deltas of sand and gravel. However, the major delta of sand and gravel formed in the area where the Colorado River emptied into the lake. Deltaic deposits prograded downstream over the silts deposited by turbidity currents.

After the dam was destroyed, most of the unconsolidated, water-saturated sediment was flushed rapidly out of the main canyon, leaving only minor remnants of sediment close to the valley walls. This material was dominantly slope wash and colluvium with silt filling the space between the larger particles. Therefore, most of the preserved sediment near the canyon walls was not composed of typical lake silts. Instead, it was composed of colluvium with intercalated laminated silt. This type of deposit is difficult to distinguish from the present accumulation of colluvium.

Table 1.

	DAM	ELEVATION	HEIGHT ABOVE RIVER	RADIOMETRIC DATE
1	Prospect	4000	2330	
2	Ponderosa	2800	1130	
3	Toroweap	3093	1443	1.2 Ma
4	Esplanade	2600	960	
5	Buried Canyon	2480	850	0.89 Ma
6	Whitmore	2500	900	
7	"D" Flows	2295	635	0.57 Ma
8	Lava Falls	2260	600	
9	Black Ledge	2033	373	
10	Gray Ledge	1813	203	
11	Layered Dbs.	1938	298	0.64 Ma
12	Massive Dbs.	1826	226	0.14 Ma

DEVELOPMENT AND DESTRUCTION OF LAVA DAMS

The hydrologic data from the Bureau of Land Management's Lake Mead Survey (1963 and 1964) provide the basic information from which we are able to calculate the rates at which the lakes behind the various lava dams were filled with water and, subsequently, sediment. These data also provide some indication of the time necessary for a lava dam to be eroded away completely (Table 1).

Rates of Formation of Dams

Although the formation of a lava dam in the Grand Canyon was a significant event that dramatically changed canyon morphology, the time needed to create a lava dam was remarkably

GRAND CANYON GEOLOGY

VOLUME OF LAVA (mi 3)	LAKE LENGTH	WATER FILL TIME	SEDIMENT FILL TIME
4.0		23 yrs.	3018 yrs.
2.5		1.5 yrs.	163 yrs.
3.7		2.62 yrs.	345 yrs.
1.8		287 days	92 yrs.
1.7		231 days	87 yrs.
3.0	100	240 days	88 yrs.
1.1	74	87 days	31 yrs.
1.2		86 days	30 yrs.
2.1	53	17 days	7 yrs.
0.3	37	2 days	10.3 mos.
0.3	42	8 days	3 yrs.
0.2		5 days	1.4 yrs.

short by any standard and certainly would be considered instantaneous in a geologic time frame. Observations of basaltic eruptions in historic times indicate that most basaltic extrusions occur in a matter of days or weeks. The major flows in the Grand Canyon, most of which were 100 to 200 feet (30 to 60 m) thick, probably moved tens of miles down the Colorado River in a matter of days.

This conclusion is supported by the fact that the upper colonnade and/or a clinkery upper surface of the flow often is preserved, essentially unmodified by erosion. This indicates that the flow was extruded in a period of time less than that required for the lake impounded behind the dam to overflow. If extrusion occurred during a longer period of time, the lake behind the dam would overflow, and erosion would modify the upper surface features of the basalts quickly.

We can calculate quite precisely the time required for the backwater behind each dam to overflow. Based on present discharge rates of the Colorado River, we know that a lake formed by backwaters behind a lava flow 100 feet (30 m) high would overflow in 17 days. The construction of a single-flow lava dam 2000 feet (600 m) high and tens of miles long could be completed within a few weeks. Therefore, it is clear that any lava flow retaining an uneroded upper surface was extruded in a matter of a few days.

Dams built from multiple flows required more time and involved cycles of partial erosion between periods of extrusion. The short time necessary for the complete erosion of a dam, however, puts definite time constraints on the development of any barrier to the Colorado River. Regardless of size and the history of eruption, the formation of every lava dam in the Grand Canyon was instantaneous from the perspective of a geologic time frame.

Rates of Reservoir Fill

The time required for reservoirs behind these dams to become filled completely with sediment also was extremely rapid. The hydrologic data for the various lakes are summarized in Table 1. These data indicate that the lakes formed behind the smaller barriers—those 150 to 400 feet (45 to 120 m) high—would overflow in 2 to 17 days. Lakes formed behind the higher dams (200 to 1000 feet or 60 to 300 m high) overflowed in a year. The highest dam, 2330 feet (699 m) high, overflowed in 22 years. Thus, the lava dams were subjected to erosion soon after they were formed, long before the interior of the lava was completely cool.

The volume of sediment carried by the Colorado River has been measured for many years by the Bureau of Reclamation. These data indicate that reservoirs behind the lava dams silted up in only a few hundred years at most. Many of the smaller reservoirs silted up in just a few months. Thus, the sediment load of the Colorado River was soon transported over the dam, causing normal erosion by abrasion of the river channel after the dam was formed. The data in Table 1 indicates that a reservoir formed behind the dam 150 feet (45 m) high would be filled with sediment in only 10.33 months. A dam 1150 feet (345 m) high would be full of sediment in only 345 years. The highest dam would be filled with sediment in 3000 years. Thus, every phase of the construc-

　　　　　　　　　　　　　GRAND CANYON GEOLOGY

tion of the dam and the formation of the reservoir or lake behind it, and the ultimate filling in of the lake with sediments, occurred in a very short time.

Erosion and Destruction of Dams

We do not know the precise way in which dams were eroded. Some boundary conditions can be established, however, based on data from various studies of waterfall migration and the rates of downcutting of major stream systems. Normal downcutting of the stream channel by abrasion was, of course, a significant process of erosion. It began as soon as water overflowed the dam and reached maximum efficiency when the lake silted up and a normal sediment load was transported over the dam. Another important process was the migration of rapids and waterfalls that initially formed on the downstream end of the flow (Rogers, J.D. and M. R. Pyles 1979). Two important characteristics of the intra-canyon flows facilitated the migration of waterfalls (Fig. 4):

(1) Intracanyon flows were deposited directly on the sand and gravel bed in the channel of the Colorado River. This layer of unconsolidated sediment would be eroded easily by undercutting at the plunge pool below the waterfall.

(2) The vertical columnar jointing in the basalt constitutes an all-pervasive structural weakness throughout the flow. The hexagonal columns produced by the jointing impart a low cohesive strength to the rock body so that the columns would readily topple into the plunge pool beneath the waterfall.

In addition, hydraulic plucking of the columns probably occurred in a river with a discharge as high as that of the Colorado. The combined effects of the unconsolidated sand and gravel substratum and columnar jointing in the basalt resulted in rapid upstream migration of the waterfalls. As the waterfalls migrated upstream and approached the head of the dam, they became higher and higher. The process of erosion then was accelerated by the increase in potential energy for undercutting and scouring. Erosion of the highest dams must have presented a spectacular scene. As the waterfalls advanced towards the highest section of the dam, a tremendous scour hole must have been generated at the base of the falls. Exceptionally high waterfalls—those over 2500 feet (759 m)—would form on the highest dams. As erosion proceeded headward, the stability of the dam almost certainly

A

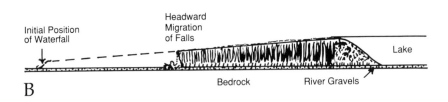

B

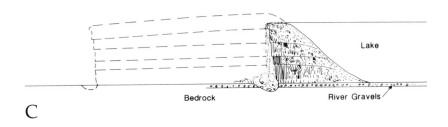

C

D

Figure 4. Diagrams showing erosion of a lava dam by headward migration of waterfalls. (A) As soon as the backwater overflowed the lava dam, a small waterfall would form at the downstream end of the lava flow. (B) Upstream migration of the falls would be accelerated by undercutting of the unconsolidated river sediments beneath the flow and by the weakness of the rock resulting from vertical columnar jointing. (C) As the waterfalls migrated headward, the stability of the dam could be jeopardized by the pore pressure in the columnar jointing. (D) Some dams may have failed catastrophically. This event could have been minimized if contemporaneous downcutting lowered the level of the overflow before the falls migrated to the head of the dam.

GRAND CANYON GEOLOGY

was jeopardized by the enormous pore pressure in the columnar joints. At some critical point, dams may have failed as catastrophic events, with a rapid discharge of a tremendous volume of water and saturated sediment that had accumulated in the lake behind the dam.

Since the rate of waterfall migration is well documented in some areas, it can provide insight into possible time frames for the destruction of the lava dams. Niagara Falls, for example, has migrated headward at the rate of three feet per year; in the last 8,000 years, then, it has migrated a distance of more than eleven miles (18 km). If this figure is typical for the migration of a waterfall on a large river, the lava dams in the Grand Canyon, which generally were less than twenty miles long, would take a maximum of 20,000 years to be completely destroyed by headward migration of waterfalls alone. From these data, we can conclude tentatively that the lava dams were destroyed in less than 20,000 years. It also seems safe to conclude that the time interval for the various phases of the buildup and destruction of the dam would be on the following orders of magnitude:

(1) Single-flow dams would be formed in a matter of several days, whereas the higher and compound lava dams would take up to several years for construction;

(2) Water would fill the reservoir behind these dams in a matter of months. At most, only seven years would be required for the water to fill the reservoir behind the highest dam completely;

(3) Sediment would fill the small reservoirs within one to seven years. In the higher reservoirs, it would take between 100 and 1,000 years to fill the lake completely with sediments. Most of the dams probably would be destroyed within 10,000 to 20,000 years after they were formed.

It is apparent from these observations that the twelve lava dams we know of occupied the canyon for a relatively brief period of time, probably no more than a total of 240,000 years.

The Prospect and Ponderosa Dams

The oldest lava dams in the Grand Canyon of which remnants are still preserved are two huge structures formed by thick, massive flows. These oldest known flows are preserved in Prospect Canyon and are referred to as the Prospect Dam. Remnants of a younger massive dam also are preserved in the Prospect area

Figure 5(a). Photograph of the Prospect Dam. Although only one remnant of this dam remains, the thick flows suggest that the dam did not extend downstream for more than a few miles. The absence of an erosional surface between flows suggests that extrusion was rapid and the barrier was completely constructed before the backwater in the lake overflowed.

and downstream at Mile 181.6. This dam is referred to as the Ponderosa Dam.

The Prospect Dam

The only remaining remnant of the Prospect Dam is a sequence of exceptionally thick flows exposed in the large alcove just east of Prospect Canyon. Here, a vertical cliff of basalt almost 2000 feet (600 m) high extends from the talus slopes near the Colorado River up to the level of the Esplanade (Fig. 5). Unfortunately, the boundaries between the flow units are difficult to see from view points on the river or from view points on the north rim near Vulcan's Throne. Observations made from a helicopter, however, reveal that the Prospect flows are extremely thick and that there are no more than three or four major flow units, each of which is more than 800 feet (240 m) thick. The Prospect flows, therefore, are some of the thickest flows in the canyon.

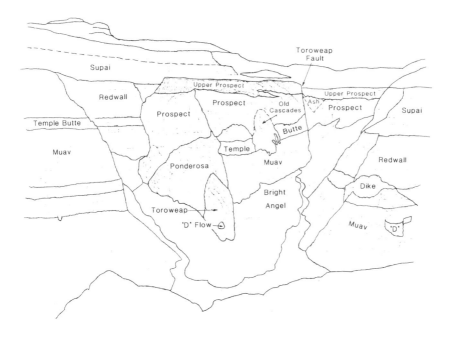

Figure 5(b). A cross-sectional drawing of the Prospect Dam. The structure of the dam consisted of several units: (1) the elliptical units formed by basalt interacting with backwater; (2) hydroexplosive tephra at the front of the dam; (3) the main flows with long, sinuous columnar jointing, and; (4) the basalt cap formed shortly after overflow of the lake. The Upper Prospect Flow consists of a horizontal unit nearly 400 feet thick which once spread across the top of Prospect Canyon and across part of the Esplanade. It formed a cap rock covering the older flows and cinder cones that filled the ancient canyon. Headward erosion along the Toroweap Fault has re-excavated Prospect Canyon and formed a steep, V-shaped gorge over a mile long. Note the offset of the Upper Prospect flow near the apex of the canyon produced by recurrent movement along the Toroweap fault. Recent movement also has displaced the alluvium covering the Upper Prospect flow.

The oldest units in Prospect Valley are a sequence of bulbous or elliptical basalt bodies with interspersed tephra. These flows are overlain by the thick, massive Prospect flow.

The elliptical structures probably were produced by the interaction of lava with the water of the Colorado River. Typical pillow structures would not be expected to form in abundance from a flow entering a river because most of the lava would not be covered completely with water. Only the upstream margins of the flows would interact directly with the river. The downstream segment of the flow would remain essentially dry until the backed-up river water overflowed the barrier.

The elliptical units are several exceptionally thick and massive flows that form a sheer vertical cliff. The internal structure of these flows consists of long, slightly sinuous columnar joints that tend to radiate out from a central point like huge shocks of wheat hundreds of feet high and only tens of feet wide. The only obvious stratigraphic break visible in this sequence of flows appears midway up the cliff at the approximate elevation of the contact between the Muav and Temple Butte limestones. A cross section showing the relationships between the flows in the Prospect alcove is shown in Figure 6.

On top of the sequence of thick flows exposed in the alcove east of Prospect Canyon is a single flow approximately 400 feet (120 m) thick that is characterized by massive columnar joints. This "Upper Prospect Flow" is not confined between the walls of the ancient Prospect Canyon like the underlying units. Instead, it spreads across the top of the canyon and over parts of the Es-

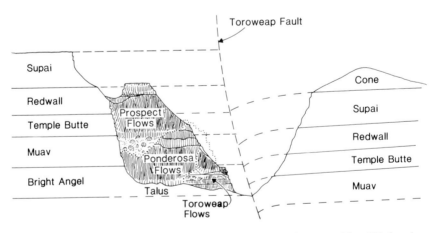

Figure 6. The flows in Prospect Canyon as seen from the west side of Vulcan's Throne. Over a mile of headward erosion has occurred in Prospect Canyon along the Toroweap fault, re-excavating a sharp, V-shaped gorge in the basalt and underlying Paleozoic rocks (Fig. 5). The thick, massive Prospect flow forms a sheer, vertical cliff almost 2000 feet high near the left margin of the drawing.

planade (Fig. 5). This flow probably blocked the river where it entered the canyon, but since the flow is eroded back from the walls of the inner gorge, we cannot determine with any certainty whether this unit formed a dam at an elevation of 4000 feet (1200 m) or if it cascaded into the inner gorge after the Prospect Dam was formed. The horizontal position of the flow at the very walls of the inner gorge, plus the considerable thickness of the flow and its massive columnar jointing, gives no suggestion that the flow cascaded into the Grand Canyon. Rather, it supports the conclusion that a lava dam once existed at this elevation.

The Prospect Dam is the oldest lava dam in the Grand Canyon for which we have found remnants. There could be older dams associated with the extrusions capping Mount Trumbull and Mount Dellenbaugh, but we found no evidence of their existence in the inner gorge. The old age of the Prospect Dam indicates the degree to which the flows in Prospect Canyon have been eroded. The steep, V-shaped gorge of Prospect canyon, which is over a mile long, has been eroded in the Prospect flows. This face shows that the Prospect flows are much older than the basalt that fills similar major tributaries to the Grand Canyon, such as Toroweap Valley and Whitmore Wash, both of which are filled with a sequence of flows only slightly modified by erosion. The flows in Toroweap Valley and Whitmore Wash, therefore, almost certainly are much younger than those that fill Prospect Canyon.

From the elevation of the Prospect Dam remnant, we are able to reconstruct the shoreline of the lake it formed. Prospect lake was the largest and deepest lake to form in the Grand Canyon. It extended all the way up through the Grand Canyon into today's Lake Powell region of Utah (Fig. 7). The rather extensive terrace deposits of gravel, sand, and silt in the Bull Frog area of Lake Powell occur at elevations of approximately 4000 feet (1200 m) and probably are remnants of the Colorado River delta formed in the lake. The waters in the lake were deep enough to inundate most of the tributaries in the Grand Canyon for a significant distance.. This meant that the configuration of the shoreline assumed a distint dendritic pattern. The shoreline throughout much of the Grand Canyon in the vicinity of the park headquarters was very close to the base of the vertical cliffs of the Redwall Formation. Calculations show that the lake behind the Prospect Dam required 22 years to fill with water and 3,018 years to fill with sediment.

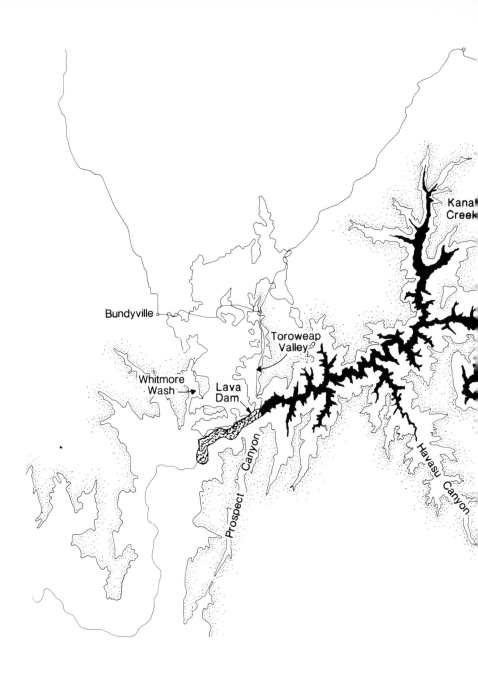

GRAND CANYON GEOLOGY

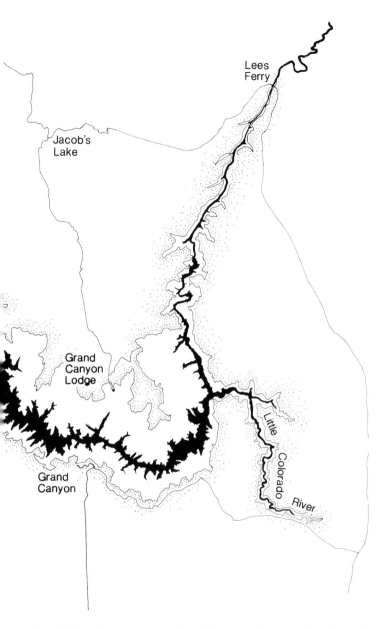

Figure 7. The Prospect Lake. The Prospect Dam was the highest lava dam formed across the Colorado River. The lake formed behind the dam that extended all the way through the Grand Canyon and up into Utah. It formed more than 1.2 million years ago.

Figure 8(a). The Ponderosa flow at Mile 181.6, as seen from the south rim. The Ponderosa in this area is 800 feet thick and extends from near river level to the base of the cascades. Much of the lower part of the Ponderosa is obscured by younger flows that are juxtaposed against it. The Ponderosa Dam was formed from a single, thick flow in which the classic three-part columnar jointing is well developed. It probably was a relatively short dam and did not extend downstream much beyond Mile 190.

The Ponderosa Dam

The Ponderosa Dam was similar to, but somewhat smaller than, the Prospect Dam. It was constructed by a single flow at least 1000 feet (300 m) thick. The top of the barrier was at an elevation of 2800 feet (840 m), or 1130 feet (339 m) above river level. The best

exposures are in the alcove east of Prospect Canyon (Fig. 8), at the
west end of the Esplanade Cascades on the north side of the river
(Mile 181.6).

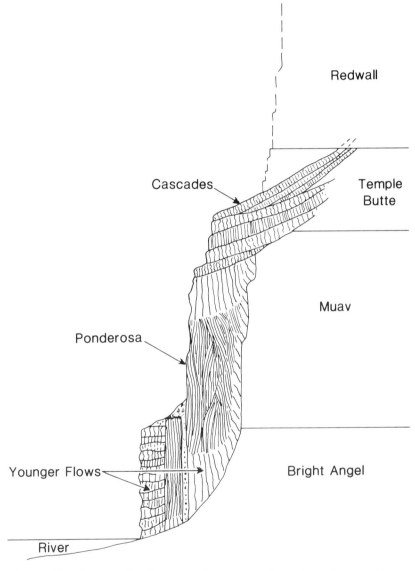

Figure 8(b). Diagram showing the juxtaposed relations of Ponderosa and
younger flows. The Ponderosa is juxtaposed against the Paleozoic rocks and
is capped by younger flows and cascades. Remnants of two younger dams
are juxtaposed against the Ponderosa.

The lake that formed behind the Ponderosa Dam extended through the park headquarters area of the Grand Canyon upstream to somewhere near Nankoweap Rapids. It took only a little more than a year (450 days) for the Ponderosa Lake to fill and overflow. Within 162 years, the lake was full of sediment.

The Toroweap, Esplanade, and Buried Canyon Dams

After the destruction of the massive barriers of the Prospect and Ponderosa dams, three smaller dams were built. These complex dams were built from multiple units and represent a different style of extrusion. They are referred to as the Toroweap, Esplanade, and Buried Canyon dams after large exposures preserved on the north side of inner gorge at Mile 179 (mouth of Toroweap Canyon), Mile 181 (below the Esplanade Cascades), and Mile 183 (the Buried Canyon).

The complex dams were built from a series of flows 50 to 200 feet (15 to 60 m) thick, interbedded with various amounts of river gravel and tephra. Most of the flows are characterized by a well-developed basal colonnade and a thick entablature consisting of slightly sinuous columns. One of the important features of complex dams is that the upper surface of each flow commonly shows evidence of erosion prior to deposition of the overlying younger unit. Locally deep channels are cut in the flows and then filled with younger basalt, gravel, and tephra. The major units of basalt, therefore, are separated frequently by lenses of gravel, sand, and, in some cases, tephra. Judging from this evidence of erosion in the upstream part of each flow during the formation of the dam, it is likely that waterfall migration destroyed a large segment of the downstream part of each flow before the extrusion of the succeeding younger flows. The downstream profile of the final complex dams, therefore, were likely steep and relatively short (Fig. 3d)

Each of the major flow units in the complex dams range from 150 to 200 feet (45 to 60 m) in thickness, and each caused an immediate barrier to form across the Colorado River. The small lakes accumulating behind these barriers rose rapidly and overflowed within a matter of a few days. (Calculations based on discharge rates of the Colorado River indicate that a barrier 100 feet high would overflow in only 2.3 days.) The overflow of the river then eroded and modified the blocky rubble on the upper surface of the flow, together with much of the upper colonnade. Small waterfalls that formed at the end of the flow immediately began to migrate upstream. Subsequent extrusion of a younger flow probably overlapped the eroded end of the older unit. The flows in the complex dams must have been formed by a fairly rapid series of volcanic extrusions with only short time intervals between extrusions—time sufficient only for the upper segment of the overflows to be modified by erosion of the Colorado River.

Based on the present rate of discharge and the current sediment load of the Colorado River, overflow of a barrier 200 feet (60 m) high probably occurred within three days. It seems most likely that the small lake was completely silted up within ten months. Without question, the small lakes formed behind each lava flow were filled completely with sediment prior to the next extrusion. Thus, by the time the complex dams were constructed, the final lake was silted in completely. Since a barrier formed by a single flow would be eroded completely within 20,000 years, the construction of the complex dams must have been completed in less than 100,000 years.

The Toroweap Dam

Remnants of the Toroweap Dam occur on the north wall of the inner gorge beneath Vulcan's Throne. They are largely obscured from viewpoints at the Toroweap Campground, but from the south rim (at the mouth of Prospect Canyon) and from the river, the thick sequence of flows can be seen adhering to the canyon wall (Fig. 9).

Excellent exposures of the upstream part of the Toroweap Dam are found on the north wall of the inner gorge, just upstream from the mouth of Toroweap Canyon (Fig. 10). Here, a detailed cross section of the internal structure of the Toroweap Dam is exposed to reveal the structure of the dam at points where the lava made contact with the waters of the Colorado River. Each of the major units show three distinct types of structures. The front of the flow is characterized by a series of billowy, elliptical structural forms, many of which are twenty or thirty feet (6 or 9 m) in diameter. The size of the elliptical structures grade down to less than three feet in diameter.

Geologists believe these features formed as a single flow entered the inner gorge and interacted with water from the Colorado River (Fig. 11). When the flows first entered the river, some of the material probably formed pillow structures after quenching. Since the flow was thicker than the depths of the Colorado River,

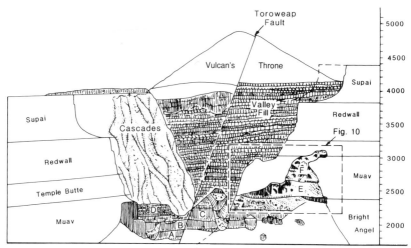

Figure 9. Diagram showing the relationships between the major flows at the mouth of Toroweap Valley. Remnants of the Toroweap Dam are exposed from river level to an elevation of approximately 3000 feet. They extend far beyond the mouth of Toroweap Valley and are juxtaposed against the Bright Angel, Muav, and Temple Butte formations. Toroweap Valley is filled with a sequence of thin flows to the level of the Esplanade. Recent cascades cover the western part of this sequence. Vulcan's Throne rests upon the flows that fill Toroweap Valley. The Toroweap fault displaces the entire sequence, including Vulcan's Throne.

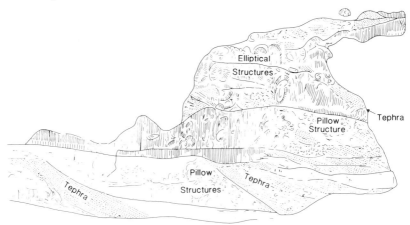

Figure 10. Sketch showing the internal structure of the Toroweap Dam, exposed in remnants high on the north wall of the inner gorge at Mile 178.4. The individual flows are 150-200 feet thick and are characterized by basal colonnade and entablature jointing. Eastward, the flows develop elliptical structures that, in turn, grade into pillow basalts and stratified tephra. The tephra is inclined upstream. These structures suggest that this area is near the head of the lava dam.

412   GRAND CANYON GEOLOGY

however, much of the flow was not covered immediately with water. This material proceeded downstream, essentially dry. Backwater of the Colorado River probably overflowed the barrier in a matter of a few days. As a result, water flowed over the top of the hot lava, greatly influencing the nature of the cooling and the resulting internal columnar jointing. Water from the overflow of the river would seep easily into large cracks in the lava—causing part of the flow adjacent to the fracture to cool very rapidly. Cooling in these cases would proceed inward from the upper surface and inward from the walls of the vertical cracks. Many geologists believe that this type of cooling produced the elliptical, pseudo-pillow structures found here.

At the point of contact with the river water, the lava was quenched, and water interacted with the basalt, causing it to contract and explode. The hydroexplosive activity resulting from the lava making contact with the water probably produced a small amount of tephra that accumulated in the steadily rising lake behind the barrier. This material was deposited at the angle of repose to form the upstream dipping layer of tephra. Undoubtedly, single-flow units developed lobes or offshoots that were diverted upstream a short distance over the elliptical structures and tephra. Thus, several minor flow lobes likely developed from the major flow unit.

In addition to the large remnant of the Toroweap Dam preserved below Vulcan's Throne, several small, isolated remnants are preserved high on the canyon walls on the south side of the inner gorge. These remnants, some 1400 to 1600 feet (420 to 480 m) above the present level of the river, can be observed easily from the Toroweap Campground.

The Toroweap Dam was at least 1443 feet (433 m) high, making it one of the highest dams in the Grand Canyon. Its history includes five major periods of extrusion, each followed by short periods of backwater overflow and erosion. Remnants of the Toroweap sequence are juxtaposed against the Ponderosa near the mouth of Prospect Canyon, indicating that it is younger than both the Ponderosa and Prospect dams. Radiometric dates obtained from the oldest units (flow A) indicate that the Toroweap Dam is 1.16 to 0.18 million years old (McKee and others 1968). This indicates that the Toroweap Dam is older than the one formed in the Esplanade and Buried Canyon.

Toroweap Lake was one of the larger lakes to form in the

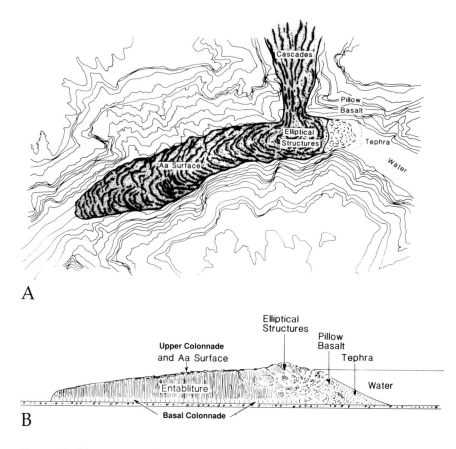

A

B

Figure 11. The structure of a hypothetical flow 200 feet thick in the Grand Canyon. Only the upstream margin of the flow would interact with water from the river. This would produce hydroexplosive tephra, which would be deposited at the upstream margin of the dam. Some pillow structures would be produced where lava came into direct contact with the river. Most of the flow would move downstream essentially dry until the backwater overflowed the barrier. The Colorado River would overflow the barrier of a simple flow within two days. It would flow over the top of the basalt and influence the cooling of the lava. The interaction of the river water with the lava would be most intense near the front of the dam and would form the elliptical structure. The overflow also may have had an effect on the development of the entablature in the downstream part of the flow.

Grand Canyon. It extended all the way through the present National Park Visitor's area and upstream into the vicinity of Lees Ferry (Fig. 12). Throughout much of the park region, the lake's depth was approximately 750 feet (225 m). In the region of the visitor's center, the lake essentially flooded all of Granite Gorge—with the shoreline occurring very close to the vertical wall of the Tapeats Sandstone.

The gravel, sand, and silt that form the terrace deposits at Lees Ferry at an elevation of 3600 feet (1080 m) probably were deposited as a delta built by the Paria and Colorado rivers where they emptied into the Toroweap Lake. Likewise, the silt deposits that form the main floor of Havasu Canyon represent a major remnant of Toroweap Lake deposits.

If the Toroweap Dam formed instantaneously, the lake behind it would overflow in 2.6 years. It would be completely full of sediment in 345 years. Since the Toroweap Dam was constructed over a period of time, the lake undoubtedly was full of sediment by the time the dam was completed. After the extrusion that built the Toroweap Dam terminated, erosion by headwater migration of the waterfall and the downcutting of the stream channel probably destroyed the dam in less than 10,000 years.

The Esplanade Dam

On the north side of the river between Miles 181 and 182, a sequence of flows that formed a complex dam at an elevation of 2600 feet (780 m), or 960 feet (288 m) above river level, is preserved beneath the Esplanade Cascades (Fig. 13). These flows can be seen from viewpoints west of the present Toroweap Campground and resemble in some respects the Toroweap sequence. Both are preserved beneath major cascades.

The Esplanade sequence also is similar in some respects to that preserved in the Buried Canyon at Mile 184. These similarities have led some geologists to consider the Toroweap, Esplanade, and Buried Canyon lavas as part of a single complex lava dam. However, the stratigraphic sequence in each dam is distinctly different, and radiometric dates indicate three separate ages for the sequences.

Several small remnants of the Esplanade Dam are preserved in tributary canyons on the south wall of the inner gorge between Mile 183.5 and Mile 184. They provide important documentation concerning the height of the Esplanade Dam. All are preserved in "hanging valleys" at an elevation of 2600 feet (780 m).

The lake that formed behind the dam was similar to, but slightly smaller than, the Toroweap Lake and extended upstream

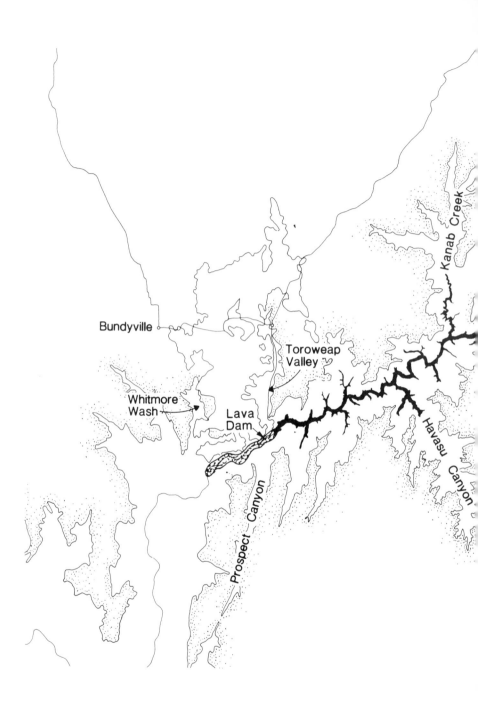

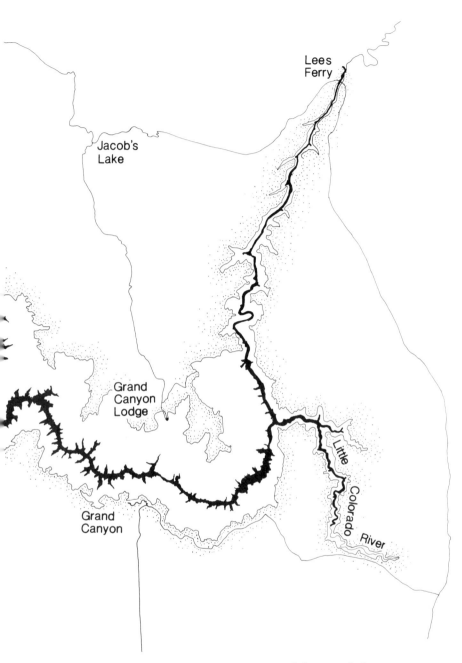

Figure 12. The Toroweap lake. The Toroweap lake extended upstream through the Grand Canyon to the area of Lees Ferry. The shoreline throughout much of the canyon was near the cliff of the Tapeats Sandstone.

LATE CENOZOIC LAVA DAMS

Figure 13. The Esplanade Dam. The lower units exposed near the lower right part of the photograph are similar to the lower flows of the Toroweap Dam. The upper units below the cascades are distinctive in that the columnar jointing consists of rough, thick columns without a basal or upper colonnade.

only to Hance Rapids. It was exceptionally narrow throughout its entire extent and did not extend any appreciable distance up the tributaries. It probably was full of sediment by the time the youngest flow was extruded. Headward migration of waterfalls undoubtedly destroyed the dam within 10,000 years.

The Buried Canyon Dam

One of the most remarkable exposures of basalt in the Grand Canyon is on the north side of the river at Mile 183, half way between Toroweap Valley and Whitmore Wash. Here, a segment of the Grand Canyon was filled with a sequence of older basaltic flows, now preserved as a buried canyon (Fig. 14). The basalts here are not remnants clinging to the wall of the canyon as is the case in most areas. In this region, the basalts fill the entire lower part of the original inner gorge, just as they did when they clogged the canyon and formed a dam. The floor and walls of the original canyon are exposed in a perfect cross section. This unique exposure is preserved because the Colorado River shifted to the outside of a slight meander bend at the time it overflowed the lava dam. Subsequent downcutting formed a completely new canyon, leaving a section of the original gorge filled with basalt preserved on the inside of the meander.

The Buried Canyon contains nine major flow units totaling 650 feet (195 m) in thickness. The top of the highest flow stands at an elevation of 2480 feet (744 m), or 860 feet (258 m) above the present river level. The flows are overlain by a deposit of river gravels 160 feet (48 m) thick and are capped by a younger volcanic cone.

In addition to the remnant of the dam preserved in the Buried Canyon at Mile 184, a number of small remnants are preserved as "hanging valleys" in small tributary valleys on both sides of the

Figure 14. The Buried Canyon Dam at Mile 183. (Note the boat near the sandbar for scale.) This segment of flows occurs in the "original" inner gorge of the canyon and is preserved because the Colorado River shifted southward around a meander bend and cut a new gorge in the Paleozoic strata instead of eroding the basalt. The Buried Canyon Dam was formed by a series of flows 150 to 200 feet thick, all of which show evidence of erosion prior to the extrusion of the next younger lava.

river between Miles 184 and 185 (Fig. 15). These remnants represent segments of the higher basalts that formed the dam and flowed from the inner gorge up the small tributaries.

Several flows in the Buried Canyon sequence were dated by Brent Dalrimple from the U.S. Geological Survey. They range in age from 0.89 to 1.14 million years old. Based on these figures, geologists calculate that the buildup of the dam took place over a period of approximately 250,000 years. Large erosional channels in the basalt sequence indicate that the dam was built and partly destroyed four times during this period of time.

The lake behind the Buried Canyon Dam was similar to that formed by the Toroweap Dam and Esplanade Dam (Fig. 12).

The Whitmore Dam

One of the most obvious expressions of lava flows damming the Colorado River is in the vicinity of Whitmore Wash from Mile 187 to 190. Large remnants of a sequence of relatively thin lava flows can be seen filling Whitmore Wash and adjacent canyons when the area is viewed from the air above Mile 190 (Fig. 16).

Figure 15. Remnants of the Buried Canyon Dam preserved downstream at Mile 184-185

GRAND CANYON GEOLOGY

Figure 16. The Whitmore Dam as seen from the air above Mile 192. View looking northeast. Large remnants of the Whitmore Dam are preserved on both sides of the river and form wide terraces. The source of the flows was from cascades into Whitmore Wash.

Downstream, this sequence is preserved as prominent terraces that outline large segments of the original dam on both sides of the canyon. What makes these exposures exceptionally striking is the way they are preserved. The remnants of the dam are not preserved in protected alcoves (which is more typical) but, instead, form large terraces—some more than one-quarter mile wide and over a mile long. Thus, they are large enough to constitute a significant geomorphic feature in the canyon, recognizable in aerial photographs and on topographic maps. This is the only dam in which relatively large segments of the original upper surface is preserved.

Whitmore Dam is distinctive in that it was composed of numerous thin-flow units, most of which range from ten to twenty feet (3 to 6 m) in thickness (Fig. 17). More than forty individual flow units fill the valley of Whitmore Wash. Little or no sediment separates the flow units in the exposures at the mouth of Whitmore Wash nor in the adjacent valleys, indicating that the flows were extruded in rapid succession.

The source of the flows that built the Whitmore Dam is very clear. As Figure 16 shows, lavas were extruded from eruptive centers near the southern tip of the Uinkaret Plateau and from eruptive centers on the Esplanade. The lavas flowed westward and cascaded into Whitmore Wash. They then moved due south down Whitmore Wash and entered the Colorado River at a point where the river makes an abrupt turn to the south. There was, therefore, no real deviation in the trends of the lava flows as they moved from Whitemore Wash into the Colorado River.

Since the dam was built up by numerous thin flows, the water of the Colorado River immediately overflowed the lava barriers. At various times, it also deposited river gravels. In this particular situation, there was a strong tendency for the river to impinge against the south wall of the canyon because the lavas were moving into the canyon from the north. Subsequent flows were

Figure 17. Remnants of the Whitmore Dam preserved high on the north canyon wall about one-half mile downstream from Whitmore Wash. The Whitmore Dam was distinctive in that it was composed of numerous, thin flows. Remnants of younger, lower dams are juxtaposed against the canyon wall in the central part of this photograph.

GRAND CANYON GEOLOGY

deposited on the river gravels near the south wall, whereas lavas deposited near the north and west walls were superimposed on one another with interlayers of gravel.

Whitmore Dam was 900 feet (270 m) high and over a half-mile wide, making it the widest lava dam in the Grand Canyon. The head of the dam was right at the confluence of the Whitmore Wash and Colorado River. If it had been constructed instantaneously, only 230 days would be required for the lake behind it to fill with water and overflow. However, the interbedded gravels in the Whitmore flows indicate that the lake filled with water and sediment and that it was overflowing at the same time that the lavas were being extruded.

Since the lake behind Whitmore Dam was close to the elevation of the Esplanade Dam, its configuration and shoreline probably were similar to that shown in Figure 12. The lake, extending upstream through the Grand Canyon to Hance Rapids, was 100 miles (160 km) long, and throughout most of its extent, it was very narrow. In Havasu Creek, the lake extended up to the base of the Mooney Falls, and the level surface of lake silts between Mooney Falls and Beaver Falls most likely are deposits formed in Whitmore Lake.

D Dam

A number of small, isolated remnants of a sequence of thin basalt flows are preserved on both sides of the canyon near Lava Falls Rapids at Mile 179.5. These flows clearly are younger than the Toroweap sequence but are older than a series of younger dams and represent a barrier across the canyon 635 feet (190 m) high. The sequence of lava originally was described by McKee and Schenk (1942) and is one of the most distinctive series of flows in the canyon. It consists of a series of flow units 5 to 15 feet (2 to 5 m) thick that are separated by thin layers of clinkers and ash. Remnants of the D flow are relatively small and inconspicuous; unless one is consciously looking for them, they may be missed entirely.

The Younger Dams

Five younger dams ranging in height from 200 to 600 feet (60 to 186 m) also were formed in the Grand Canyon. Most of these dams were constructed from a single flow that extended downstream several tens of miles. One was formed from multiple, thin

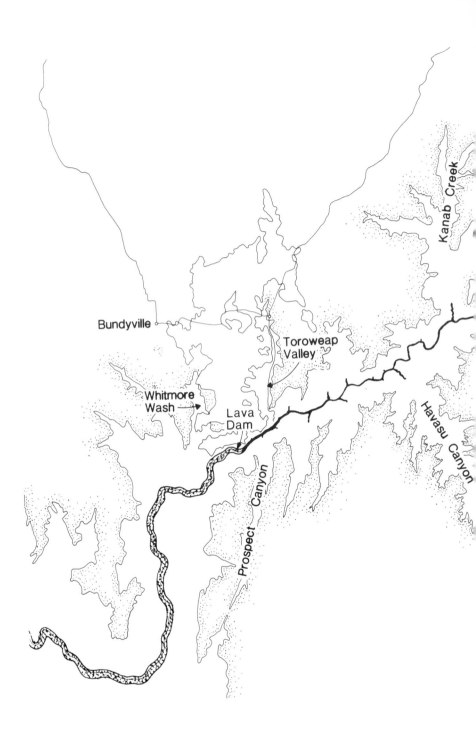

GRAND CANYON GEOLOGY

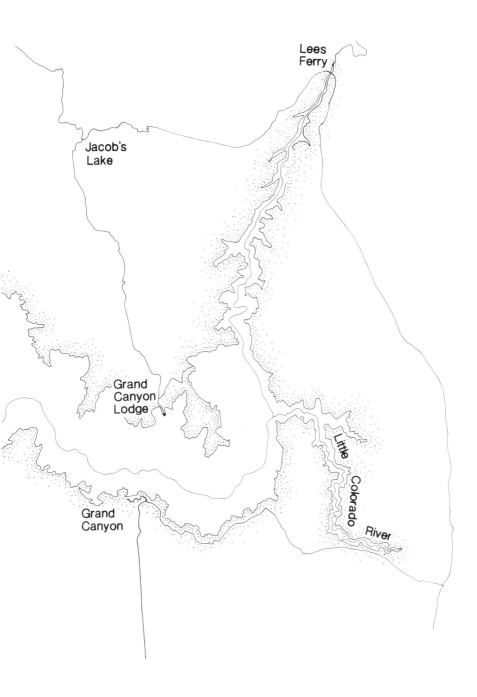

Figure 18. Lakes formed in the Grand Canyon by younger, single-flow dams

diabase flows. Numerous remnants of these younger dams are preserved near the walls of the canyon in the areas of least vigorous erosion. The size of the remnants varies from quite small to more than a mile long. Their geomorphic and stratigraphic relationships, for the most part, are indicated by juxtaposition. The best exposures for demonstrating the age sequence of the younger intracanyon flows are in the vicinity of Whitmore Wash, where five major intracanyon flows are exposed (Fig. 17).

The lakes formed behind the younger dams were barely more than 1/8-mile wide, occupying little more than the present river channel (Fig. 18). They extended upstream to the middle Granite Gorge but did not inundate any of the land in the Grand Canyon Visitors Area. The lakes were long, narrow, and deep channels. The largest filled with water in 17.5 days and was silted up completely in 6.5 years.

Lava Cascades

Two major lava cascades, the "Toroweap Cascades" and the "Esplanade Cascades" (Fig. 1), can be seen from the vicinity of Vulcan's Throne. In addition, smaller "frozen lava falls" are found in the inner gorge east of Whitmore Wash.

The Toroweap Cascades, which spill over the rim of the inner gorge just west of Vulcan's Throne at the mouth of Toroweap Valley, originated from centers of intrusion high on the southern tip of the Uinkaret Plateau just south of Mt. Emma. They flowed over the outer rim of the canyon, across the Esplanade, and then cascaded over the rim of the inner gorge into the Colorado River 3000 feet (900 m) below. The cascades represent some of the most recent volcanic activity within the canyon and clearly are equivalent to Stage IV flows on the Uinkaret Plateau. They are younger than Vulcan's Throne and likely represent volcanic eruptions within the last ten to twenty thousand years. The cascades are relatively thin flows, rarely more than thirty feet thick, and on the Esplanade, they retain many of their original surface features such as pressure blisters. On the steep slopes, the cascades are jumbled and poorly defined. Near the western margins of the Toroweap Cascades, a relatively thick sequence of steeply inclined flows indicates that a considerable volume of lava entered the Grand Canyon from Toroweap Valley.

The Esplanade Cascades, which cap the intracanyon flows, are similar to the Toroweap Cascades in that they represent the

Figure 19. Vulcan's Forge. Vulcan's Forge is a volcanic neck in the middle of the Colorado River—a short distance upstream from Vulcan's Throne.

most recent volcanic activity within the area. The source of the cascades clearly was from centers of eruption on the Esplanade. The lava spread over the relatively flat surface of the Esplanade and spilled over the rim of the inner gorge at several places, both in this area and to the west of it.

Intrusions

The lava cascades between Toroweap Valley and Whitmore Wash are so striking when seen from the air or from the vicinity of Vulcan's Throne that there is a tendency to conclude that the cascades were the source of all intracanyon flows. There are, however, numerous dikes and cones within the inner gorge—indicating that much of the lava forming the lava dams may have been extruded within the inner gorge of the canyon. One of the largest, and certainly the most spectacular, evidences of volcanic events within the canyon is a huge mass of basalt referred to as Vulcan's Forge or Lava Pinnacle. It is located near the center of the Colorado River at Mile 177.9 (Fig. 19). Another volcanic neck similar in size to Vulcan's Forge is exposed on the south wall of the canyon at Mile 180.2. It is exceptional because the face of the canyon wall dissects the neck and exposes a vertical cross section nearly 700 feet (210 m) high. The neck is surrounded by a white alteration zone that makes it exceptionally conspicuous from viewpoints on the Esplanade about a mile west of Vulcan's Throne.

Several dike systems are exposed on both the north and south walls of the Grand Canyon. These systems, however, are relatively obscure. The dikes are three to five feet (1 to 2 m) wide and follow a vertical joint system to the top of the Esplanade. They are thin, tabular bodies that extend upward from the bottom of the canyon with little variation in width. Several additional dikes intrude the older intracanyon flows and laminated ash in the vicinity of Lava Falls Rapids and downstream at Mile 182.5.

The most prominent dike, however, is located high on the south rim of the Grand Canyon near the mouth of Prospect Valley (Fig. 20). This dike is thirty to forty feet (10 to 12 m) thick and forms a high wall projecting above the surrounding country rock. In contrast to the smaller dikes, the Prospect dike trends in an east-west direction, parallel to the Colorado river. A dike of similar size, but with a much less imposing topographic expression, is exposed along the rough trail descending the cascades from near Vulcan's Throne to the river below. It also trends in an east-west direction but is only slightly more than ten feet (3 m) thick.

Cones

Most of the cinder cones found within the inner gorge are associated with the Toroweap fault. In addition to Vulcan's Throne and the large cone on the canyon rim at the mouth of Prospect Canyon, remnants of five additional cones are found down in the canyon along the Toroweap fault zone. Two of these occur below Vulcan's Throne, and two are found along the trail to the Colorado river. Also, a large cinder cone, located along the fault, was buried beneath the lavas in Prospect Canyon and now is exposed at the head of the new valley excavated along the trace of the Toroweap fault.

Several large remnants of cinder cones also adhere to the wall of the inner gorge in the vicinity of the Esplanade Cascades. The largest of these, called Bill's Cone, is a very recent feature as large as Vulcan's Throne. Although much of this cone has sloughed off into the canyon, the upper part of the cone and crater appears to be relatively unmodified by erosion. Small remnants of older cones adhere to the canyon wall in the same vicinity.

Figure 20. Large dike just west of the mouth of Prospect Canyon

RATES OF EROSION

The series of lava flows within the Grand Canyon provide an unusually well-documented record of rates of erosion, slope retreat, and the adjustment of streams to equilibrium. This is possible because the relative ages of the twelve lava dams have been determined by juxtaposition, and radiometric dates provide a series of time benchmarks from the earliest extrusions to the latest. In addition, geologists have measured the thickness, length, and volume of the lava dams so that they can calculate the rates at which erosion has cut through the dams. What we found is that the Colorado River was able to erode through the lava dams at an astounding rate. When we consider the total thickness of all the rock through which the river has cut its channel in the last million years or so, it is clear that the river has the capacity to erode through any rock type almost instantaneously and that the profile of the Colorado River essentially is at equilibrium. In addition, slopes of the Grand Canyon appear to be in quasi-equilibrium, and when equilibrium is upset, adjustment is extremely rapid. The rationale is as follows.

RATES OF DOWNCUTTING

All available evidence indicates that prior to the extrusions of lava into the Grand Canyon some one to two million years ago, the Colorado River had cut down to its present gradient and stratigraphic position. The size and shape of the canyon walls at the time the basalts were extruded were essentially the same as that which we see today. One of the most significant conclusions of our studies, with respect to processes of erosion and canyon cutting, is that after each lava dam was formed, the Colorado River eroded through the dam, down to its original profile *but no farther*. This process of reexcavating the canyon took place at least twelve times during the last 1.5 million years. The cumulative thickness of basalt in these twelve lava dams was 11,300 feet (3390 m). Thus, the Colorado River actually has cut through a cumulative vertical thickness of nearly two miles (3 km) of rocks. In all probability, the actual downcutting through the basalts took place in less than a million years because there were large time intervals between periods of dam destruction when the Colorado River was not influenced by the presence of lava. The best estimates (based on the time necessary for the destruction of the lava dams) would be that actual downcutting through the dams took place in approximately 200,000 years. During the intervening time, the Colorado River was flowing at its normal gradient with neither large-scale deposition nor erosion taking place.

The fundamental question of the age of the Grand Canyon (how long has it taken the Colorado River to cut through a sequence of Paleozoic strata one mile deep) is not just a question of how fast could the river erode but how fast was the region uplifted. Based on the rates at which the river has been able to cut through the sequence of lava dams (or the Paleozoic strata when its channel was displaced beyond the lava flows), it appears that the Colorado River has the capacity to cut down faster than tectonic processes can produce uplift.

This conclusion is supported by the fact that recurrent movement on the Toroweap fault has displaced the strata more than 150 feet in slightly more than one million years. There is no indication of rapids or waterfalls associated with the fault scarp, indicating that the escarpment produced by recurrent movement along the fault is erased by erosion immediately after it forms.

RATES OF SLOPE RETREAT

In addition to providing an insight into the capacity of a river to downcut and to maintain a profile of equilibrium, the remnants of the lava dams provide important documentation concerning rates of slope retreat and the relationship between slope profiles, stream gradient, and tectonic uplift.

Throughout the sections of the western Grand Canyon where remnants of lava dams still remain, one fact is completely clear: the remnants of lava dams are preserved only in the more protected parts of the canyon; ie., on the insides of meander bends, in protected alcoves, and as hanging valleys in the mouths of minor tributaries. Those remnants clinging to the canyon walls are preserved only as thin slivers, commonly only a few tens of feet wide. Except for the youngest flows, which may be 200 to 400 feet (60 to 120 m) wide in the broader sections of the canyon below Whitmore Wash, the remnants of the dams remain as thin sheets plastered against the canyon walls.

Actual downcutting of the river by abrasion along the river channel would produce a vertical gorge only 200 to 250 feet (60 to 75 m) wide (the average width of the Colorado River channel). Slope retreat has been responsible for widening the rest of the inner gorge. At the present time, the inner gorge at the level of Temple Butte Limestone is roughly 2000 feet (600 m) wide. This means that approximately 900 feet (270 m) of slope retreat occurred on each side of the canyon after each lava dam was formed, whereas the downcutting produced by abrasion of the channel has been restricted to a narrow zone 200 to 250 feet (60 to 75 m) wide. At the base of the Muav Limestone, the canyon is approximately 1000 feet (300 m) wide; at that level, 400 feet (120 m) of slope retreat on each side othe river has occurred.

With the destruction of each lava dam and the rapid re-establishment of the river gradient to its original profile, there also was a rapid and contemporaneous retreat of the slopes back to their original profile. The process occurred so fast that the original slope profile of the canyon was re-established before the formation of the next lava dam.

The sequence of juxtaposed remnants near the southern end of the Esplanade sequence serves as an important example. Here, remnants of four lava dams are in juxtaposition (Ponderosa, Esplanade, and two younger dams). Remnants upstream and downstream from this site indicate that five additional units were once present in this area. After each lava dam was breached, there was an average distance of 700 feet (210 m) of slope retreat back to the original canyon wall. The cumulative distance of slope retreat that actually occurred in this area of the canyon was 4900 feet (1470 m), a rate of nearly one mile (1.6 km) per million years. If we consider all of the late Cenozoic lava dams in the canyon, the cumulative distance of slope retreat at the level of the Redwall Limestone would be 11,600 feet (3480 m). At the top of the Bright Angel, the distance would be 5500 feet (1650 m).

In each case, after a lava dam was eroded, the basalt retreated to within a few feet of the original canyon wall. Then, the processes of slope retreat essentially stopped. In many places, the processes of slope retreat completely removed the basaltic flows, *but the process of slope retreat did not enlarge the canyon and go back beyond the original canyon walls.*

Several important conclusions are obvious from these facts: (1) the energy system that causes slope retreat in the Grand Canyon has the capacity to erode canyon walls at a rate of one to two miles per million years or more. (2) The slopes recede rapidly to a profile of quasi-equilibrium. They then recede at a very slow rate. (3) Slope retreat is balanced delicately with the stream gradient. Renewed downcutting of the stream channel is accompanied by renewed slope retreat. (4) The profile of the Grand Canyon apparently is in a state of quasi-equilibrium. (5) Periods of major rapid slope retreat are initiated by tectonic uplift.

SUMMARY

We can conclude from these studies that erosion of the Grand Canyon did not take place at an imperceptibly slow and constant rate. Instead, it was governed by the style and degree of tectonic uplift. Inasmuch as movement on the major faults in the Colorado Plateau occurs in pulses, the actual processes of erosion also must occur in pulses; and, inasmuch as a tectonic disturbance produces only a few feet of displacement at a time, erosion back to equilibrium occurs in a series of small pulses separated by relatively long periods of quiescence.

CHAPTER 18

EARTHQUAKES AND SEISMICITY OF THE GRAND CANYON REGION

David S. Brumbaugh

INTRODUCTION

The Grand Canyon is located in a seismically active part of the earth's crust. This region, shown in Figure 1, also contains a highly complex structure. The only faults represented in Figure 1 are those that have been active in the last four million years (m.y.). The seismicity of this area suggests that some of these faults may be active at present.

Physiographically, the Grand Canyon lies within the Colorado Plateau, which really is a series of plateaus rising eastward across the major boundary faults: Grand Wash, Hurricane, Toroweap, and West Kaibab (Fig. 1). The Grand Canyon lies at the southern end of the Intermountain Seismic Belt, which stretches north-south across central Utah. The continuation of this belt of seismicity into Arizona has been the topic of debate; nevertheless, there is no argument that the Grand Canyon region is active seismically. This seismicity may represent a tectonic boundary for the Colorado Plateau, which therefore lies inside of the accepted physiographic boundary at the Grand Wash Cliffs (Brumbaugh 1987).

Little previous work of any detail exists on the seismicity of the Grand Canyon region. Earlier workers surveyed the seismicity of the entire state of Arizona but placed little or no emphasis on the Grand Canyon region (Sturgul and Irwin 1971; DuBois and others 1982). While studies of the northern part of the state, from the Mogollon Rim to the Utah border, do exist, the Grand Canyon region is not the focus of this work.

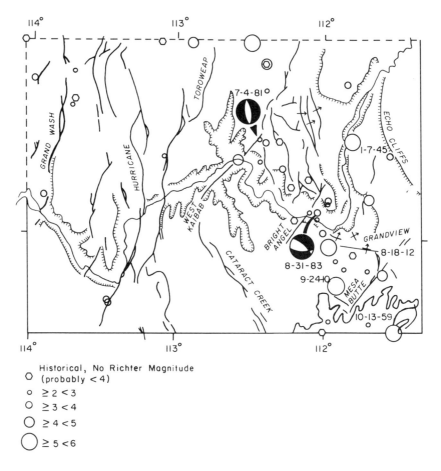

Figure 1. Seismicity of the Grand Canyon region. Epicenters are indicated by circles and hexagons. Faults active in the last 4 m.y. shown by solid lines, monoclines by ⇀. Grand Canyon escarpment = ⊥⊥⊥⊥⊥, except where noted (Echo Cliffs). Fault plane solutions (focal mechanisms) are shown for two events: 7/4/81 and 8/31/83.

HISTORIC SEISMICITY

Geologists know that 45 earthquakes have occurred in the Grand Canyon region in this century (Table 1, Fig. 1). These quakes range in size from 5.75 M_L down to 2.00 on the Richter scale. Events smaller than 2.00 undoubtedly have occurred, but their numbers cannot be estimated reliably since geologists do not have instrumental coverage of sufficient density to detect these small events unless they occur close to established seismograph stations.

Five events of magnitude 5.0 or greater on the Richter scale occurred in the Grand Canyon region in this century. On September 24, 1910, an event that probably registered in the 5-6 magnitude range occurred just northwest of the Mesa Butte fracture system. Minor damage was reported at Cedar Wash and at Flagstaff, Arizona.

The next moderately sized event in northern Arizona occurred on August 18, 1912. The magnitude of this tremor was estimated to be 5.5 (DuBois and others 1982). The epicenter was located near the south rim of the Grand Canyon (Fig. 1). Houses were damaged as far away as Williams, Arizona, some 59 miles (95 km) to the south. Fortunately, damage was minimal because of the sparse population base existing in northern Arizona at the time.

The event of January 7, 1945, had a Richter scale magnitude of 5.1, and its epicenter was placed in the eastern Grand Canyon, just west of the Echo Cliffs (Fig. 1). It was felt by very few and did no damage.

During 1959, two events of 5.0 or greater on the Richter scale occurred in the Grand Canyon region. On July 21, 1959, Fredonia, Arizona, located on the Arizona-Utah border, was shaken by the largest instrumentally recorded earthquake in Arizona for which we have a magnitude. This event had a calculated Richter magnitude of 5.75 and caused minor damage at Fredonia. It was felt over a 24,400 square-mile (62,500 sq-km)-area of Arizona and Utah. This event was followed on October 13 by a smaller shock of magnitude 5.0 that was located about 34 miles (55 km) southeast of the event of August 1912. It was felt strongly in Flagstaff, but no damage was associated with it.

The rate of occurrence of earthquakes in this century in the Grand Canyon region is comparable to that in the Intermountain

Seismic Belt of Utah or, on a broader scale, that of the Western Mountain Region (Fig. 2). It should be understood that the time span for which data is available is rather small for the moderate-to-large events (MM VI-IX). Therefore, our information may not be as reflective of true rates of recurrence as we would like. Nevertheless, despite effects caused by periodicity, it is clear that recurrence rates in northern Arizona are lower, on the average, than those for the Intermountain Seismic Belt. The rate varies from a maximum north of the Grand Canyon, which is comparable to that of the Intermountain Seismic Belt in Utah (Kruger-Kneupfer and others 1985), to much lower rates south of the canyon. The slope of the recurrence curve for the Grand Canyon data (Fig. 2) suggests the occurrence of an MMVII (5.5-6.0) event twice every fifty years or so.

TECTONICS

The Grand Canyon region of the southern Colorado Plateau is dominated structurally by high-angle, normal faults that cluster into large, northeast, north-south, and northwest-trending fracture systems. Several of these fracture systems are well exposed where they cross the Grand Canyon (Fig. 1). Crossing the canyon from west to east are the Grand Wash, Hurricane, and Toroweap faults, all predominantly north-south trending. The West Kaibab and Bright Angel fracture systems are prominent northeast-trending groups of normal faults. The northwest-trending Grand-view-Phantom system intersects the walls of the eastern Grand Canyon. Two prominent systems, the Mesa Butte (northeast) and Cataract Creek (northwest), do not appear as canyon-crossing fault systems (Fig. 1).

The level of earthquake activity in this century in the Grand Canyon region, as well as the clustering of activity in the vicinity of some of the fracture systems, suggests some of these faults may be active today. Shoemaker and others (1974) concluded that the distribution of epicenters indicated that the Bright Angel and Mesa Butte fault systems are active. Both of these, as previously noted (Fig. 1), are prominent, well-expressed, northeast-trending fracture systems. No surface scarps of historic age exist to indicate that any of the fracture systems are active. Nor have the 5.0 or greater events resulted in any surface rupture. Geomorphic mapping suggests the possibility that many of the northeast, north-

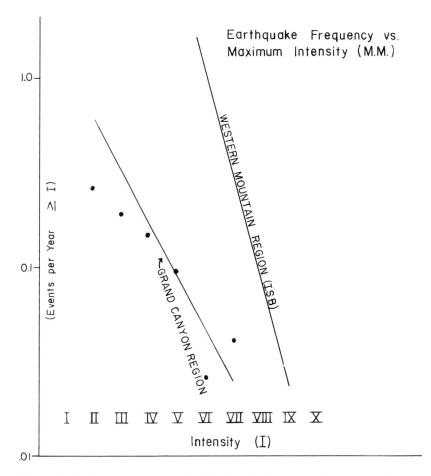

Figure 2. Earthquake frequency for the Grand Canyon region compared to the Intermountain Seismic Belt (ISB). Recurrence curves are represented by a solid line for each region. Western Mountain Region curve is from Podolny and Cooper (1974).

south, and northwest-trending fracture systems may have been active in late Pliocene or Quaternary time (Scarborough and others 1986).

If we ignore the mapped surface traces of fracture systems and concentrate only on the pattern displayed by epicenter locations, the most prominent trend of activity is a northwest trend stretching from the West Kaibab fracture system to the Mesa Butte system. Furthermore, two fault plane solutions exist (7/4/81, 8/31/83), neither one of which indicates movement on northeast-trending fault surfaces. The 8/31/83 event shows two potential fault surfaces, N40°W and N86°E. Only the N40°W trend comes close to paralleling the trends of any of the three major fracture systems. The 7/4/81 event lies at the junction of the West Kaibab and the Crazy Jug fracture systems (Fig. 1). The potential fault surface orientations have trends of N24°W and N30°W (Kruger-Kneupfer and others 1985). It should be noted that the solution is a composite one—that is, it combines the 7/4/81 event with others close to it in time and space. There are those who feel that composite solutions are less reliable than single event ones. Finally, examination of the distribution of epicenters of just the largest events [M≥4.0], which undoubtedly are controlled by tectonic stresses, clearly shows a northwestern alignment as well.

A tectonic boundary may be defined by changes in tectonic elements. This includes parameters like structural style, crustal thickness, and seismicity. The concentrated band of seismicity in the Grand Canyon region marks the locus of a change in crustal thickness and a change in structural style. The West Kaibab fault zone marks the eastern limit of Basin and Range type faults on the southern Colorado Plateau. Crustal thicknesses in the western Grand Canyon region are more like those found beneath the Basin and Range terrane.

Brumbaugh (1987) has suggested that the belt of concentrated seismicity in the Grand Canyon region represents the presentday tectonic boundary of the Colorado Plateau—a boundary, furthermore, that has migrated with time, as suggested by earlier workers (Best and Hamblin 1978). This migration has expanded the area of the Basin-Range from the southwest and encroached into the Colorado Plateau.

SUMMARY

The seismicity of the Grand Canyon region of northern Arizona is contiguous with the Intermountain Seismic Belt of Utah. This, plus a similar recurrence rate north of the canyon (and a common style of faulting), suggests that the seismicity of the Grand Canyon region represents an extension of the Intermountain Seismic Belt of Utah. An analysis of the distribution of epicenters (Fig. 1) indicates that the majority of the events are concentrated in a rather narrow, northwest-trending band between the Mesa Butte and the West Kaibab fracture systems. This

Table 1. Historic Seismicity, Grand Canyon Region

DATE (Mo-D-Yr)	ORIGIN TIME(GMT)	EPICENTER	MAGNITUDE	INTENSITY
9-24-10	0405	35.75 x 111.50	(5.5?)	VII
8-18-12	211200	35.95 x 111.95	$5.5M_L$	VII
8-17-38	090806	36.7 x 113.7		
12-28-38	043736	37.0 x 114.0		
1-31-44	042458	36.9 x 112.4		
1-07-45	222532	36.5 x 111.8	5.1	VI
10-27-47	041540	35.5 x 112.0		
7-21-59	173929	37.0 x 112.5	$5.75M_L$	VII
10-13-59	081500	35.5 x 111.5	5.00	VI
2-15-62	071243	36.9 x 112.4	$4.50M_L$	V
2-15-62	090645	37.0 x 112.9	$4.40M_L$	V
8-28-64	065046	37.0 x 113.1		
6-07-65	142801	36.1 x 112.2		
9-03-66	075320	36.5 x 112.3		
10-03-66	160350	35.8 x 111.6	$4.4M_b$	V
12-01-66	092040	36.2 x 113.9	$3.7M_b$	IV
5-26-67	074842	36.421 x 111.557	3.1	II
7-20-67	135110	36.3 x 112.1		
8-07-67	162444	36.5 x 112.4	$3.86M_L$	IV
8-07-67	164032	36.4 x 112.6	$4.00M_L$	IV
9-04-67	232744	36.2 x 111.7	$4.60M_L$	V
11-05-70	094557	36.265 x 112.227		
11-24-70	164756	36.357 x 112.273	$3.00M_L$	II
12-03-70	034724	35.874 x 111.906	$2.80M_L$	
12-16-70	134419	36.844 x 113.715	$2.60M_L$	
12-16-70	134647	36.679 x 113.748	$2.20M_L$	
3-27-71	043911	36.672 x 112.393	$2.60M_L$	
5-06-71	165718	36.419 x 113.084	$2.20M_L$	
12-15-71	125814	36.791 x 111.824	$3.00M_L$	II
2-28-76	205358	35.91 x 111.788		
8-02-76	21:28	35.826 x 111.720	$2.10M_L$	
8-07-76	08:48	35.571 x 111.979	$2.00M_L$	
8-08-76	12:44	35.836 x 111.856	$2.70M_L$	
8-09-76	21:43	35.541 x 111.678	$2.90M_L$	
8-05-79	191015	36.796 x 113.984	$3.70M_L$	IV
1-12-81	085913	35.658 x 113.469	$3.50 M_L$	III
7-04-81	195945	36.509 x 112.441	2.60	
11-19-82	205734	36.027 x 112.006	$3.00M_L$	II
8-31-83	081009	36.135 x 112.037	3.30	III
9-13-83	152246	36.13 x 112.09	$2.20M_L$	
9-13-83	162905	36.096 x 112.045	$2.65 M_L$	
9-13-83	202920	36.175 x 111.965	$2.20M_L$	
7-18-84	142933	36.192 x 111.958	$3.00M_L$	II
9-09-85	020409	36.39 x 112.438	2.30	

trend of activity, plus the northwestern trend of potential fault planes from the 7/4/81 event and one from the 8/31/83 event, suggest that it is the northwest-trending fracture systems, such as the Grandview-Phantom, which contain presently active faults.

It seems likely that the seismicity of the Grand Canyon region is an expression of a tectonic boundary that is marked by a change in structural style as well as in crustal thickening. This seismicity is the product of an active stretching or extension of the crust in the Grand Canyon region. This crust, moreover, is not homogeneous, but prefractured in northeastern, north-south, and northwestern directions by large fault systems. A preferential reactivation of only northwest-trending faults, such as the Grandview-Phantom, suggests that the stretching of the crust must be occurring perpendicular to this trend, that is, in a northeastern-southwestern direction. If this extension continues to affect the crust of the Colorado Plateau in the future, the tectonic boundary will continue to migrate into the interior of the plateau—eventually leaving the Grand Canyon region a seismically quiet area.

CHAPTER 19

ROCK MOVEMENT
AND MASS WASTAGE
IN THE GRAND CANYON

Richard Hereford
Peter W. Huntoon*

INTRODUCTION

In this chapter, we discuss the processes by which the Grand Canyon is widened and enlarged—rock movement and mass wastage. Although the evolution of the Grand Canyon has been the subject of numerous studies (e.g., McKee and others 1967; Hunt 1969; Lucchitta, this volume), these works have been concerned with the history of the course of the river or a general description of topographic features. Little attempt has been made to record features or events of significance in interpreting rock movement and mass wastage. Exceptions are the works of McKee (1933a), Péwé (1968), Ford and Breed (1970), Huntoon (1973), Hereford (1984), and Webb and others (1987). Other studies, though directly applicable to the Grand Canyon, have been concerned with geologic processes in general. These include studies by Reiche (1937), Strahler (1940), Koons (1955), Ahnert (1960), Bradley (1963), and Schumm and Chorley (1966).

*Adapted from Ford, T. D., P. W. Huntoon, G. H. Billingsley, and W. J. Breed, "Rock Movement and Mass Wastage in the Grand Canyon." In: *Geology of the Grand Canyon*. Flagstaff: MNA Press, 1977.

Erosive processes such as water and wind obviously operate in the Grand Canyon. But since they show no specific distinctness (except for flash floods), we need not discuss them here. Flash floods have been considered by Péwé (1968) and Leopold (1969).

The features described herein result from small-scale processes such as slab failure, rock avalanche, rock fall, talus slide, and debris flow—as well as large features such as gravity faults, valley anticlines, and rotational landslides. All these processes provide a means of supplying the river with sediment for transport out of the Grand Canyon (Leopold 1969).

SMALL SCALE WASTAGE

Geologists define slab failure as the sudden collapse of a cliff face cut in solid rock (Fig. 1). McKee (1933a) described many slab failures, including seven observed by Mr. Emory Kolb that occurred within sight of his house on the south rim over a period of twenty years. Other rock falls were heard frequently during the night by Mr. Kolb. Bradley (1963) and Schumm and Chorley (1966) have described numerous examples in Glen Canyon and adjacent parts of the Colorado Plateau.

Conversation with two Grand Canyon boatmen during a recent trip on the Colorado River revealed that they had noticed an average of two major slab falls per year during their trips on the river. Many individuals floating the river during rainy season report hearing small falls. A list of known falls and possible dates of occurrence are presented in Table 1. These examples demonstrate that slab falls are relatively common in the Grand Canyon.

The causes of slab failure range from trivial to catastrophic. The latter could include earthquakes, but there is no record of a slab failure resulting from earth tremor in the Grand Canyon. The atmospheric shock waves from thunder or sonic booms also may cause dislodgement, but geologists have recorded no incident to date. However, damage from sonic booms has been observed elsewhere in the Southwest in recent years.

A common feature of fine-grained sandstone in the canyon walls of the Southwest is large-scale conchoidal fracture surfaces. Several of the buttes in Monument Valley exhibit these features, and some of the cliff dwellings have conchoidal fracture surfaces as their rear walls (e.g., Betatakin). Bradley (1963) has referred to

Figure 1. Recent rock fall, south wall of Marble Canyon at mile 7.3. Rock fall occurred early spring, 1970.

Table 1. Recorded Rock Falls in Grand Canyon, 1968-1973

Date of Fall	Formation	Est. weight and Length of Fall
est. 1968 or later	Kaibab	___
est. 1969-70	Kaibab	fell ca. 200'
est. winter 1970 or 1971	Kaibab	at least 40 tons:450'
3-20-69	Permian formations, esp.Kaibab and Toroweap	numerous small rocks
est. late Nov. 1971	Toroweap boulder rolled down Hermit slope	ca. 15 tons; 800'
1-11-69	Coconino	150 lb.; 200'
est. 6/69	Coconino	slab measuring about 300' x 20' x 20' fell about 300 ft.
est. 11/70	Supai	several tons; 30-40 ft.
7-25-67	Redwall	300 lb; 450'
1-1-70	Redwall	3 boulders at 150 lbs. each; fell ca. 1400'
7-9-73	Redwall	400 lb.; 500'
est. winter or spring, 1971	Redwall	ca. 30 tons; 1100'
est. March, 1973	Redwall	35 tons; 240'
est. winter 1972 or 1973	Lower Redwall	total slide ca. 300-400'
2-23-74	Shinumo Qtzite	450 lb.; 90'
6-22-72	Vishnu Schist	several boulders equal at least 1 ton; 200' or less
est. winter, 1973	talus	1 ton; 30-40'

Location	Source	Comments
1/2 mile below Navajo Bridge, Colo. River	Museum river trip, May 1973	——
right side; Colo. River mile 5.5	Museum river trip, May 1973	300' high talus fan spreads to river's edge
south side, Colo. River, mile 7 1/4 Marble Canyon	Museum river trip, May 1973	talus covers Hermit exposure to river
South Canyon, in Marble Canyon	Billingsley	high winds gusting 65 mph
Rider Canyon in Marble Canyon	Billingsley	very fresh scars some 4' deep in the shale
Salt Trail Canyon	Billingsley	windy day
small point directly west of Powell Point	seen by Nat. Pk. Serv. ranger—reported by Ford and Breed	——
Cottonwood spring in Sowats Canyon	Billingsley	"very fresh" smashed cottonwoods
Little Colo. River gorge, 8 mi upstr. from Colo. River	Billingsley	fell into Little Colo. River
Colo. River, mile 167	Billingsley	fell into river
Colo. River, mile 50; Marble Canyon	Billingsley	——
Colo. River, mile 155 1/2; north side	Billingsley on Museum river trip, 1973	fell into Colo. River
President Harding rapids: Colo. River mile 43 3/4	Billingsley	fell to edge of river
Colo. River, mile 44	Museum river trip, May 1973	talus covering Muav
1/2 mi. from Colo. River in 75-mi. Canyon	Billingsley	——
Colo. River, mile 79	Billingsley	thunderstorm in progress
Tanner Wash, Marble Canyon	Billingsley	snow and ice present

them as exfoliation fractures. They are caused by the release of compressive stress as canyons are cut down, and he has noted examples in Glen Canyon and Monument Valley. To the writers' knowledge, there has been no discussion of these fractures in the Grand Canyon literature though they are a common feature. They are best developed in the Coconino Sandstone and some of the thicker sandstones of the Supai Group. The conchoidal marks commonly are twenty feet across, with the curved lines convex up, indicating that the point of yielding has been the base of a bed responding to both vertical pressure and to the expansive release of compression. This results in a flat, vertical slab spalling along a fracture plane—where no joint existed previously. Carson and Kirkby (1972) classed this phenomenon as slab failure. In the Grand Canyon, the stress apparently has been sufficient to spall large slabs of sandstone as the river cut down. Some dislodge as slab falls, while others remain in position—ready to fall when loosened by rain, wind, frost wedging, or other effects.

Geologists define a rock avalanche as the collapse of the rock face through the failure of a random pattern of micro-joints, which may or may not be superimposed on a larger joint network (Carson and Kirkby 1972). Although we have recorded no specific example from the Grand Canyon, it seems likely that this is a common process wherever there are inherent small-scale fractures in a rock. The small joints and stylolites in the Redwall Limestone are in this category. Solution weakening of these features as water seeps through could be responsible for the great alcoves so common in the Redwall. In the Coconino Sandstone and the thicker sandstones of the Supai Group, the combination of weaknesses along cross bedding and exfoliation fractures probably causes many rock avalanches.

Unless a fall is observed to be specifically a slab failure or a rock avalanche, it may not be possible to distinguish from the debris which of the two has occurred. The presence of a conchoidal scar, however, would suggest the former. Some rock avalanches occur when seepage weakens the underlying shale.

As far as we know, most recent falls have been from the Coconino Sandstone and the Kaibab Limestone, with a few from the Redwall Limestone. In the Coconino Sandstone, falls seem to occur when the underlying Hermit Shale becomes weakened and is no longer able to support the weight of sandstone above. This may be caused by the slow seepage of water from the base of the sandstone above the impervious shale, even though the rock face may have appeared dry immediately before the fall. The rapid rain-wash erosion of shale slopes below scarps noted by Schumm and Chorley (1966) also must contribute to the undermining process.

Falls have taken place at all seasons, at all times of the day, in any weather. Since the weakening process is a long one, perhaps only a little extra heating on a hot day, or a light shower, or a touch of frost is the critical factor. McKee (1933a) noted that falls took place often during spring snow melt and during summer storms.

Carson and Kirkby (1972) define a rock fall as the fall of single, small blocks from a cliff face—generally derived from a thin, superficial zone of the rock face by the forces of atmospheric attack. Such rock falls are common in the Grand Canyon, and it is not unusual for single rocks to be heard bounding down the cliffs. The most common causes of dislodgement are diurnal expansion and contraction, trickling rainwater during a shower, and frost heave.

A large, ancient deposit of probable rock fall origin is present at the mouth of Nankoweap Creek in Marble Canyon. The age of the deposit, which is 180 feet (60 m) above the river, is Pleistocene. This is indicated by a uranium-trend date of 210,000 $\pm$ 25,000 years from a calcic soil horizon at the surface of the deposit (Machette and Rosholt, in press). The geology of the deposit suggests that it resulted from a rock fall originating near the top of the steep east wall of Marble Canyon (Hereford 1984). Large blocks fell nearly 3200 feet (1 km) into the Colorado River channel, damming the water flow.

Talus slides may be defined as the sudden downward movement of loose talus, much of which may have been derived from earlier falls of any of the above categories. Such slides seldom are observed, and even though they leave distinctive scars, few have been recorded. Perhaps the most obvious examples are the slides that have wiped out sections of unmaintained trails and, in a few cases, parts of the Bright Angel and Kaibab trails.

The best examples of talus slides on a large scale are seen in the western Grand Canyon (mile 270). The high level of Lake Mead several decades ago enabled water to seep into the base of several talus slopes along the river and weaken the slope base by lubrication. Scars over ten feet (3 m) high developed along the tops of the slopes as the whole mass gradually slid into the river.

The causes of talus slides are similar to those of rock falls. Seepage of rainwater along the top of shales underlying a talus slope may weaken the shales to the point where they no longer can support the weight above them. An intense rain shower, a frost, or a rise or fall in temperature may weaken the adhesion of rocks in the talus. The passage of an animal or a hiker across the talus, earth tremors, or sonic booms are other factors that could cause talus movement.

Debris flows are mixtures of rock particles of varying sizes and water that move as viscous masses. In comparison, flash floods have much more water in proportion to sediment. An example of a mudflow near Lees Ferry has been described briefly by Péwé (1968, p. 15), who observed that "such mudflows on a larger scale must have occurred in the past in all of the tributary canyons of the Colorado River." It is difficult to recognize the deposits of debris flows in many cases because subsequent streamflow over them winnows out the finer sediments, leaving only the gravel and boulders.

A study by Webb and others (1987) found recent debris flow deposits in thirty-six Grand Canyon tributaries. They suggested that debris flow is an important sediment transport process in these streams. Coarse sediment carried by debris flows to the Colorado River channel forms debris fans and rapids.

Two large debris flows were noted by Billingsley in the vicinity of the Little Colorado and the Colorado River confluence a few days after their occurrence. Trees were sheared off at their base, and boulders weighing over several tons were rafted along on top of very coarse material (Fig. 2).

LARGE-SCALE FEATURES AND GRAVITY TECTONICS

Cutting of the Grand Canyon during the Pliocene produced enormous topographic relief. This created an environment that facilitates near-surface, gravity tectonic deformation of the rocks in and adjacent to the deep canyons. If we consider the elevation of the Colorado River to be a local energy datum, the rocks that lie above the river have potential energy that is directly proportional to their elevation. All the rocks above the river release energy by moving downslope to the river. The result is that large, gravity-

Figure 2. Mudflow near confluence of the Little Colorado and Colorado rivers

induced stress gradients exist in the rocks adjacent to the canyon. These stress gradients increase in intensity toward the canyon walls. Occasionally, the potential energy of the elevated rocks is converted into kinetic energy—causing mechanical deformation of the rocks as they move to lower, more stable positions closer to the river.

Geologists have identified several gravity tectonic processes in the Grand Canyon. The rate of these processes ranges from slow to rapid, and mechanically, the processes involve both fracture and flowage failures. Only gravity tectonic processes that produce large-scale features will be treated here; these are valley anticlines, high-angle gravity faults, and rotational landslides.

Valley Anticlines

A valley anticline whose axis closely parallels the Colorado River is present between Fishtail Canyon (Fig. 3) and Parashant Canyon (Billingsley and Huntoon 1983). Identical valley anticlines also occur in the principal tributary canyons along this sixty-mile reach—including Kanab and Tuckup canyons. The

Figure 3. Valley anticline with strike parallel to the Colorado River upstream from the mouth of Havasu Canyon. Note dip of strata away from the river. View is downstream.

occurrence of the anticlines along the sinuous trends of the canyons reveals a genetic link between the two. Huntoon and Elston (1980) deduced that the anticlines are the product of a lateral squeezing toward the canyons of the saturated shaly parts of the Cambrian Muav Limestone and the underlying Bright Angel Shale. The driving mechanism for this deformation is the huge difference in lithologic load between the heavily loaded rocks under the 2100-foot-high (640 m) canyon walls and the unloaded canyon floors (Sturgul and Grinshpan 1975).

The limbs of a typical valley anticline dip away from the canyon at angles of up to sixty degrees, and the folding extends over 800 feet (270 m) into the canyon walls. The anticlinal bulge is a horizontal shortening structure across the river in which the shortened beds arch upward into the void provided by the canyon. Deformation in the anticlines ranges from ductile to brittle.

Similar valley anticlines are missing in the easternmost Grand Canyon and Marble Canyon, where the same rocks occur at river level. The lack of anticlines is explained by the absence in this section of shaly rock that would deform when saturated.

Numerous sets of minor, low-angle, conjugate thrust faults that parallel the axes of the anticlines occur in the Muav Limestone. The intersecting faults in these sets dip, respectively, toward and away from the river. They are particularly numerous and have larger offsets in locations where the Muav Limestone is steeply folded. They also occur, however, in beds dipping as little as two degrees (Huntoon and Elston 1980). The thrust faults accommodate horizontal shortening across the anticlines by causing structural thickening and shortening of the deformed strata.

Steeply dipping kink bands also parallel the anticlines. These have formed in response to shortening across the fold axis (Reches and Johnson 1978). In outcrop, such bands can be mistaken for small displacement reverse faults. Most of the observed kinks displace the rocks upward toward the river, and the bands containing the kinked strata dip toward the river. The kink bands usually cause deformation in zones no wider than six feet (2 m).

Gravity Faults

High-angle gravity faults in the eastern Grand Canyon are brittle failures unrelated to deep-seated tectonic processes (Huntoon 1973). Faults in this area occur on narrow ridges flanked by deep canyons that expose the Bright Angel Shale. These faults represent brittle failure accompanying rather minor ductile deformation within the Bright Angel Shale (Huntoon 1973). They develop as the overlying column of rock (which in some cases is 4000 feet [1200 m] thick) settles into the ductile shale base. As Figure 4 shows, the faults generally are restricted to Paleozoic strata above the base of the Bright Angel Shale, and they range from very steeply dipping to nearly vertical. Displacements are greatest at the top of the sections and die out downward. The faults are laterally discontinuous and seldom extend more than three miles.

High-angle gravity faults are important in the morphologic evolution of Grand Canyon scenery. Once the faults form, erosion proceeds rapidly along them, segmenting what formerly was a ridge between two canyons into a string of buttes. As erosion progresses, the buttes become isolated from the plateau. Ironically, the faults that initiate the butte-forming process vanish in time because they erode. Consequently, we do not find faults around some deeply eroded old buttes, such as Solomon Temple in the eastern Grand Canyon (Huntoon and others 1986).

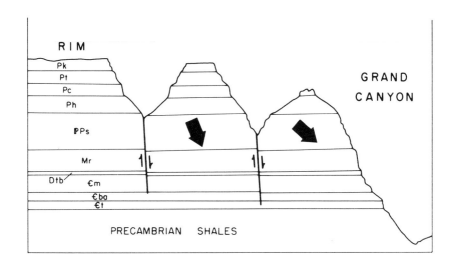

Pk – Kaibab Formation	Mr – Redwall Limestone
Pt – Toroweap Formation	Dtb – Temple Butte Formation
Pc – Coconino Sandstone	€m – Muav Limestone
Ph – Hermit Shale	€ba – Bright Angel Shale
PPs – Supai Formation	€t – Tapeats Sandstone

Heavy arrow shows relative movement of the butte

Figure 4. Erosion along high-angle gravity faults that develop across ridges allows buttes to separate from the ridges. Faults develop as buttes settle into ductile Cambrian and Precambrian shales.

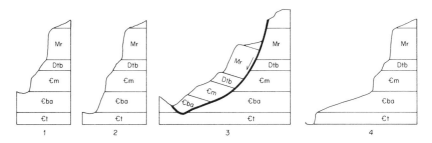

Figure 5. Stabilization of cliff profiles through rotational landsliding, Grand Canyon, Arizona. (1) Channel is downcut to top of Bright Angel Shale, profile is stable. (2) Channel has cut through Bright Angel Shale, shale slope is too steep to buttress overburden load. (3) Profile fails through rotational sliding. (4) Erosion of debris results in stable cliff profile. Mr - Redwall Limestone; Dtb - Temple Butte Formation; C - Muav Limestone; C -Bright Angel Shale; G -Tapeats Sandstone

454   GRAND CANYON GEOLOGY

Rotational Slides

The most important gravity tectonic structures in terms of canyon widening are rotational slides. These slides are massive blocks or rows of blocks (Fig. 5) that detach from the canyon wall and glide into the canyon along low-angle, concave upward normal faults. Because these faults are concave-upward, the descending mass is forced to rotate backward against the glide plane. The displaced blocks commonly are called "toreva blocks" (Reiche 1937).

Rotational slides most frequently occur in two settings in the Grand Canyon. The largest involve the Redwall cliff, which is inclusive of the Pennsylvanian Esplanade Sandstone and the Cambrian Bright Angel Shale. This section commonly is 1600 feet (490 m) or more thick, depending on how much of the Bright Angel Shale is involved in the slide. Much smaller rotational slides and calving involve the Cambrian Tapeats Sandstone, 250 to 400 feet (75 to 120 m) thick, which detaches above the ductile Precambrian Galeros Formation in the eastern Grand Canyon (Huntoon and others 1986). Rotational slides can develop wherever cliffs form above ductile strata, so other cliffed sections also spawn slides in the Grand Canyon.

Large rotational slides involving the entire Redwall cliff above the Bright Angel Shale develop because steep slopes on the Bright Angel Shale do not provide sufficient buttressing to support the lithostatic loads imposed on the shale at the base of the cliff. Rotational slides currently are forming along the Colorado River between Deer Creek and Kanab Canyon. In this reach, the river has just begun to erode through the Muav Limestone into the underlying Bright Angel Shale. The result is a sheer-walled, narrow canyon less than 0.3 miles (0.5 km) wide but over 1600 feet (490 m) high. As the river incises into the Bright Angel Shale, the foot of the cliff rests on a progressively heightening shale cliff. Without lateral buttressing, the section above the shale fails, and toreva blocks slide into the canyon. Progressive tiers of blocks calve off the canyon wall until the slope on the Bright Angel Shale is sufficiently wide to support the cliffed section above it.

Huntoon (1975) proposed that a stable canyon profile is attained through a series of catastrophic rotational slides shortly after the Bright Angel Shale is exhumed by the downcutting of the river. Saturation of the shale undoubtedly facilitates the process. Because of regional structural undulations, incision of the Bright

Angel Shale has not occurred uniformly over time along the length of the Grand Canyon. Consequently, the inception of rotational sliding has varied from location to location. It has yet to occur between Hundred Forty Mile and Stairways canyons because the Bright Angel Shale has not been exposed or is just beginning to be incised along that reach.

The region between Tapeats and Fishtail canyons provides spectacular evidence that canyon widening results from rotational sliding above the Bright Angel Shale. The most recent slides occurred west of Deer Canyon. Here, a tier of huge blocks has calved off the north wall of the canyon and completely filled a one-mile reach of the Colorado River channel. The blockage forced the river out of its channel and cut a parallel channel 0.3 miles (0.5 km) to the south (Huntoon, 1975, p. 5). The youthfulness of this slide is revealed by the fact that the floor of the blocked channel is about the same elevation as the present channel and that through-flowing tributary drainage has not been reestablished through the slide debris. Sediments ponded in the lake behind the blockage

Figure 6. Former channel of the Colorado River that was blocked by a Pliocene(?) rotational slide from the south side of Cogswell Butte (right) between Tapeats and Deer canyons. Notice that the river excavated a new canyon to the south of the blockage and that the floor of the filled channel lies 500 feet above the modern channel, attesting to the antiquity of the slide. View is downstream to the west.

Figure 7. Row of rotational slide blocks from the Redwall cliff, Surprise Valley. A second, closer row of blocks lies buried by alluvium in the center of the valley. View is toward the northeast.

still are preserved in thick sections on the south side of the Colorado River two miles upstream from Deer Canyon.

Halfway between Deer and Tapeats canyons is a second blocked channel shown in Figure 6. The floor of the channel is 240 feet (75 m) above the present river level, which attests to its antiquity. A blockage of the Colorado River caused its course to be shifted southward, where it cut a new channel.

Surprise Valley, north of Cogswell Butte, contains two or more rows of old, probably Pliocene, toreva blocks (shown in Fig. 7). When these blocks fell into and blocked a former tributary to Tapeats Canyon is unknown.

Carbon Butte is the most unusual toreva block in the Grand Canyon. It is described in detail by Ford and Breed (1970). The butte (Fig. 8) includes rocks from the Bright Angel Shale through the Redwall Limestone and occupies a position between the east and west forks of Carbon Canyon in the eastern Grand Canyon. The butte detached from the Redwall cliff on a listric slip surface bottoming between the Cambrian Tapeats Sandstone and the underlying Precambrian Kwagunt Formation. The detached mass slid southward, trailing behind it large chunks of Tapeats Sandstone that were torn from its base. The detached block came to rest one mile (1.6 km) south and 1800 feet (550 m) below its starting point. The track it followed was a valley eroded on resistant Precambrian strata along the south-plunging axis of a Precambrian syncline.

Large, young landslides and rotational slides are common in the western Grand Canyon downstream from Whitmore Wash. As in the eastern Grand Canyon, failure of the Bright Angel Shale is responsible for most of the slides, and the overlying strata up through the Esplanade Sandstone typically comprises the detached mass. Numerous examples of blocked stream channels exist in the region. The most notable is a huge detachment 1.2 miles long that blocked the Colorado River near Two Hundred Five Mile Canyon. This slide displaced the Colorado River eastward along its entire length. Several large slides occurred in tributary canyons in the western Grand Canyon.

Figure 8. Carbon Butte (small, left-dipping butte in center) detached from the ridge to the left and glided one mile down a shallow valley eroded along the axis of a syncline in Precambrian rocks, trailing in its wake blocks of Tapeats Sandstone torn from their base. Notice that the Redwall section subsequently eroded from the ridge from which Carbon Butte fell. View is toward the northeast, with Chuar Butte in the background.

SUMMARY

Today, the processes of slab failure, rock avalanche, rock fall, talus slide, debris flow, landslide, and gravity tectonics are actively widening and enlarging the Grand Canyon. Evidence indicates that these processes were more active in the past. Since we have not been able to conduct complete quantitative estimates of the rates or to determine the relative importance of these processes, the topic requires additional research. Nonetheless, the large-scale processes of landslides, debris flows, and gravity tectonics probably have been the principal means of canyon enlargement. These processes move debris downslope to the Colorado River, which transports the debris out of the Grand Canyon.

CHAPTER 20

SIDE CANYONS OF THE COLORADO RIVER, GRAND CANYON

Andre R. Potochnik
Stephen J. Reynolds

INTRODUCTION

The majesty of the Grand Canyon emanates from the magnitude of its impressive dimensions. The Colorado River has cut an exquisite gorge; its tributaries are no less spectacular. Trails from the rim are few, the hike is lengthy, and access to much of the canyon from above is limited; however, a river trip through the Grand Canyon, with ample time for hiking, enables one to explore many side canyons with relative ease. Unlike a view from the rim, a view from the river provides an inside-out perspective (Fig. 1). The rim view paints an overall picture of the grand-scale geological scene. In the side canyons, it is possible to examine the details of geological features and relationships that tell a more complete story.

The sequence of rocks in the Grand Canyon can be divided into four groups, each separated by a major unconformity (Fig. 2). The oldest group includes Proterozoic igneous and metamorphic rocks (e.g., Vishnu Group) that were formed during a major episode of deformation and metamorphism about 1.8 to 1.7 billion

Figure 1. The Colorado River within Marble Canyon, as seen from Saddle Canyon. Photograph by S. Reynolds

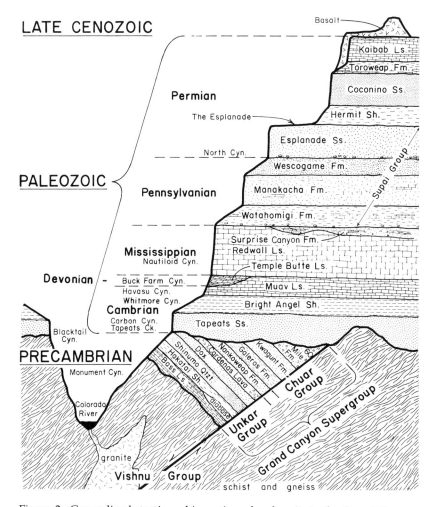

Figure 2. Generalized stratigraphic section of rock units in the Grand Canyon

years ago. These rocks are overlain by the Grand Canyon Supergroup, a series of tilted sedimentary rocks and interlayered mafic flows and sills that formed between 1.3 and 0.8 billion years ago (Elston, personal communication, Elston and McKee 1982). Extensive erosion of these Proterozoic rocks formed a conspicuous, regionally planar unconformity on which the canyon's characteristic, horizontally stratified Paleozoic rocks were deposited between 550 and 150 million years ago. The fourth group of rocks in the canyon includes a veneer of late Cenozoic sediments and volcanic rocks that were deposited mostly since six million years ago, during and after the main period of canyon cutting. The classic physiography of the canyon and surrounding Colorado Plateau is largely a signature of Cenozoic erosion in response to post-mid-Cretaceous uplift of the region to a height of about two miles (3.2 km) above sea level.

From its source in north-central Colorado to its mouth at the Gulf of California, the Colorado River crosses four major physiographic provinces: the Rocky Mountains, the Colorado Plateau, the Transition Zone, and the Basin and Range Province. On the final leg of its long journey across the Colorado Plateau, the river follows a sinuous course in northern Arizona through the Grand Canyon. The Grand Canyon, the Transition Zone, and the Colorado Plateau end abruptly at the Grand Wash Cliffs, where the river flows into the Basin and Range Province at Lake Mead. The canyon is 277 miles (444 km) long, up to 18 miles (30 km) wide, and nearly one mile (1.6 km) deep. Although many rivers in the world follow the trend of pre-existing structural weaknesses, a peculiar feature of the Colorado River in the Grand Canyon is that it generally cross-cuts major north/south faults and folds and flows against the regional dip of strata through most of its length. It parallels the major structures only twice and for a mere one-sixth of its length (Fig. 3).

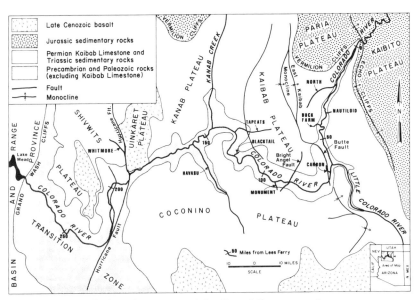

Figure 3. Simplified geologic map of the Grand Canyon region

In contrast, side canyons are influenced more strongly by structural and geomorphic controls—such as faults, regional dip of strata, and rock hardness. Below, we summarize these controls and show how they have resulted in the formation of segments of the Grand Canyon, each having a distinctive side-canyon morphology. Finally, we discuss geological features of some of the more interesting side canyons. This discussion is, in part, modified from Potochnik and Reynolds (1986).

CONTROLS ON THE MORPHOLOGY OF SIDE CANYONS

The morphology of side canyons reflects various combinations of stratigraphic and structural controls, especially the regional dip of strata, folds and faults, and differential erosional susceptibility of the various rock units. Other factors influencing side-canyon morphology include climate and amount of time since the beginning of side-canyon incision.

A first-order control of side-canyon morphology is the regional or subregional dip of Paleozoic strata. Side canyons that flow down the regional dip tend to be longer and wider than those draining into the river against the dip. This is exemplified along the north rim of the canyon near Bright Angel Creek. Here, we see a more branched network of longer streams flowing with the southward dip, compared to those flowing against it from the south rim (Fig. 4).

A more local control of side-canyon morphology is displayed along folds and faults. Fault-controlled canyons, such as Bright Angel Creek, are recognized by distinct linear trends and longer-than-normal lengths (Fig. 4), attributes that reflect the easily eroded character of faulted and brecciated rock along the faults. Regional folds, such as the structurally low sag along Havasu Creek (Fig. 4), control the general location of some major tributaries. More structurally abrupt folds, such as the East Kaibab monocline (Fig. 3), cause rapid downstream changes in the morphology of a single side canyon as the stream cuts across a wide variety of rock units.

The character of side canyons is influenced strongly by differences in the erosional susceptibility of different rock units. Canyons cut in hard, resistant rocks, such as the Proterozoic Vishnu

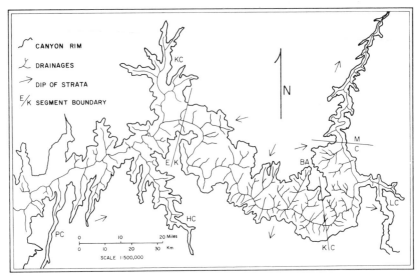

Figure 4. Simplified drainage-network map of the eastern Grand Canyon, showing the four segments of the canyon discussed in the text. Segments are as follows: M - Marble Canyon; C - Chuar; K - Kaibab; and E - Esplanade. Labeled tributaries are as follows: BA - Bright Angel Creek; HC - Havasu Creek; KC - Kanab Creek; LC - Little Colorado River; and PC - Prospect Canyon

Group or the Redwall Limestone, have steep walls and narrow, V-shaped or vertical profiles. In contrast, canyons cut entirely in soft rocks, especially the easily eroded Dox Sandstone and Chuar Group, are broad and bowl-shaped, with rounded, receding walls. Most side canyons display both characteristics because of the successive alternation of relatively hard and soft rocks. A rapid lateral retreat of resistant cliffs probably begins as the underlying, soft rock is exposed. As a consequence, side canyons commonly become wider once the stream has incised down through a cliff-former and into an underlying weak unit.

Climatic variation also causes different side-canyon morphologies. Precipitation varies considerably, according to changes in elevation across the region. Greater precipitation in uplifted regions, such as the Kaibab Plateau, provides more erosive power to streams originating in those areas. Frost action and chemical weathering also are greater in those same regions. Climatic control is best illustrated by comparing the relatively unincised

466   GRAND CANYON GEOLOGY

Desert Facade (6000 ft [1829 m] elevation) near the mouth of the Little Colorado River to the topographically higher (7300 ft [2225 m] elevation) and more incised, pine-covered, south rim to the west. Climatic changes with time also have affected the types and magnitudes of processes responsible for incising and enlarging side canyons. For example, a change from a warmer and more humid climate in the Eocene to a colder and drier one in the Oligocene (Frakes and Kemp 1973) probably affected the rate of denudation of Mesozoic rocks from the plateau prior to any canyon cutting. Pleistocene climatic fluctuations likewise may have affected the recurrence interval of high-discharge flow events and the rates at which lava dams were scoured from the river.

Base-level control is of fundamental importance when considering the causative factors in canyon development. The local base level that controls Grand Canyon tributaries is the Colorado River. An interesting example of this effect is the lower gorge of the Little Colorado River. Upstream from the gorge, the Little Colorado River meanders across northeastern Arizona in a broad, open valley cut in Mesozoic strata. In its last thirty miles (48 km), however, the river gradient increases by about 500 percent and cuts a gorge through the entire 2900-foot (884-m)-thick Paleozoic section. Such an abrupt change in gradient probably was initiated by integration of the Little Colorado River into the main Colorado River system. Regional considerations indicate this occurred about four million years ago, when Lake Bidahochi, centered in the present Little Colorado River Valley, was drained (Scarborough 1987). Through the process of headwater erosion, the ephemeral Little Colorado will be adjusting its gradient to this change for thousands of years to come.

MORPHOLOGICAL DIVISIONS OF THE GRAND CANYON

The Grand Canyon can be subdivided into four segments based on the general morphology of side canyons: (1) Marble Canyon; (2) Chuar; (3) Kaibab Plateau; and (4) Esplanade. The differences between these four segments can be observed readily on the Geologic Map of Arizona (Wilson and others 1969) and in Figure 4.

Marble Canyon Segment

This segment includes all canyons cut into the Marble Platform from below Lees Ferry to Little Nankoweap Creek (mile 52). It is characterized by fairly short, narrow, relatively unbranched gorges—such as North and Nautiloid canyons. These streams flow in accord with the regional northeast dip, while the river flows against it, resulting in distinctively barbed tributaries (Fig. 4). Despite fairly large watersheds, the canyons do not incise a large area of the Marble Platform and have the appearance of being newly formed.

Chuar Segment

From Little Nankoweap Creek to the beginning of the Upper Granite Gorge, canyon development is influenced strongly by the East Kaibab monocline and the Butte fault. The most distinctive feature in this segment is the abrupt appearance of enormous tributaries from the west and the virtual absence of lateral gorges from the east (Fig. 4). The river flows south, paralleling the base of the East Kaibab monocline, then swings broadly westward as it cuts across the monocline, deep into the core of the Kaibab Plateau. This segment is the structurally and topographically deepest part of the Grand Canyon. The large tributaries from the Kaibab uplift owe their size to the greater runoff and more significant relief of the lofty upland region. Moreover, broad, open valleys in their upper reaches are a consequence of incision into the soft Chuar Group shales and the Dox Formation. On the east rim, the absence of canyons along the Desert Facade is the result of low rainfall, resistant rock, and the regional dip of Paleozoic strata away from the river. The Little Colorado River is the only tributary from the east along this entire 48-mile (80-km)-long segment, yet it is the largest of the tributaries entering the Grand Canyon. Not far below its confluence with the Little Colorado River, the main Colorado River crosses a splay of the East Kaibab monocline, where the easily eroded Upper Proterozoic shales are exposed on both sides of the river. Here, small canyons begin to appear on the east side because of the weak shales, despite the continued dip of strata away from the river.

Kaibab Segment

This segment includes all canyons controlled by the Kaibab upwarp between Red Canyon (mile 77) and Kanab Creek (mile

143) (Fig. 4). Through the uplifted Kaibab Plateau, the river cuts to its deepest stratigraphic level, nearly 1800 feet (547 m) into the Proterozoic crystalline basement. Through most of this distance, the main river channel is incised in a narrow, steep-walled canyon called the Inner Granite Gorge. Snowmelt from the Kaibab Plateau produces several large, perennial streams that flow radially toward the canyon off the southern and western flanks of the plateau. The north rim tributaries commonly are long, fault controlled, and display a well-developed dendritic pattern. In contrast, southern tributaries flow against the dip of the strata and form short, steep, and generally unbranched canyons (Fig. 4). Outwash from these small, south-side tributaries produces many of the Grand Canyon's most tyrannical rapids (e.g., Hance, Sockdolager, Horn Creek, and Granite Falls).

Esplanade Segment

This segment is named for the characteristically broad topographic bench formed on the Esplanade Formation by the erosional retreat of the Kaibab-Coconino cliffs. It includes the entire western Grand Canyon downstream from Kanab Creek (mile 143) (Fig. 4). The Esplanade surface actually becomes a prominent feature rather abruptly at the lower end of the Upper Granite Gorge (mile 114) but does not become fully developed until the river leaves the Kaibab upwarp near Kanab Creek. Many side canyons in this segment are uncommonly long and linear, reflecting a strong control by north-trending faults that break the western Colorado Plateau into a series of fault blocks. Such faults control the position and trend of Prospect Creek, Peach Springs Canyon, and other side canyons, in addition to the overall course of the main Colorado River along the Hurricane fault (Fig. 3). The great width of these canyons is confined to the portion above the level of the Esplanade. Much narrower gorges are cut in the underlying Pennsylvanian-Cambrian rocks. The ten-fold increase in width of the Esplanade west of the Kaibab Plateau suggests that canyon-cutting into the Kaibab Limestone began much earlier here than in areas to the east.

The Esplanade segment also includes two major tributaries—Kanab Creek and Havasu Creek—and side canyons with long and complex geological histories. For example, the present mouth of Prospect Canyon is a mere remnant of a wider, deeper canyon that was filled with basalt flows after its incision (Hamblin 1976).

Other canyons, such as Milkweed and Peach Springs canyons, originally were formed by north-northeast-flowing drainages that predated cutting of the present Grand Canyon (Young 1982); the walls of these presentday side canyons contain remnants of sedimentary and volcanic rocks deposited during this earlier drainage regime.

GEOLOGICAL HIGHLIGHTS OF
SELECTED SIDE CANYONS

A variety of factors has influenced the size, orientation, and morphology of side canyons. Some of these factors are local features, such as faults, whereas others are regional in extent, resulting in the four canyon segments discussed above. To illustrate the attributes of side canyons and the processes that control them, the geological highlights of nine side canyons (Fig. 3) are presented below. These nine canyons have been chosen for discussion because they display a wide range of geological features in a variety of settings. We have included brief descriptions of the aesthetics of most canyons in an attempt to convey the natural wonder of the Grand Canyon as viewed from both river level and from the vantage point of short hikes up the side canyons.

North Canyon

In the first twenty miles (32 km) below Glen Canyon Dam, the Colorado River cuts through the red-colored Mesozoic rocks of the Vermilion and Echo cliffs into the underlying, light-colored Permian rocks that form the familiar rim of the Grand Canyon. By North Canyon (mile 20 from Lees Ferry), the river also has deeply incised the underlying, rust-colored Permian Hermit Shale and the Pennsylvanian-Permian Supai Group. North Canyon, typical of the Marble Canyon segment, is a narrow cleft cut in sculptured formations of the Supai Group—including a gray conglomerate that marks the base of the Esplanade Sandstone. The canyon walls are steep and high, with the warm colors, soft textures, and acoustics of a cathedral. Parts of the canyon have the feeling of a sanctuary because of smooth, concave walls formed by dramatic, curved fractures in the sandstone (Fig. 5). The unusual fractures presumably reflect exfoliation of the sandstone walls that was caused by canyon incision.

Figure 5. North Canyon. Curved fractures and bedding surfaces occur in the Permian Esplanade Sandstone of the Supai Group. Photograph by A. Potochnik

Nautiloid Canyon

Not far downstream from North Canyon, gray- and cream-colored strata appear at river level and form a small cliff along the river's shores. Through the next twelve miles (19 km), the river's course becomes confined within the towering walls of the canyon's single most formidable barrier to river-rim hiking— the Redwall Limestone. This fossiliferous, carbonate rock was deposited in a vast, shallow sea that inundated much of western North America during Mississippian time, more than 300 million years ago. In 1869, John Wesley Powell named Marble Canyon for the beautiful, marblelike polish of the Redwall Limestone flanking the river. The limestone's true ivory and gray colors, exposed by running water or recent rockfalls, usually are concealed by a red iron-oxide stain derived from the overlying Supai Group and Hermit Shale.

The 500-foot (152-m) vertical walls of nearly pure limestone and dolomite are highly resistant to erosion in the semiarid climate of the inner canyon. Numerous solution caverns, however, were formed by groundwater dissolution of the limestone during the geological period between deposition of the Redwall and the overlying Supai Formation (McKee 1976). Many large caverns developed along vertical joints or cracks in the limestone and are conspicuous—particularly near mile 35, where a large set of joints bisects the canyon.

In this vicinity, a narrow cleft called Nautiloid Canyon enters from the east (Fig. 3). A short climb into this canyon reveals numerous fossilized remains of the chambered nautiloid (Fig. 6). These fossils were the first of their kind to be found in the Grand Canyon (Breed 1969). The polished limestone floor of Nautiloid Canyon bears many cross sections of these ancient, cone-shaped creatures, whose size can be up to twenty inches (55 cm) in length. The tentaclelike appendages on some specimens may be relics of

Figure 6. Fossil nautiloid on stream-polished outcrop of Mississippian Redwall Limestone. Photograph by S. Reynolds

GRAND CANYON GEOLOGY

soft parts. These fossils, found in the cherty Thunder Springs Member of the Redwall Limestone, are but one of several types of preserved marine life.

Buck Farm Canyon

Small rapids and riffles occasionally punctuate the shady serenity of Marble Canyon as the river cuts into older Paleozoic strata downstream from Nautiloid Canyon. Only the keenest observer will notice where the river intersects the inconspicuous horizontal contact between the Mississippian Redwall Limestone and the underlying Cambrian Muav Limestone. The parallel layering and similar appearance of these two red-stained lime- stones belies the staggering difference in their ages. Rocks repre- senting Late Cambrian through Early Mississippian time (nearly 180 million years of earth history) are missing at this stratigraphic boundary. A clue to this unrecorded period is evident in the canyon walls upstream from Buck Farm Canyon (mile 41), where purplish lenses of Devonian Temple Butte Limestone fifty feet (15 m) high and hundreds of feet long are present between the Redwall and Muav limestones (Fig. 7). Especially well-exposed examples of these lenses are present in Buck Farm Canyon.

The bowl-shaped cross section of these lenses suggests the following evolutionary scenario. After deposition of the Cam-

Figure 7. Lens of Devonian Temple Butte Limestone in a channel cut into the underlying Cambrian Muav Limestone. The lens is overlain by Mississippian Redwall Limestone. Photograph by S. Reynolds

brian Muav Limestone, the continent emerged above sea level, and streams carved channels into the landscape. Subsequent transgression of a Devonian sea filled these channels with impure limestone, which accumulated to even greater thicknesses in the western Grand Canyon (Beus 1987). Both Temple Butte Limestone and Muav Limestone then were eroded down to a peneplain as the landmass once again emerged from the sea in Late Devonian time. This second erosional period so thoroughly leveled the landscape that the Temple Butte Limestone in the eastern Grand Canyon was preserved only at the bottom of channels in which it first had accumulated. A third marine advance from the west in middle Missisippian time blanketed the region with the Redwall Limestone. This activity preserved the channel infillings in the geological record. The relatively recent cutting of the Grand Canyon affords cross-sectional views of these channels.

How did the landscape appear during the erosional intervals, and what lived in the Devonian sea? The flat-lying, undisturbed nature of the Paleozoic rocks tells us that dynamic crustal activity, such as mountain building, faulting, and volcanism, was absent. During the first erosional episode, the area was a bleak, featureless landscape of Muav Limestone that was incised by local stream channels. Probably no plants or animals lived on land, but early, bony-plated fishes first made their appearance during the transgression of the Devonian sea. The landmass subsequently reemerged, and the earliest land plants took root. The subsequent inundation by the Mississippian sea brought a much greater abundance and diversity of marine life—including corals, sponges, shellfish, echinoderms, and nautiloids.

Carbon Canyon

The Marble Canyon segment of Grand Canyon ends about eleven miles (18 km) downstream of Buck Farm Canyon, where the river enters the Chuar segment. The configurations of the side canyons change dramatically in response to the uplift of the broad Kaibab Plateau. Carbon Canyon, a lateral gorge entering the river from the west at mile 65, provides a fascinating structural and geomorphical perspective of the eastern boundary of the uplifted Kaibab Plateau.

The canyon walls in the lower portion of Carbon Canyon consist of purple sandstone and siltstone of the Proterozoic Dox Formation and buff-colored, coarse-grained arkose of the overlying Cambrian Tapeats Sandstone. The latter exhibits striking examples of honeycomb-weathering and colorful Liesegang banding. The purple- and rust-colored, parallel banding forms graceful, curving patterns throughout the rock (Fig. 3). It was caused by precipitation of iron oxides as groundwater migrated through the saturated matrix of the porous sandstone.

Evidence of the forces that uplifted the Kaibab Plateau relative to areas to the east is displayed dramatically a short distance up Carbon Canyon. Here, bedding planes in the Tapeats Sandstone become gently tilted upward as one walks up the canyon until a place is reached where the beds abruptly bend vertically. At this point, the narrow gorge opens into a wide valley of rolling hills with many small tributaries that drain the soft shales of the Proterozoic Chuar Group. The Kaibab Plateau, underlain by the entire Paleozoic sequence, is visible on the western skyline five miles (8 km) away and 2000 feet (610 m) higher than the elevation of the same formations along the river. The sharp upturn in the strata in Carbon Canyon is a local fold caused by the Butte fault, which parallels the East Kaibab monocline, the eastern boundary of the Kaibab Plateau. The Butte fault was a normal fault (west side down) in the Proterozoic but was reactivated as a reverse fault (west side up) during the Late Cretaceous- to-early-Tertiary uplift of the Kaibab Plateau. Cenozoic erosion has breached the monocline, causing removal of the entire Paleozoic sequence and exposure of the underlying Chuar Group shales.

Figure 8. Liesegang banding in Cambrian Tapeats Sandstone, Carbon Canyon. Bedding dips to the left. Photograph by S. Reynolds

Monument Creek

Near mile 93, within the Kaibab segment, Monument Creek enters the Colorado River from the south side in the deepest section of the Upper Granite Gorge. Within the walls of this side canyon are exposures of Early Proterozoic metamorphic and granitic rocks, whose resistance to erosion is responsible for the steep-walled, "V" shape of the inner gorge (Fig. 9). The steep metamorphic layering and numerous convoluted folds within the rocks were formed during metamorphism and deformation that occurred during amalgamation and suturing of terranes along a continental margin to the southeast (Anderson 1986). The mountains that formed during this tectonic episode were eroded away before the Cambrian sea encroached on the landscape and buried it beneath the beach sands of the Tapeats Sandstone. The "Great Unconformity" between the steeply dipping metamorphic rocks and the overlying, gently inclined Tapeats Sandstone is well exposed and represents over one billion years of time.

In the more recent geological past, boulders carried down the present canyon of Monument Creek and deposited at its mouth

Figure 9. Upper Granite Gorge. Dark-colored walls are Proterozoic metamorphic and igneous rocks. Photograph by S. Reynolds

GRAND CANYON GEOLOGY

have partially dammed the Colorado River, creating the major rapid of Granite Falls (Fig. 13). Monument Canyon, which contains remnants of debris deposited in 1984, has the characteristics that a side canyon needs to form a large rapid in the river. These include a steep stream gradient, a narrow canyon with a flat floor, a sufficient supply of large boulders, and a source of fine-grained material to generate mudflows or debris flows (large boulders are more easily transported in a medium that is thicker than water). Monument Creek does not have a large drainage area, compared to other canyons; apparently, this factor is of secondary importance in creating a large rapid.

Blacktail Canyon

The Upper Granite Gorge ends in the vicinity of Blacktail Canyon (mile 120), where the regional westward dip of the strata causes the Tapeats Sandstone to descend to river level. Blacktail Canyon is a narrow, somewhat tubelike notch cut along the Great Unconformity between the sandstone and underlying schist of the Proterozoic Vishnu Group. The details of the unconformity are incredibly well exposed along the polished walls of the canyon (Fig. 10). The vertical metamorphic layering in the schist is over-

Figure 10. The Great Unconformity between Tapeats Sandstone and the underlying Proterozoic Vishnu Group, Blacktail Canyon. Material from the light-colored layer within the schist has been incorporated into the thin, basal conglomerate of the overlying sandstone. The unconformity represents more than one b.y. missing from the geologic record. Photograph by A. Potochnik

lain by sandstone and conglomerate derived from weathering and erosion of the schist. Although thin, vertical quartz veins in the schist were somewhat resistant to weathering, they finally were eroded into the small quartz pebbles now found in the basal sandstone. When standing here, it is easy to imagine the waves of the Cambrian sea 550 million years ago crashing onto jagged hills of schist and churning the metamorphic rock into sandy beaches as the sea advanced across the barren landscape.

Tapeats Creek

A few miles downstream, near mile 134, a cold and clear-flowing perennial stream called Tapeats Creek joins the Colorado River. The lowest formations of the Middle Proterozoic Unkar Group are exposed near river level (Fig. 11). These formations have a characteristic tilt that readily distinguishes these rocks from the more flat-lying Paleozoic rocks. A creekside path through ancient ruins and garden sites of the Anasazi Indians traverses upward through the Unkar Group into the overlying Paleozoic rocks. Here, Thunder Springs bursts from a cavern high in the Muav Limestone wall.

Figure 11. Proterozoic diabase (dark ledge at river level) and slope-forming units of the Proterozoic Unkar Group along the Colorado River near Tapeats Creek. Tilted units of Unkar Group are overlain unconformably by Cambrian Tapeats Sandstone. Photograph by J. Blaustein

The Bass Formation, the oldest unit in the Unkar Group, contains wavy, fossilized algal mats, the oldest preserved evidence of life revealed in the Grand Canyon. Bright, reddish orange shales and siltstones of the overlying Hakatai Shale contain ripplemarks and mudcracks, features that suggest deposition in a tidal-flat environment. A gradational contact between these shales and the underlying Bass Limestone indicates that Hakatai mudflats gradually displaced the algal marine environment as the Bass sea retreated from the area. The tidal flats, in turn, were covered by a thick sequence of sand deposited near the shoreline of a sea. Consolidation of these sands formed the overlying Shinumo Quartzite, a cliff-forming unit that constitutes the steep-walled, narrow canyon of upper Tapeats Creek. A sill of dark-colored diabase approximately one billion years old occurs within the Bass Formation. As the layers in the Bass Formation were pushed apart to accommodate the magma, they reacted with this magma to form thin layers of green serpentine and fibrous, chrysotile asbestos.

An upcanyon view from high on the Thunder River switchbacks reveals the angular unconformity between the Shinumo Quartzite and the overlying Cambrian rocks (Fig. 12). The Tapeats Sandstone, a beach sand of the advancing Cambrian sea, was deposited on the shore of a Shinumo Quartzite island that stood as a large remnant of late Proterozoic erosion. As the sea deepened, the island became submerged, and offshore muds of the Bright Angel Shale lapped across the top of the former island.

Havasu Creek

Downstream from Tapeats Creek, the river turns west and quickly passes out of the Kaibab segment and into the Esplanade segment. The regional tilt of the rock layers causes the cliff-forming Paleozoic limestones, once again, to appear at river level. The confluence of Havasu Creek with the Colorado River near mile 157 is easily missed. A narrow Muav Limestone gorge obscures the enormity of this large tributary, second only to the Little Colorado River in size. Havasu Creek is known for its spectacular waterfalls. The verdant banks of this perennial, aqua-blue stream are lined with velvet ash, cottonwood, and wild grape.

Travertine deposits are perhaps the most fascinating geological feature of Havasu Creek. The travertine is formed by the

Figure 12. Tilted Proterozoic Shinumo Quartzite (Unkar Group) overlain unconformably by flat-lying Cambrian strata, Tapeats Creek. The resistant quartzite was once a craggy island. Along its flanks, Cambrian seas deposited the sands of the Tapeats, which forms the dark ledge shown in the left center of the photograph. The island later was buried by marine muds, which were compacted into the overlying, slope-forming units of the Cambrian Bright Angel Shale. Photograph by S. Reynolds

precipitation of calcium carbonate as the creek waters warm and evaporate during the long flow to the Colorado. The travertine tends to encrust the surface and to take the form of any object over which the water passes.

Distinctive features of travertine cementation are the flat-topped and sinuous dams so commonly seen in the creek. These dams form by a self-enhancing process. An obstruction tends to catch sticks and leaves, which become encrusted with calcium carbonate, thereby increasing the size of the obstruction. When the obstruction becomes large enough, mosses colonize it and provide an additional substrate that increases its width and size. Eventually, a dam forms across the channel with perhaps one or two spillways through which the stream flows. The process becomes self-restricting in the spillways because water velocity is sufficient to prevent accumulation of debris, growth of moss, and precipitation of travertine.

Whitmore Wash

Near mile 188, below the notorious Lava Falls rapids, Whitmore Wash preserves evidence of a time even more tumultuous than that experienced by the river traveler while navigating the rapids. Whitmore Wash and the area around Lava Falls contain remnants of dark, basalt lava flows that once filled the Grand Canyon to a depth of more than 1400 feet (427 m). The lava erupted from volcanic vents, such as Vulcan's Throne, that pierced the Esplanade surface approximately 3000 feet (910 m) above the canyon bottom (Hamblin 1976 and chapter 17).

A number of volcanic vents occur near the Hurricane fault, a major, recently active, north-south fault that may have served as a conduit for the ascending lavas. Upon eruption, the flows of molten lava cascaded over the walls of the Grand Canyon into the Colorado River 3000 feet (910 m) below, creating enormous clouds of steam and filling tributary canyons that drained into the Grand Canyon prior to volcanism. A cross-sectional view of the lava-filled canyon of "old" Whitmore Wash is visible from river level where the Whitmore Trail climbs the north wall of the Grand Canyon into Whitmore Valley. Less than one-half mile downstream from the trail is the "new" Whitmore Wash, a narrow side canyon that drains the same extensive watershed as the former wash. The new wash, however, has cut into the Paleozoic limestones instead of excavating the more erosionally resistant lava that fills the old channel. In the main Grand Canyon, the dams formed by lava flows were more transient, probably surviving less than 10,000 years (McKee and others 1968, Damon and others 1967, Hamblin 1976).

SUMMARY

The canyons described above represent just a small sample of the geological features and natural beauty found within side canyons of the Colorado River. Each canyon is unique, both in scenery and in the array of exposed geological features. Short hikes within these canyons complement the river running, which alternates between the relaxing tranquility of long, slow-moving stretches and the burst of apprehension, excitement, and chaos within the rapids. The entire experience is difficult to describe, but impossible to forget.

BIBLIOGRAPHY

Adams, J.E. and M.L. Rhodes. 1960. Dolomitization by Seepage Refluxion. *American Association of Petroleum Geologists Bulletin*, v. 44, pp. 1912-1921.

Ahlbrandt, T.S., S.S. Andrews, and D.T. Gwynne. 1978. Bioturbation in Eolian Deposits. *Journal of Sedimentary Petrology*, v. 48, pp. 839-848.

Ahnert, 1960. The Influence of Pleistocene Climates upon the Morphology of Cuesta Scarps on the Colorado Plateau. Association of American Geographers *Annals*, v. 50, pp. 139-156.

Aitken, J.D. 1978. Revised Models for Depositional Grand Cycles, Cambrian of the Southern Rocky Mountains, Canada. *Bulletin of Canadian Petroleum Geology*, v. 26, pp. 515-542.

————. 1967. Classification and Significance of Cryptalgal Limestones and Dolomites, with Illustrations from the Cambrian and Ordovician of Southwestern Alberta. *Journal of Sedimentary Petrology*, v. 37, pp. 1163-1178.

Akers, J.P., J.H. Irwin, P.R. Stevens, and N.E. McClymonds. 1962. Geology of the Cameron Quadrangle, Arizona. *U.S. Geological Survey Map* GQ 162.

Aldrich, L. T., G.W. Wetherill, and G.L. Davis. 1957. Occurrence of 1350 Million-Year-Old Granitic Rocks in the Western United States. *Geological Society of America Bulletin*, v. 68, pp. 655-656.

Alf, R.M. 1968. A Spider Trackway from the Coconino Formation, Seligman, Arizona. *Southern California Academy of Sciences Bulletin*; v. 67(2), pp. 125-128.

Allmendinger, R.W. and others. 1987. Overview of the COCORP 40°N Transect, Western United States, the Fabric of an Orogenic Belt. *Geological Society of America Bulletin*, v. 98, pp. 308-319.

Altany, R.M. 1979. Facies of the Hurricane Cliffs Tongue of the Toroweap Formation, Northwestern Arizona. *In:* Baars, D.L., ed., *Permianland. Four Corners Geological Society Guidebook*, 9th Field Conference, pp. 101-104.

Anderson, J.J., P.D. Rowley, R.J. Fleck, and A.E.M. Nairn. 1975. Cenozoic Geology of the Southwestern High Plateaus of Utah. *Geological Society of America Special Paper* 160. 88 pp.

Anderson, Phillip. 1986. Summary of the Proterozoic Plate Tectonic Evolution of Arizona from 1900 to 1600 m.y.a. *In:* Beatty, Barbara and P.A.K. Wilkinson, eds., Frontiers in Geology and Ore Deposits of Arizona and the Southwest. *Arizona Geological Society Digest*, v. 16, pp. 5-11.

Anderson, R.E. 1978. Geologic Map of the Black Canyon 15-minute Quadrangle, Mohave County, Arizona, and Clark County, Nevada. *U.S. Geological Survey Geologic Quadrangle Map* GQ 1394.

Anderson, R.E. and P.W. Huntoon. 1979. Holocene Faulting in the Western Grand Canyon, Arizona, Discussion and Reply. *Geological Society of American Bulletin*, v. 90, pp. 221-224.

Anderson, R.Y. 1982. A Long Geoclimatic Record from the Permian. *Journal of Geophysical Research*, v. 8, pp. 7285-7294.

Anderson, R.Y., W. E. Dean, D.W. Kirkland, and H.I. Snider. 1972. Permian Castile Varved Anhydrite Sequence, West Texas and New Mexico. *Geological Society of America Bulletin*, v. 83, pp. 59-86.

Armstrong, A. K. and B. L. Mamet. 1974. Biostratigraphy of the Arroyo Penasco Group, Lower Carboniferous (Mississippian), North-central New Mexico. *In:* Guidebook, New Mexico Geological Society, Ghost Ranch, Central Northern New Mexico, 25th field conference, pp. 145-158.

Armstrong, A. K. and J. E. Repetski. 1980. The Mississippian System of New Mexico and Southern Arizona. *In:* Fouch, T. D. and E. R. Magathan, eds., *Paleogeography of West Central United States, Rocky Mountain Paleogeography Symposium I*, Rocky Mountain section, Society of Economic Paleontologists and Mineralogists. Denver, Co., pp. 82-100.

Babcock, R. S., E.H. Brown, M.D. Clark, and D.E. Livingston. 1979. Geology of the Older Precambrian Rocks of the Grand Canyon. Part II. The Zoroaster Plutonic-Complex and Related Rocks: *Precambrian Research,* v. 8, pp. 243-275.

Babenroth, D.L. and A.N. Strahler. 1945. Geomorphology and Structure of the East Kaibab Monocline, Arizona and Utah. *Geological Society of America Bulletin,* v. 56, pp. 107-150.

Badiozamani, K. 1973. The Dorag Dolomitization Model—Application to the Middle Ordovician of Wisconsin. *Journal of Sedimentary Petrology,* v. 43, pp. 965-984.

Bagnold, R.A. 1980. An Empirical Correlation of Bedload Transport Rates in Flumes and Natural Rivers. Royal Society of London, *Royal Society of London Proceedings,* Series A, v. 372, no. 1751, pp. 453-473.

————. 1966. An Approach to the Sediment Transport Problem from General Physics. U.S. Geological Survey Professional Paper 422-I, 37 pp.

Baird, Donald. 1952. Revision of the Pennsylvanian and Permian Footprints *Limnopus, Allopus, and Baropus. Journal of Paleontology,* v. 26, pp. 832-840.

Baker, V.R. 1984. Flood Sedimentation in Bedrock Fluvial Systems. *In:* Koster, E.H. and R.J. Steel, R.J., eds., *Sedimentology of Gravels and Conglomerates,* Canadian Society Petroleum Geologists, Memoir 10, pp. 87-98.

————. 1973. Paleohydrology and Sedimentology of Lake Missoula Flooding in Eastern Washington. *Geological Society of America Special Paper* 144, 69 pp.

Baker, V.R. and G. Pickup. 1987. Flood Geomorphology of the Katherine Gorge, Northern Territory, Australia. *Geological Society of America Bulletin,* v. 98, pp. 635-646.

Bakhmeteff (Bakhmetev), B. A. 1932. *Hydraulics of Open Channels.* New York: McGraw-Hill, 329 pp.

Barnes, C.W. 1987. Geology of the Gray Mountain Area, Arizona. *Geological Society of American Centennial Field Guide, Rocky Mountain Section,* pp. 379-384.

Bassler, R. S. 1941. A Supposed Jellyfish from the Precambrian of the Grand Canyon. *U. S. National Museum Proceedings,* v. 89, pp. 519-522.

Basu, A. 1981. Weathering before the Advent of Land Plants: Evidence from Unaltered Detrital K-feldspars in Cambro-Ordovician Arenites. *Geology,* v. 9, pp. 132-133.

Beer, Bill. 1988. We Swam the Grand Canyon. *The Mountaineers,* Seattle, 169 pp.

Belden, W. 1954. The Stratigraphy of the Toroweap Formation, Aubrey Cliffs, Coconino County, Arizona. M.S. thesis, University of Arizona, Tucson, 87 pp.

Bennett, V. C. and D.J. DePaolo. Proterozoic Crustal History of the Western United States as Determined by Neodymium Isotopic Mapping. *Geological Society of American Bulletin* (in press).

Best, M.G. and W.H. Brimhall. 1970. Late Cenozoic Basalt Types in the Western Grand Canyon Region. *In:* Hamblin, W.K. and M.G. Best, eds., *The Western Grand Canyon District: Guidebook to the Geology of Utah.* Utah Geological Society, pp. 57-74.

Best, M. G. and W.K. Hamblin. 1978. Origin of the Northern Basin and Range Province: Implications from the Geology of its Eastern Boundary. *In* Smith, R. B. and G.P. Eaton, eds., *Cenozoic Tectonics and Regional Geophysics of the Western Cordillera.* Geological Society of America, Memoir 152, pp. 313-340.

Beus, S. S. 1989. Devonian and Mississippian Geology of Arizona. *In:* Jenney, J.P. and S.J. Reynolds, eds., Geologic Evolution of Arizona. *Arizona Geological Society Digest,* v. 17.

————. 1987. Devonian and Mississippian Geology of Arizona. *In* Jenny, J. P. and S. J. Reynolds, eds. *Geologic Evolution of Arizona.* Arizona Geological Society Digest, v. 17.

————. 1986. A Geologic Surprise in the Grand Canyon. Arizona Bureau of Geology and Mineral Technology, *Fieldnotes,* v. 16 (3), pp. 1-4.

Beus, S.S. 1980. Devonian Serpulid Biotherms in Arizona. *Journal of Paleontology,* v. 54 (5), pp. 1125-1128.

Beus, S.S. 1980. Late Devonian (Frasnian) Paleography and Paleoenvironments in Northern Arizona. *In* Fouch, T. O. and R. Magathan, eds., *Paleozoic Paleogeography of West-Central United States, Rocky Mountain Paleography Symposium 1.* Rocky Mountain Section, Society of Economic Paleontologists and Mineralogists. pp. 55-69.

———. 1978. Late Devonian (Frasnian)Invertebrate Fossils from Jerome Member of the Martin Formation, Verde Valley, Arizona. *Journal of Paleontology, ,* Vol. 52, pp. 40-54.

Beus, S. S. and W. Breed. 1968. A New Nautiloid Species from the Toroweap Formation in Arizona. *Plateau ,* v. 40 (4), pp. 128-135.

Beus, S.S. 1964. Fossils from the Kaibab Formation at Bee Springs, Arizona. Contributions to the Geology of Northern Arizona. *MNA Bulletin No. 40 ,* pp 59-64.

Beus, S. S., R.R. Rawson, R.O. Dalton, Jr., G. M. Stevenson, V. S. Reed, and T. M. Daneker. 1974. Preliminary Report on the Unkar Group (Precambrian) in Grand Canyon, Arizona. *In:* Karlstrom, T.N.V., G. A. Swann, and R.L. Eastwood, eds., *Geology of Northern Arizona, Part 1 - Regional Studies.* Geological Society of America Field Guide, Rocky Mountain Section. Flagstaff, Arizona, pp. 34-64.

Billingsley, G.H. 1989. Mesozoic Strata at Lees Ferry, Arizona. *In:* S. Beus, ed., *Centennial Field Guide Volume 2.* Boulder: Rocky Mountain Section of the Geological Society of America, pp. 67-71.

Billingsley, G. H. and S.S. Beus. 1985. The Surprise Canyon Formation, an Upper Mississippian and Lower Pennsylvanian(?) Rock Unit in the Grand Canyon, Arizona. *U.S. Geological Survey Bulletin* 1605A, pp. A27-A33.

Billingsley, G. H. and John D. Hendricks. 1989. Physiographic Features of Northwestern Arizona. *In:* T.L. Smiley, J.D. Nations, T.L. Péwé, and J.P. Schafer, eds., *Landscapes of Arizona: The Geological Story.* Lantham: University Press of America, pp. 67-71.

Billingsley, G. H. and P.W. Huntoon. 1983. Geologic Map of Vulcan's Throne and Vicinity, Western Grand Canyon, Arizona. Grand Canyon Natural History Association.

Billingsley, G. H. and E. D. McKee. 1982. Pre-Supai Buried Valleys. *In:* McKee, E. D., ed., *The Supai Group of Grand Canyon.* U.S. Geological Survey Professional Paper 1173, pp. 137-154.

Bissell, H.J. 1969. Permian and Lower Triassic Transition from the Shelf to Basin (Grand Canyon, Arizona to Spring Mountains, Nevada). *In:* Geology and Natural History of the Grand Canyon Region. *Four Corners Geological Society Guidebook,* pp. 135-169.

Blackwelder, Eliot. 1934. Origin of the Colorado River. *Geological Society of America Bulletin,* v. 45, pp. 551-566.

Blackwell, D.D. 1978. Heat Flow and Energy Loss in the Western United States. *In:* Smith, R.B. and G.P. Eaton, eds., *Cenozoic Tectonics and Regional Geophysics of the Western Cordillera.* Geological Society of America Memoir 152, pp. 175-208.

Blair, W.N., 1978. Gulf of California in Lake Mead Area of Arizona and Nevada during Late Miocene Time.*The American Association of Petroleum Geologists Bulletin,* v. 62 (7), pp. 1159-1170.

Blakey, R.C. 1988. Basin Tectonics and Erg Response. *Sedimentary Geology,* v. 56, (1-4), pp.127-152.

———. 1980. Pennsylvanian and Early Permian Paleogeography, Southern Colorado Plateau and Vicinity. *In:* Fouch, T.D. and E. R. Magathan, eds., *Paleozoic Paleogeography of West-Central United States: Rocky Mountain Section,* S.E.P.M., Denver, pp. 239-257.

Blakey, R.C. 1979. Oil Impregnated Carbonate Rocks of the Timpoweap Member, Moenkopi Formation, Hurricane Cliffs Area, Utah and Arizona. *Utah Geology,* v. 6 (1), pp. 45-54.

Blakey, R.C. and Rex Knepp. 1988. Pennsylvanian and Permian Geology of Arizona. *In:* Jenney, J. P. and S. J. Reynolds, eds., *Geologic Evolution of Arizona,* Arizona Geological Society Digest, v. 17.

———. 1983. Permian Shoreline Eolian Complex in Central Arizona. Dune Changes in Response to Cyclic Sea Level Changes. *In:* Brookfield, M.E. and T.S. Ahlbrandt, eds., *Eolian Sediments and Processes.* Amsterdam: Elsevier Science Publishers, pp. 551-581.

Blakey, R.C., Fred Peterson, and Gary Kocurek. 1988. Synthesis of Late Paleozoic and Mesozoic Eolian Deposits of the Western Interior of the United States. *Sedimentary Geology,* v. 56 (1-4), pp. 3-126.

485

Bloeser, B. 1985. *Melanocyrillium*, a New Genus of Structurally Complex Late Proterozoic Microfossils from the Kwagunt Formation (Chuar Group), Grand Canyon, Arizona. *Journal of Paleontology*, v. 59, pp. 741-765.

Bloeser, B., J. W. Schopf, R. J. Horodyski, and W. J. Breed. 1977. Chitinozoans from the Late Precambrian Chuar Group of the Grand Canyon, Arizona. *Science*, v. 195, pp. 676-679.

Bohannon, R.G. 1984. Nonmarine Sedimentary Rocks of Tertiary Age in the Lake Mead Region, Southeastern Nevada and Northwestern Arizona. *U.S. Geological Survey Professional Paper* 1259, 72 pp.

Bowers, W.E. 1972. The Canaan Peak, Pine Hollow, and Wasatch Formations in the Table Cliff Region, Garfield County, Utah. *U.S. Geological Survey Bulletin*, v. 1331B, pp. 707-730.

Boyce, Joseph Michael. 1972. The Structure and Petrology of the Older Precambrian Crystalline Rocks, Bright Angel Canyon, Grand Canyon, Arizona. M. S. thesis, Northern Arizona University, Flagstaff, 88 pp.

Bradley, William C. 1963. Large Scale Exfoliation in Massive Sandstones of the Colorado Plateau. *Geological Society of America Bulletin*, v. 74 (5) , pp. 519-528.

Brady, L.F. 1961. A New Species of *Paleohelcura Gilmore* from the Permian of Northern Arizona. *Journal of Paleontology*, v. 35, pp. 201-202.

————. 1949. *Onischoidichnus*, A New Name for *Isopodichnus Brady* 1947 not Bornemann 1889. *Journal of Paleontology*, v. 23, p. 573.

————. 1947. Invertebrate Tracks from the Coconino Sandstone of Northern Arizona. *Journal of Paleontology*, v. 32, pp. 466-472.

————. 1939. Tracks in the Coconino Sandstone Compared with Those of Small Living Arthropods. *Plateau*, v. 12, pp. 32-34.

Brand, L.R. 1979. Field and Laboratory Studies on the Coconino Sandstone (Permian) Vertebrate Footprints and Their Paleoecological Implications. *Palaeogeography, Palaeoclimatology, Palaeoecology*, v. 28, pp. 25-38.

————. 1978. Footprints in the Grand Canyon; *Origins*, v. 5, pp. 64-82.

Brathovde, J.R. 1986 Stratigraphy of the Grand Wash Dolomite (Upper? Cambrian), Western Grand Canyon, Mohave County, Arizona. M.S. thesis. Northern Arizona University, Flagstaff, 140 pp.

Breed, W.S. 1968. The Discovery of Orthocone Nautiloids in the Redwall Limestone-Marble Canyon, Arizona. Four Corners Geological Society, 5th Field Conference, p. 134.

Bremner, J.A.C. 1986. Microfacies Analysis of Depositional and Diagenetic History of the Redwall Limestone in the Chino and Verde Valleys. M. S. thesis. Northern Arizona University, Flagstaff, 161 pp.

Brenckle, P.L. 1973. Smaller Mississippian and Lower Pennsylvanian Calcareous Foraminifers from Nevada. Cushman Foundation for Foraminiferal Research, *Special Publication 11*, pp. 1-82.

Brookfield, M.E. 1977. The Origin of Bounding Surfaces in Ancient Eolian Sediments. *Sedimentology*, v. 24, pp. 303-332.

Brown, E. H., R.S. Babcock, M.D. Clark, and D.E. Livingston. 1979. Geology of the Older Precambrian Rocks of the Grand Canyon. Part I. Petrology and Structure of the Vishnu Complex. *Precambrian Research*, v. 8, pp. 219-241.

Brown, Silas C. and Robert E. Lauth. 1958. Oil and Gas Potentialities of Northern Arizona. *In*: Roger Anderson and John Harshbarger eds., *Guide Book of the Black Mesa Basin*. Albuquerque: New Mexico Geological Society, pp. 153-160.

Brumbaugh, D.S. 1987. A Tectonic Boundary for the Southern Colorado Plateau. *Tectonophysics*, v. 136, pp. 125-136.

Carson M. and M. Kirkby. 1972. *Hillslope Form and Process*. Cambridge: Cambridge University Press, pp. 448 & 449.

Chan, Marjorie and Gary Kocurek. 1988. Complexities in Eolian and Marine Interactions: Processes and Eustatic Controls on Erg Development. *Sedimentary Geology*, v. 56(1-4), pp. 283-300.

Chapin, C.E. 1983. An Overview of Laramide Wrench Faulting in the Southern Rocky Mountains with Emphasis on Petroleum Exploration. *In*: Lowell, J.D., ed., *Rocky Mountain Foreland Basins and Uplifts*. Rocky Mountain Association of Geologists, pp. 169-179.

486

Chapin, C.E. and S. M. Cather. 1983. Eocene Tectonics and Sedimentation in the Colorado Plateau-Rocky Mountain Area. *In*: Lowell, J.D. ed., *Rocky Mountain Foreland Basins and Uplifts*. Rocky Mountain Association of Geologists, pp. 33-56.

Cheevers, C.W. 1980. Stratigraphic Analysis of the Kaibab Formation in Northern Arizona, Southern Utah, and Southern Nevada. M.S. thesis, Northern Arizona University, Flagstaff, 144 pp.

Cheevers, C.W. and R.R. Rawson. 1979. Facies Analysis of the Kaibab Formation in Northern Arizona, Southern Utah, and Southern Nevada. *Permianland: Four Corners Geological Society Source Book*, pp. 105-133.

Chow, V.T. 1959. *Open-Channel Hydraulics*. New York: McGraw-Hill, 680 pp.

Chronic, H. 1983. *Roadside Geology of Arizona*. Missoula: Mountain Press Publishing Co., 314 pp.

————. 1952. Molluscan Fauna from the Permian Kaibab Formation, Walnut Canyon, Arizona. *Geological Society of America Bulletin*, v. 63, pp. 95-166.

Clark, M.D. 1979. Geology of the Older Precambrian Rocks of the Grand Canyon. Part III. Petrology of the Mafic Schists and Amphibolites. *Precambrian Research*, v. 8, pp. 277-302.

————. 1976. The Geology and Petrochemistry of the Precambrian Metamorphic Rocks of the Grand Canyon, Arizona. Ph.D. dissertation, University of Leicester, Leicester, England, 216 pp.

Clark, R. 1980. Stratigraphy, Depositional Environments, and Petrology of the Toroweap and Kaibab Formations (lower Permian), Grand Canyon Region, Arizona. Ph.D. dissertation, State University of New York at Binghampton, 395 pp.

Clemmensen, Lars B., H. Olsen, and R. Blakey. 1989. Erg-margin Deposits in the Lower Jurassic Moenave Formation and Wingate Sandstone, Southern Utah. *Geological Society of America Bulletin*, v. 101, pp. 759-773.

Cloud, P. E. 1968. PreMetazoan Evolution and the Origins of the Metazoa. *In*: E. T. Drake, ed., *Evolution and the Environment*. Yale University Press, New Haven and London, pp. 1-72.

————. 1960. Gas as a Sedimentary and Diagenetic Agent. *American Journal of Science*, v. 258A, pp. 35-45.

Cloud, P. E. and M. A. Semikhatov. 1969. Proterozoic Stromatolite Zonation. *American Journal of Science*, v. 267, pp. 1017-1061.

Coney, P.J. 1981. Accretionary Tectonics in Western North America. *Arizona Geological Society Digest*, v. 14, pp. 23-37.

Cooley, M.E., J.W. Harshbarger, J.P. Akers, and W.F. Hardt. 1969. Regional Hydrogeology of the Navaho and Hopi Indian Reservations, Arizona, New Mexico, and Utah. *U.S. Geological Survey Professional Paper 521A*, 61 pp.

Cooley, M. E., B. N. Aldridge, and R. C. Euler. 1977. Effect of the Catastrophic Flood of December 1966, North Rim Area, Eastern Grand Canyon, Arizona. *U.S. Geological Survey Professional Paper 980*, 43 pp.

Cooper, G. A. and R. E. Grant. 1972. Permian Brachiopods of West Texas. *Smithsonian Contributions to Paleobiology*, v. 14, pp. 1-23.

Cotter, E. 1978. The Evolution of Fluvial Style, with Special Reference to the Central Appalachian Paleozoic. *In*: Miall, A.E., ed., *Fluvial Sedimentology*. Canadian Society of Petroleum Geologists Memoir 5, pp. 361-384.

Crimes, T.P. 1975. The Stratigraphic Significance of Trace Fossils. *In*: Frey, R.N., ed., *The Study of Trace Fossils*. New York: Springer-Verlag, pp. 109-130.

————. 1970. The Significance of Trace Fossils in Sedimentology, Stratigraphy, and Paleoecology with Examples from Lower Paleozoic Strata. *In*: Crimes, T.P. and J.C. Harper, eds., *Trace Fossils*. Geological Journal Special Issue 3, pp. 101-126.

Cudzil, M.R. and S.G. Driese. 1987. Fluvial, Tidal, and Storm Sedimentation in the Chilhowee Group (Lower Cambrian), Northeastern Tennessee, U.S.A. *Sedimentology*, v. 34, pp. 861-883.

Dalton, R.O., Jr. 1972. Stratigraphy of the Bass Formation. M.S. thesis, Northern Arizona University, Flagstaff, 140 pp.

Damon, P. E. 1968. Application to the Dating of Igneous and Metamorphic Rocks with the Basin Ranges. *In*: Hamilton, E.I. and F.B. Farquhar, eds., *Radiometric Dating for Geologists*. New York: Interscience, pp. 36-44.

487

Damon, P.E., M. Shafiqullah, and R.B. Scarborough. 1978. Revised Chronology for Critical Stages in the Evolution of the Lower Colorado River [abs.]. *Geological Society of America Abstracts with Programs*, v. 10 (3), p. 101.

Damon and others 1967 . Correlation and Chronology of Ore Deposits and Volcanic Rocks. U.S. Atomic Energy Commission Annual Progress Report No. C00-689-76 (Contract AT(11-1)-689.

Daneker, T. M. 1974. Sedimentology of the Precambrian Shinumo Quartzite, Grand Canyon, Arizona. (Abstract.) Geological Society of America. *Abstracts with Programs*, v. 6 (5), p. 438.

Darton, N.H. 1910. A Reconnaissance of Parts of Northwestern New Mexico and Northern Arizona. *U.S. Geological Survey Bulletin*, v. 435, p. 27.

Davidson, E.S. 1967. Geology of the Circle Cliffs area, Garfield and Kane Counties, Utah. *United States Geological Survey Bulletin*, v. 1229, 140 pp.

Davis, G.A., J.L. Anderson, E.G. Frost, and T.J. Schackelford. 1980. Mylonitization and Detachment Faulting in the Whipple-Buckskin-Rawhide Mountains Terrane, Southeastern California and Western Arizona. *In*: Crittenden, M.D., P.J. Coney, and G.H. Davis, eds., *Cordilleran Metamorphic Core Complexes*. Geological Society of America Memoir 153, pp. 79-129.

Davis, W.M. 1903a. An Excursion to the Plateau Province of Utah and Arizona. *Harvard College Museum of Comparative Zoology Bulletin*, v. 42, pp. 1-50.

_____. 1901. An Excursion to the Grand Canyon of the Colorado. *Bulletin of the Museum of Comparative Zoology* (Harvard), v. 38, pp. 107-201.

Dawson, W. 1897. Note on Cryptozoon and other Ancient Fossils. *Canadian Records of Science,*. v. 8, p. 208.

DeCourten, F.L. 1976. Trace Fossils of the Kaibab Formation (Permian) of Northern Arizona. M. S. thesis, University of California, Riverside, 72 pp.

Denison, R.G. 1951. Late Devonian Freshwater Fishes from the Western United States. Chicago Natural History Museum, *Fieldiana*. Geology, v. 11 (5), pp. 221-261.

Dew, E.A. 1985. Sedimentology and Petrology of the DuNoir Limestone (Upper Cambrian) Southern Wind River Range, Wyoming. M.S. thesis, Northern Arizona University, Flagstaff, 193 pp.

Dickinson, W.R. 1981. Plate Tectonic Evolution of the Southern Cordillera. *Arizona Geological Society Digest*, v. 14, p. 113-135.

Dolan, R.A. and Howard and D. Trimble. 1978. Structural Control of the Rapids and Pools of the Colorado River in the Grand Canyon. *Science*, v. 202, pp. 629-631.

Dott, R.H. 1974. Cambrian Tropical Storm Waves in Wisconsin. *Geology*, v. 2, pp. 243-246.

Douglas, R. J. W., H. Gabrielse, J.O.Wheeler, D.F. Stott, and H.R. Belyea. 1970. Geology of Western Canada. *In*: R. J. W. Douglas, ed., *Geology and Mineral Resources of Canada: Geological Survey of Canada*, Economic Geology Report No. 1, pp. 365-488.

DuBois, S.M., A.W. Smith, N.K. Nye, and T.A. Nowak. 1982. Arizona Earthquakes, 1776-1980. *Arizona Bureau of Geology and Mineralogy, Geological Survey Branch Bulletin*, v. 193, 456 pp.

Dutton, C.E. 1882. Tertiary History of the Grand Canyon District. *U.S. Geological Survey Monograph* 2, 264 pp.

Edzwald, J.K. and C.R. O'Melia. 1975. Clay Distributions in Recent Estuarine Sediments: *Clays and Clay Minerals*, v. 23, pp. 39-44.

Elliott, D.K. 1987. *Chancelloria*, an Enigmatic Fossil from the Bright Angel Shale (Cambrian) of Grand Canyon, Arizona. *Journal of the Arizona-Nevada Academy of Science*, v. 21, pp. 67-72.

Elliott, D.K. and D.L. Martin. 1987. A New Trace Fossil from the Cambrian Bright Angel Shale, Grand Canyon, Arizona. *Journal of Paleontology*, v. 61, pp. 641-648.

Elston, D.P. 1989. Grand Canyon Supergroup: Northern Arizona. Stratigraphic Summary and Preliminary Paleomagnetic Correlations with Parts of Other North American Proterozoic Successions. *In*: Jenney, J. P. and S. J. Reynolds, eds., *Geologic Evolution of Arizona*. Arizona Geological Society Digest, v. 17.

————. 1986. Magnetostratigraphy of Late Proterozoic Chuar Group and Sixtymile Formation, Grand Canyon Supergroup, Northern Arizona: Correlation with Other Proterozoic Strata of North America. (Abstract.) Geological Society of America. *Abstracts with Programs* 1986, (Rocky Mountain Section), v. 18, p. 353.

Elston, D.P. 1979. Late Precambrian Sixty Mile Formation and Orogeny at the Top of the Grand Canyon Supergroup, Northern Arizona. *U.S. Geological Survey Professional Paper* 1092, 20 pp.

Elston, D.P. and S.L. Bressler. 1977. Paleomagnetic Poles and Polarity Zonation from Cambrian and Devonian Strata of Arizona. *Earth and Planetary Science Letters*, v. 36, pp. 423-433.

Elston, D.P. and C. S. Grommé. 1974. Precambrian Polar Wandering from Unkar Group and Nankoweap Formation, Eastern Grand Canyon, Arizona. Geological Society of America. *Abstracts with Programs*, v. 6, pp. 440-441.

Elston, D.P. and E. H. McKee. 1982. Age and Correlation of the Late Proterozoic Grand Canyon Disturbance, Northern Arizona. *Geological Society of America Bulletin*, v. 93, pp. 681-699.

Elston, D.P. and G. B. Scott. 1976. Unconformity of the Cardenas-Nankoweap Contact (Precambrian), Grand Canyon Supergroup, Northern Arizona. *Geological Society of America Bulletin*, v. 87, pp. 1763-1772.

————. 1973. Paleomagnetism of Some Precambrian Basaltic Lava Flows and Red Beds, Eastern Grand Canyon, Arizona. *Earth and Planetary Science Letters*, v. 28, pp. 235-265.

Elston, D.P. and Richard A. Young. 1989. Development of Cenozoic Landscape of Central and Northern Arizona: Cutting of Grand Canyon. *In*: Elston, D., G. Billingsley, and R. Young, eds., *Geology of Grand Canyon, Northern Arizona with Colorado River Guides*. Washington, D.C.: American Geophysical Union, pp. 145-153.

Finks, R.M. 1960. Late Paleozoic Sponge Faunas of the Texas Region: the Siliceous Sponges, *American Museum of Natural History Bulletin*, v. 120, pp. 1-160.

Finks, R.M., E.L. Yochelson, and R.P. Sheldon. 1961. Stratigraphic Implications of a Permian Sponge Occurrence in the Park City Formation of Western Wyoming: *Journal of Paleontology*, v. 35, pp. 564-568.

Finnell, T.L. 1966. Geologic Map of the Cibecue Quadrangle, Navajo County, Arizona. *U.S. Geological Survey Geologic Quadrangle Map* GQ-545.

————. 1962. Recurrent Movement on the Canyon Creek Fault, Navajo County, Arizona. *In*: *Short Papers in Geology, Hydrology, and Topography*. U.S. Geological Survey Professional Paper 450-D, pp. D80-D81.

Ford, T.D. , W. J. Breed, and J. S. Mitchell, 1972. Name and Age of Upper Precambrian Basalts in the Eastern Grand Canyon.*Geological Society of America Bulletin*, v. 83 (1), pp. 223-226.

————. 1972a. The Chuar Group of the Proterozoic, Grand Canyon, Arizona. 24th International Geological Congress (Montreal), Proceedings, Section I, *Precambrian Geology*, pp. 3-10.

————. 1972b. The Problematical Precambrian Fossil *Chuaria*. 24th International Geological Congress (Montreal), Proceedings, Section I, *Precambrian Geology*, pp. 11-18.

————. 1970. Carbon Butte, An Unusual Landslide. *In*: *Plateau*, v. 43, pp. 9-15.

————. 1969. Preliminary Geologic Report of the Chuar Group, Grand Canyon (with an appendix on palynology by C. Downie). *In*: *Geology and Natural History of the Grand Canyon Region*. Four Corners Geological Society 5th Field Conference Handbook, pp. 114-122.

Ford, T.D. , W. J. Breed, and J. S. Mitchell, 1972. Name and Age of Upper Precambrian Basalts in the Eastern Grand Canyon.*Geological Society of America Bulletin*, v. 83 (1), pp. 223-226.

Ford, T.D. , P. W. Huntoon, G. H. Billingsley, and W. J. Breed. 1974. Rock Movement and Mass Wastage in the Grand Canyon. *In*: Breed, W.J. and E. Roat, eds., *Geology of the Grand Canyon*. Museum of Northern Arizona Press, Flagstaff, pp. 116-128.

Frakes, L.A. and E.M. Kemp. 1973. Paleogene Continental Positions and Evolution of Climate. *In*: Tarling, D. H. and S.K. Runcorn, eds., *Implications of Continental Drift to the Earth Sciences*. London: Academic Press, v. 1, pp. 539 - 558.

Fryberger, S.G., A.M. Al-Sari, and T.J. Clisham. 1983. Eolian Dune, Interdune, Sand Sheet, and Siliciclastic Sabkha Sediments of an Offshore Prograding Sand Sea, Dhahran Area, Saudi Arabia. *American Association of Petroleum Geologists* , v. 67, pp. 280-312.

489

Gardner, L.S. 1941. The Hurricane Fault in Southwest Utah and Northwest Arizona. *American Journal of Science*, v. 239, pp. 241-260.

Gilbert, G.K. 1875. Report on the Geology of Portions of Nevada, Utah, California, and Arizona. *U.S. Geographical and Geological Survey West of the 100th Meridian* (Wheeler), v. 3 (1), pp. 17-187 and pp. 503-567.

————. 1874. On the Age of the Tonto Sandstone (abs.).*Philosophical Society of Washington*, Bulletin 1, p. 109.

Giletti, B. J. and P.E. Damon. 1961. Rubidium-strontium Ages of Some Basement Rocks from Arizona and Northwestern Mexico. *Geological Society of America Bulletin*, v. 72, pp. 639-644.

Gilmore, C.W. 1928. Fossil Footprints from the Grand Canyon; Third Contribution. *Smithsonian Miscellaneous Collections*, v. 80 (8), 16 pp.

————. 1927. Fossil Footprints from the Grand Canyon; Second Contribution. *Smithsonian Miscellaneous Collections*, v. 80 (3), 78 pp.

————. 1926. Fossil Footprints from the Grand Canyon. *Smithsonian Miscellaneous Collections*, v. 77 (9), 41 pp.

Gilmour, E.H. and I.D. Vogel. 1978. Bryozoans of the Toroweap Formation, Southern Nevada. Geological Society of America. *Abstracts with Programs* 10 (3), p. 107.

Glaessner, M.F. 1984. *The Dawn of Animal Life*. Cambridge: Cambridge University Press, 244 pp.

————. 1969. Trace-fossils from the Precambrian and Basal Cambrian. *Lethaia*, v. 2, pp. 369-393.

————. 1966. Precambrian Palaeontology. *Earth Science Reviews*, v. 1, pp. 29-50.

Glazner, A.F., J.E. Nielson, K.A. Howard, and D.M. Miller, 1986. Correlation of the Peach Springs Tuff, a Large-volume Miocene Ignimbrite Sheet in California and Arizona. *Geology*, v. 14 (10), pp. 840-843.

Gordon, M., Jr. 1984. Biostratigraphy of the Chainman Shale. *In*: Lints, J., Jr., ed., Western Geology Excursions. *Geological Society of America Annual Meeting Guidebook*, v. 1, Department of Geological Sciences, Mackay School of Mines, Reno, Nevada, pp. 74-77.

Graf, W. L. 1980. The Effect of Dam Closure on Down Stream Rapids. *Water Resources Research*, v. 16, pp. 129-136.

————. 1979. Rapids in Canyon Rivers. *Journal of Geology*, v. 87, pp. 533-551.

Gregory, H.E. 1950. Geology and Geography of the Zion Park Region, Utah and Arizona. *U.S. Geological Survey Professional Paper 220*, 200 pp.

Grover, P.W. 1987. Stratigraphy and Depositional Environment of the Surprise Canyon Formation, an Upper Mississippian Carbonate-Clastic Estuarine Deposit, Grand Canyon, Arizona. M.S. thesis, Northern Arizona University, Flagstaff, 166 pp.

Gutschick, R.D. 1943. The Redwall Limestone (Mississippian) of Yavapai County, Arizona. *Plateau* , 16 (1), pp. 1-11.

Hamblin, W.K. 1976. Late Cenozoic Volcanism in the Western Grand Canyon. *In*: Breed, W.S. and E. Roat, eds. *Geology of the Grand Canyon*. Museum of Northern Arizona Press, pp. 142-169.

————. 1965. Origin of Reverse Drag on the Downthrown Side of Normal Faults. *Geological Society of America Bulletin* 76, pp. 1145-1164.

Hamilton, W. 1982. Structural Evolution of the Big Maria Mountains, Northeastern Riverside County, Southeastern California. *In*: Forst, E.G. and D.L. Martin, eds. *Mesozoic-Cenozoic Tectonic Evolution of the Colorado River Region, California, Arizona, and Nevada*. Cordilleran Publishers, pp. 1-27.

Häntzschel, W. 1975. Trace Fossils and Problematica. *In*: Teichert, C., ed., *Treatise on Invertebrate Paleontology, Part W, Miscellanea, Supplement 1*. Geological Society of America and University of Kansas Press, 269 pp.

————. 1962. Trace Fossils and Problematica. *In*: R. C. Moore, ed., *Treatise on Invertebrate Paleontology, Part W, Miscellanea*. Geological Society of America and University of Kansas Press, 229 pp.

Haq, B.U. and F.W.B. Van Eysinger, 1987. *Geologic Time Table*, 4th edition. Amsterdam: Elsevier Science Publishers.

Hardie, L.A. 1987. Dolomitization—A Critical View of Some Current Views. *Journal of Sedimentary Petrology*, v. 57, pp. 166-183.

Harland, W.B. and others, eds. 1982. *A Geologic Time Scale.* Cambridge Earth Science Series, University Press, Cambridge, England, 131 pp.

Harrison, J. E. and Z. E. Peterman. 1971. Windermere Rocks and Their Correlatives in the Western United States. Geological Society of America. *Abstracts with Programs,* v. 2 (7), pp. 592-593.

Harshbarger, J.W., C.A. Repenning, and J.H. Irwin. 1957. Stratigraphy of the Uppermost Triassic and the Jurassic Rocks of the Navaho Country. *U.S. Geological Survey Professional Paper* 291, 74 pp.

Haubold, Harmut. 1984. *Saurier fährten.* A. Ziemsen Verlag (Wittenberg Luthekalt, East Germany), 231 pp.

Hayes, P.T. and G.C. Cone. 1975. Cambrian and Ordovician Rocks of Southern Arizona and New Mexico and Westernmost Texas. *U.S. Geological Survey Professional Paper* 873, 98 pp.

Helley, E.J. 1969. Field Measurement of the Initiation of Large Bed Particle Motion in Blue Creek near Klamath, California. *U.S. Geological Survey Professional Paper 562-G.* 19 pp.

Hendricks, J.D., and Lucchitta, Ivo. 1974. Upper Precambrian Igneous Rocks of the Grand Canyon, Arizona, In: Karlstrom, T.N.V., Swann, G.A., and Eastwood, R.L. eds., Geology of Northern Arizona, pt. 1 - Regional Studies: Geological Society of America Field Guide, Rocky Mountain Section, Flagstaff. pp. 65-86.

Hereford, R. 1984. Driftwood in Stanton's Cave: The Case for Temporary Damming of the Colorado River at Nankoweap Creek in Marble Canyon, Grand Canyon National Park, Arizona. In: Euler, R. C., ed., *The Archaeology, Geology, and Paleobiology of Stanton's Cave.* Grand Canyon National History Association, Monograph 6, pp. 99-106.

————. 1977. Deposition of the Tapeats Sandstone (Cambrian) in Central Arizona. *Geological Society of America Bulletin,* v. 88, pp. 199-211.

————. 1975. Chino Valley Formation (Cambrian?) in Northwestern Arizona. *Geological Society of America Bulletin,* v. 86, pp. 677-682.

Hinds, N.E.A. 1938. An Algonkian Jellyfish from the Grand Canyon of the Colorado. *Science,* v. 88, pp. 186-187.

Hintze, Lehi F. 1988. *Geologic History of Utah.* Geology Studies Special Publication 7. Provo: Brigham Young University, 202 pp.

Hjülstrom, E.J. 1935. Studies of the Morphological Activity of Rivers as Illustrated by the River Fyris. *University of Uppsala [Sweden] Geological Institute Bulletin,* v. 25, pp. 221-527.

Hoffman, P. 1989. Speculations on Laurentia's First Gigayear (2.0 to 1.0 Ga): *Geology,* v. 17, pp. 135-138.

————. 1977. The Problematical Fossil *Chuaria* from the Late Precambrian Uinta Mountain Group, Utah. *Precambrian Research,* v. 4, pp. 1-11.

Hofmann, H. J. and J. D. Aitken. 1979. Precambrian Biota from the Little Dal Group, Mackenzie Mountains, Northwestern Canada. *Canadian Journal of Earth Sciences,* v. 16, pp. 150-166.

Holm, R.F. 1987. San Francisco Mountain: a Late Cenozoic Composite Volcano in Northern Arizona. In: Beus S., ed., *Centennial field Guide Volume 2.* Boulder: Rocky Mountain Section of the Geological Society of America, pp. 389-392.

Holm, Richard F. and Richard B. Moore. 1987. Holocene Scoria Cone and Lava Flows at Sunset Crater, Northern Arizona. In: Beus, S., ed., *Centennial Field Guide Volume 2.* Boulder: Rocky Mountain Section of the Geological Society of America, pp. 393-397.

Homma, M. and S. Shima. 1952. On the Flow in a Gradually Diverged Open Channel. *Japan Science Review,* v. 2 (3), pp. 253-260.

Hopkins, R.L. 1986. Depositional Environments and Diagenesis of the Fossil Mountain Member of the Kaibab Formation (Permian), Grand Canyon, Arizona. M.S. thesis, Northern Arizona University, Flagstaff, 244 pp.

Horodyski, R. J. 1986. Paleontology of the late Precambrian Chuar Group, Grand Canyon, Arizona. Geological Society of America. *Abstracts with Programs* (Rocky Mountain Section), v. 18, p. 362.

Howard, A. and R. Dolan. 1981. Geomorphology of the Colorado River in the Grand Canyon. *Journal of Geology,* v. 89 (3), pp. 269-298.

491

————. 1976. Changes in Fluvial Deposits of the Colorado River in the Grand Canyon caused by Glen Canyon Dam. *In*: Lin, R.M., ed., *First Conference on Scientific Research in the National Parks, Transactions and Proceedings No. 5*, New Orleans, Nov. 9-12, 1975, U.S. National Park Service, pp. 845-851.

Hubbert, M.K. 1951. Mechanical Basis for Certain Familiar Geologic Structures. *Geological Society of American Bulletin*, v. 62, pp. 355-372.

Hunt, C.B. 1969. Geological History of the Colorado River. *In*: The Colorado River Region and John Wesley Powell. *U.S. Geological Survey Professional Paper* 669, 145 pp.

Hunter, R.E. 1977. Basic Types of Stratification in Small Eolian Dunes. *Sedimentology*, v. 24, pp. 361-388.

Huntoon, P.W. 1981. Grand Canyon Monoclines, Vertical Uplift or Horizontal Compression? *Contributions to Geology*, v. 19, pp. 127-134.

————. 1975. The Surprise Valley Landslide and Widening of the Grand Canyon. *Plateau*, v. 48 (1 and 2) pp. 1-12.

————. 1973. High Angle Gravity Faulting In the Eastern Grand Canyon, Arizona. *Plateau*, v. 45 (3) pp. 117-127.

Huntoon, P.W., G.H. Billingsley, and M.D. Clark. 1982. *Geologic Map of the Lower Granite Gorge and Vicinity, Western Grand Canyon, Arizona*. Grand Canyon Natural History Association.

————. 1981. *Geologic Map of the Hurricane Fault Zone and Vicinity, Western Grand Canyon, Arizona*. Grand Canyon Natural History Association.

Huntoon, P.W. and D. P. Elston. 1980. Origin of the River Anticlines, Central Grand Canyon, Arizona. *U.S. Geological Survey Professional Paper* 1126A, 9 pp.

Huntoon, P.W. and J.W. Sears. 1975. Bright Angel and Eminence Faults, Eastern Grand Canyon, Arizona. *Geological Society of America Bulletin*, v. 86, pp. 465-472.

Huntoon, P.W. and others. 1986. *Geological Map of the Eastern Part of the Grand Canyon National Park, Arizona*. Grand Canyon Natural History Association.

Illing, L. V., A. J. Wells, and J. C. M. Taylor. 1965. Pennecontemporaneous Dolomite in the Persian Gulf. *In*: Dolomitization and Limestone Diagenesis—a Symposium. *Society of Economic Paleontologists and Mineralogists Special Paper* 13, pp. 89-111.

Ippen, A.T. 1951. Mechanics of Supercritical Flow. *Transactions American Society of Civil Engineers*, v. 116, pp. 268-295.

Ippen, A.T. and J.H. Dawson. 1951. Design of Channel Contractions. *Transactions American Society of Civil Engineers*, v. 116, pp. 328-346.

Irwin, C.D. 1971. Stratigraphic Analysis of Upper Permian and Lower Triassic Strata in Southern Utah. *American Association of Petroleum Geologists Bulletin*, v. 55, pp. 1976-2007.

Irwin, M.L. 1965. General Theory of Epeiric Clear Water Sedimentation. *American Association of Petroleum Geologists Bulletin*, v. 49, pp. 224-459.

Johnson, H.D. 1977. Shallow Marine Sandbar Sequences: An Example from the Late Precambrian of North Norway. *Sedimentology*, v. 24, pp. 245-270.

Kauffman, E. G. and J. R. Steidtmann. 1981. Are these the Oldest Metazoan Trace Fossils? *Journal of Paleontology*, v. 55, pp. 923-947.

Keller, G.R., R.B. Smith, and L.W. Braile. 1975. Crustal Structure Along the Great Basin-Colorado Plateau Transition from Seismic Reflection Studies. *Journal of Geophysical Research*, v. 80, pp. 1093-1098.

Kendall, C.G. St. C. and W. Schlager. 1981. Carbonates and Relative Changes in Sea Level, *Marine Geology*, v. 44, pp. 181-212.

Kent, W.N. and R.R. Rawson. 1980. Depositional Environments of the Mississippian Redwall Limestone in Northeastern Arizona. *In*: Fouch, T.D. and E.R. Magathon, eds., *Paleozoic Paleogeography Symposium 1, Rocky Mountain Section*, Society of Economic Paleontologists and Mineralogists. Denver, Colorado, pp. 101-109.

Keyes, C.R. 1938. Basement Complex of the Grand Canyon. *Pan-American Geologist*, v. 70 (2), pp. 91-116.

Kieffer, S.W., J.B. Graf, and J.C. Schmidt. 1989. Hydraulics and Sediment Transport of the Colorado River. *In*: Elston, D.P., George H. Billingsley, and Richard A. Young, eds., *Geology of the Grand Canyon, Northern Arizona (with Colorado River Guides)*, FieldTrip Guidebook T115/315 for the 28th International Geological Congress, 239 pp.

492

Kieffer, S.W. 1988. Hydraulic Maps of Major Rapids, Grand Canyon, Arizona. U. S. Geological Survey Miscellaneous Investigations Map Series I-1897A-J.

————. 1987. The Rapids and Waves of the Colorado River, Grand Canyon, Arizona. U.S. Geological Survey Open-File Report 87-096, 97 pp.

————. 1986. The Rapids of the Colorado River (a 20-minute VHS video). U. S. Geological Survey Open File Report 86-503.

————. 1985. The 1983 Hydraulic Jump in Crystal Rapid: Implications for River-running and Geomorphic Evolution in the Grand Canyon. *Journal of Geology*, v. 93, pp. 385-406.

King, Philip B. 1977. *The Evolution of North America*. Princeton: Princeton University Press, 197 pp.

Kirkland, P. 1962. Permian Stratigraphy and Stratigraphic Paleontology of the Colorado Plateau. M.S. thesis. University of New Mexico, Albuquerque, 245 pp.

Knight, R. L. and J. R. Cooper. 1955. Suggested Changes in Devonian Terminology in the Four Corners Area: *Four Corners Geological Society First Field Conference Guidebook*, pp. 56-58.

Kocurek, Gary. 1986. Origins of Low-angle Stratification in Aeolian Deposits. *In*: Nickling, W.G., ed., *Aeolian Geomorphology*. Proceedings of the 17th Annual Binghamton Symposium, pp. 177-193.

————. 1981b. Significance of Interdune Deposits and Bounding Surfaces in Aeolian Dune Sands. *Sedimentology*, v. 28, pp. 753-780.

Koons, E.D. 1955. Cliff Retreat in the Southwestern United States. *American Journal of Science*, v. 253, pp. 44-52.

————. 1948. High-level Gravels of Western Grand Canyon. *Science*, v. 197, pp. 475-476.

————. 1945. Geology of the Uinkaret Plateau, Northern Arizona. *Geological Society of America Bulletin*, v. 56, pp. 151-180.

Kruger-Kneupfer, J.L., M.L. Sbar, and R.M. Richardson. 1985. Microseismicity of the Kaibab Plateau, Northern Arizona, and Its Tectonic Implications. *Bulletin of the Seismological Society of America*, v. 75 (2), pp. 491-506.

Landau, L.D. and E.M. Lifshitz. 1959. *Fluid Mechanics*. Oxford: Pergamon Press, 536 pp.

Lapinski, P.W. 1976. The Gamma Member of the Kaibab Formation (Permian) in Northern Arizon. M.S. thesis, University of Arizona, Tucson, 138 pp.

Larson, E. E., P. E. Patterson, M. H. Amini, and J. Rosenbaum. 1986. Petrology, Chemistry and Revised Age of the Late Precambrian Cardenas Lava, Grand Canyon, Arizona. Geological Society of America. *Abstracts with Programs* (Rocky Mountain Section),v. 18, p. 353.

Leopold, L. 1969. The Rapids and the Pools--Grand Canyon. *U.S. Geological Survey Professional Paper* 669, pp. 131-145.

Lingley, W. S. 1976. Some Structures in Older Precambrian Rocks of Clear Creek—Cremation Creek Area, Grand Canyon National Park, Arizona. Arizona Geological Society Digest, v. 10, pp 27-37.

Lingley, W. S. 1973. Geology of the Older Precambrian Rocks in the Vicinity of Clear Creek and Zoroaster Canyon, Grand Canyon, Arizona. M.S. thesis, Western Washington University, Bellingham, 78 pp.

Lochman-Balk, C. 1971. The Cambrian of the Craton of the United States. *In*: Holland, C.H., ed., *Cambrian of the New World*. New York: John Wiley and Sons, Inc., pp. 79-167.

————. 1970. Upper Cambrian Faunal Patterns on the Craton. *Geological Society of America Bulletin*, v. 81, pp. 3197-3224.

Longwell, C.R. 1946. How Old is the Colorado River? *American Journal of Science*, v. 244 (12), pp. 817-835.

————. 1936. Geology of the Boulder Reservoir floor, Arizona-Nevada. *Geological Society of America Bulletin*, v. 47, pp. 1393-1476.

————. 1921. Geology of the Muddy Mountains, Nevada, with a Section to the Grand Wash Cliffs in Western Arizona. *American Journal of Science*, v. 1, pp. 39-62.

Longwell, C.R., E. H. Pampeyan, B. Bower, and R. J. Roberts. 1965. Geology and Mineral Deposits of Clark County, Nevada. *Nevada Bureau of Mines Bulletin* 62, 218 pp.

Lovejoy, E.M.P. 1980. The Muddy Creek Formation at the Colorado River in Grand Wash—The Dilemma of the Immovable Object. *Arizona Geological Society Digest*, v. 12, pp. 177-192.

Lucchitta, Ivo and S. S. Beus. 1987. Field Trip Guide for Marble Canyon and Eastern Grand Canyon. In: Davis, G. H. and G. M. VandenDolder, eds., *Geologic Diversity of Arizona and Its Margins: Excursions to Choice Areas*. Arizona Bureau of Geology and Mineral Technology, Special Paper 5, pp. 3-19.

Lucchitta, Ivo and J. D. Hendrick. 1983. Characteristics, Depositional Environment, and Tectonic Interpretations of the Proterozoic Cardenas Lavas, Eastern Grand Canyon, Arizona. *Geology*, v. 11(3), pp. 77-181.

Lucchitta, Ivo. 1984. Development of the Landscape in Northwest Arizona—the Country of Plateaus and Canyons. In: Smiley, T.L. and others, eds., *Landscapes of Arizona—the Geological Story*. University Press of America, pp. 269-302.

———. 1979. Late Cenozoic Uplift of the Southwestern Colorado River Region. *Tectonophysics*, v. 61, pp. 63-95.

———. 1975. The Shivwits Plateau. In: Application of ERTS Images and Image Processing to Regional Geologic Problems and Geologic Mapping in Northern Arizona. *Jet Propulsion Laboratory Technical Report* 32-1597, pp. 41-72.

———. 1972. Early History of the Colorado River in the Basin and Range Province. *Geological Society of America Bulletin*, v. 83, pp. 1933-1948.

———. 1967. Cenozoic Geology of the Upper Lake Mead Area Adjacent to the Grand Wash Cliffs, Arizona. Ph.D. dissertation, Pennsylvania State University, 218 pp.

Lucchitta, Ivo and S. S. Beus. 1987. Field Trip Guide for Marble Canyon and Eastern Grand Canyon. In: Davis, G. H. and G. M. VandenDolder, eds., *Geologic Diversity of Arizona and Its Margins: Excursions to Choice Areas*. Arizona Bureau of Geology and Mineral Technology, Special Paper 5, pp. 3-19.

Lucchitta, Ivo and J. D. Hendrick. 1983. Characteristics, Depositional Environment, and Tectonic Interpretations of the Proterozoic Cardenas Lavas, Eastern Grand Canyon, Arizona. *Geology*, v. 11 (3), pp. 77-181.

Lucchitta, Ivo and E.H. McKee. 1975. New Chronological Constraints on the History of the Colorado River and the Grand Canyon [abs.]. Geological Society of America. *Abstracts with Programs*, v. 9, pp. 746-747.

Lucchitta, Ivo and R.A. Young. 1986. Structure and Geomorphic Character of Western Colorado Plateau in the Grand Canyon-Lake Mead region. In: Nations, J.D., C.M. Conway, and G.A. Swann, eds., *Geology of Central and Northern Arizona*. Geological Society of America, Rocky Mountain Section, Field Trip Guidebook, pp. 159-176.

Lull, R.S. 1918. Fossil Footprints from the Grand Canyon of the Colorado. *Smithsonian Miscellaneous Collections*, v. 80, pp. 1-16.

Mamet, B.L. and B. Skipp. 1970. Preliminary Foraminiferal Correlations of Early Carboniferous Strata in the North American Cordilleran. *Les Congres et Colloques di l'Universite di Liege*, v. 55, pp. 237-348.

Marcou, J. 1856. Resume and Field Notes: United States Pacific Railroad Exploration, *Geological Report*, v. 3 (4), pp. 165-171.

Martin, D.L. 1985. Depositional Systems and Ichnology of the Bright Angel Shale (Cambrian), Eastern Grand Canyon, Arizona. M.S. thesis, Northern Arizona University, Flagstaff, 365 pp.

Martin, D.L., L.T. Middleton, and D.K. Elliott. 1986. Depositional Systems of the Middle Cambrian Bright Angel Shale, Grand Canyon, Arizona. Geological Society of America. *Abstracts with Programs*, v. 18, p. 394.

Mather, T. 1970. Stratigraphy and Paleontology of Permian Kaibab Formation, Mogollon Rim Region, Arizona. Ph.D. dissertation, University of Colorado, Boulder, 164 pp.

Matthews, J. J. 1961. *Geologic Map of the Bright Angel Quadrangle, Grand Canyon National Park, Arizona*. Grand Canyon Natural History Association.

Maxson, J.H. 1968. Geologic Map of the Bright Angel Quadrangle, Grand Canyon National Park, Arizona (revised). Grand Canyon Natural History Association.

McKee, E.D. 1982. The Supai Group of Grand Canyon. *U.S. Geological Survey Professional Paper* 1173, pp. 504.

McKee, E.D. 1979. Ancient Sandstone Considered to be Eolian. *In*: McKee, E.D., ed., A Study of Global Sand Seas. *U.S. Geological Survey Professional Paper 1052*, pp. 187-238.
————. 1976. Paleozoic Rocks of the Grand Canyon. *In*: Breed, W.J. and E. Roat, eds., *Geology of the Grand Canyon*. Museum of Northern Arizona Press, pp. 42-64.
————. 1975. The Supai Group, Subdivision and Nomenclature. *U.S. Geological Survey Bulletin* 1395-J, 11 pp.
————. 1974. Paleozoic Rocks of Grand Canyon. *In*: Breed, W.J. and E. Roat , eds., *Geology of the Grand Canyon*. Museum of Northern Arizona Press, pp. 42-75.
————. 1963. Lithologic Subdivisions of the Redwall Limestone in Northern Arizona: Their Paleogeographic and Economic Significance. *U.S. Geological Survey Professional Paper* 400-B.
————. 1947. Experiments on the Development of Tracks in Fine Cross-bedded Sand. *Journal of Sedimentary Petrology*, v. 17, pp. 23-28.
————. 1945. Small-scale Structures in Coconino Sandstone of Northern Arizona. *Journal of Geology*, v. 53, pp. 313-325.
————. 1944. Tracks That Go Uphill in Coconino Sandstone, Grand Canyon, Arizona. *Plateau*, v. 16, pp. 61-72.
————. 1939. Studies on the History of Grand Canyon Paleozoic Formations. Washington, D.C. *Carnegie Institute*, Yearbook 38, pp. 313-314.
————. 1938. The Environment and History of the Toroweap and Kaibab Formations on Northern Arizona and Southern Utah. Washington D.C.: Carnegie Institute , *Publication 492*, 268 pp.
————. 1937. Research on Paleozoic Stratigraphy in Western Grand Canyon. Washington, D. C.: Carnegie Institute, *Yearbook 36*, pp. 340-343.
————. 1933. The Coconino Sandstone - Its History and Origin. Washington D.C.: Carnegie Institute, *Publication 440*, pp. 77-115.
————. 1933a. Landslides and Their Part in Widening the Grand Canyon. *Grand Canyon Nature Notes*, v. 8, pp. 158-161.
————. 1932. Some Fucoids from the Grand Canyon. *Grand Canyon Nature Notes*, v. 7, pp. 77-81.
McKee, E.D. and J.J. Bigarella. 1972. Deformational Structures in Brazilian Coastal Dunes. *Journal of Sedimentary Petrology,* v. 42, pp. 670-681.
McKee, E.D. and W. Breed. 1969. The Toroweap Formation and Kaibab Limestone. *New Mexico Geological Society Special Publication* 3, pp. 12-26.
McKee, E.D., J.R. Douglas, and S. Rittenhouse. 1971. Deformation of Lee-side Laminae in Eolian Dunes. *Geological Society of America Bulletin*, v. 82, pp. 359-378.
McKee, E.D. and R.C. Gutschick. 1969. History of the Redwall Limestone of Northern Arizona. *Geological Society of America Memoir*, v. 114, 726 pp.
McKee, E.D., W.K. Hamblin, and P.E. Damon. 1968. K-Ar Age of Lava Dam in Grand Canyon. *Geological Society of America Bulletin*, v. 79 (1), pp. 133-136.
McKee, E.D and E.H. McKee. 1972. Pliocene Uplift of the Grand Canyon Region: Time of Drainage Adjustment. *Geological Society of America Bulletin*, v. 83 (7), pp. 1923-1932.
McKee, E.D and R.J. Moiola. 1975. Geometry and Growth of the White Sands Dune Field, New Mexico. *U.S. Geological Survey, Journal of Research*, v. 3, pp. 59-66.
McKee, E.D and C.E. Resser. 1945. Cambrian History of the Grand Canyon Region. Washington D.C.: Carnegie Institute, *Publication 563*, 232 pp.
McKee, E.D and E.T. Schenk. 1942. The Lower Canyon Lavas and Related Features at Toroweap in Grand Canyon. *Journal of Geomorphology*, v. 5, pp. 245-273.
McKee, E.D., R.F. Wilson, W.J. Breed, and C.S. Breed. 1967. Evolution of the Colorado River in Arizona. *Museum of Northern Arizona Bulletin* 44, 67 pp.
McKee, E.H. and D.C. Noble. 1976. Age of the Cardenas Lavas, Grand Canyon, Arizona. *Geological Society of America Bulletin*, v. 87, pp. 1188-1190.
McKee, E.H. and D. C. Noble. 1974. Radiometric Ages of Diabase Sills and Basaltic Lava Flows in the Unkar Group, Grand Canyon.Geological Society of America. *Abstracts with Programs*, v. 6, p. 458.
McNair, 1951. Paleozoic Stratigraphy of Part of Northwestern Arizona. *American Association of Petroleum Geologists Bulletin* 35, pp. 503-541.

495

Metzger, D.G. 1968. The Bouse Formation (Pliocene) of the Parker Blythe-Cibola Area, Arizona and California. *In*: Geological Survey Research 1968. *U.S. Geological Survey Professional Paper*600-D, pp. D126-D136.

Middleton, L.T. 1988. Cambrian and Ordovician Depositional Systems in Arizona. *In*: Jenney, J.P. and S.J. Reynolds, eds., *Geological Evolution of Arizona*. Arizona Geological Society Digest, v. 17.

Middleton, L.T. and R. Hereford. 1981. Nature and Controls on Early Paleozoic Fluvial Sedimentation along a Passive Continental Margin: Examples from the Middle Cambrian Flathead Sandstone (Wyoming) and Tapeats Sandstone (Arizona). International Association of Sedimentologists Special Congress, Keele, United Kingdom, Modern and Ancient Fluvial Systems. *Sedimentology and Processes*, p. 83.

Middleton, L.T., J.R. Steidtmann, and D. DeBoer. 1980. Stratigraphy and Depositional Setting of Some Middle and Upper Cambrian Rocks, Wyoming. *Wyoming Geological Association*, 32nd Annual Field Conference, pp. 23-35.

Miller, H. and W. Breed. 1964. *Metococeras bowmani*, A New Species of Nautiloid from the Toroweap Formation (Permian) of Arizona. *Journal of Paleontology*, v. 38, pp. 877-880.

Mullens, R. 1967. Stratigraphy and Environment of the Toroweap Formation (Permian) North of Ashfork, Arizona. M.S. thesis, University of Arizona, Tucson, 101 pp.

National Research Council. 1987. *River and Dam Management*: A Review of the Bureau of Reclamation's Glen Canyon Environmental Studies. Washington, D.C., National Academy Press, 203 pp.

Nielson, R.L. 1981. *Stratigraphy and Depositional Environments of the Toroweap and Kaibab Formations, Southwestern Utah*, Ph.D. dissertation, University of Utah, Salt Lake City, 1015 pp.

Nitecki, M. H. 1971. Pseudo-organic Structures from the Precambrian Bass Limestone in Arizona. *Geology*, v. 23 (1), pp. 1-9.

Noble, L.F. 1928. A Section of the Kaibab Limestone in Kaibab Gulch, Utah, *United States Geological Survey Professional Paper* 150, pp. 41-60.

—————. 1922. A Section of Paleozoic Formations of the Grand Canyon at the Bass Trail. *U.S. Geological Survey Professional Paper* 131-B, pp. 23-73.

—————. 1914. The Shinumo Quadrangle, Grand Canyon District, Arizona. *U.S. Geological Survey Bulletin*, v. 549, 100 pp.

Noble, L.F. and Hunter, J. F. 1916. Reconnaissance of the Archean Complex of the Granite Gorge, Grand Canyon, Arizona. *U.S. Geological Survey Professional Paper* 98-I, pp. 95-113.

O'Conner, J.E., R.H. Webb, and V.R. Baker. 1986. Paleohydrology of Pool-and-Riffle Pattern Development: Boulder Creek, Utah. *Geological Society of America Bulletin*, v. 97, pp. 410-420.

Pasteels, P. and L. T. Silver. 1965. Geochronological Investigations in the Crystalline Rocks of the Grand Canyon, Arizona (abstract). Geological Society of America Annual Meeting Program, Kansas City, 122 pp.

Peirce, H.W. 1984. The Mogollon Escarpment. *Arizona Bureau of Geology and Mineral Technology Fieldnotes*, v. 14 (2), pp. 8-11.

Peirce, H.W. and J.D. Nations. 1986. Tectonic and Paleogeographic Significance of Tertiary Rocks of the Southern Colorado Plateau and Transition Zone. *In*: Nations, J.D., C.M. Conway, and G.A. Swann, eds., *Geology of Central and Northern Arizona*. Geological Society of America, Rocky Mountain Section, Field Trip Guidebook, pp. 159-176.

Peirce, H.W., M. Shafiqullah, and P.E. Damon. 1979. An Oligocene (?) Colorado Plateau Edge in Arizona. *Tectonophysics*, v. 61, pp. 1-24.

Peterson, Fred. 1988. Pennsylvanian to Jurassic Eolian Transportation Systems in the Western United States, *Sedimentary Geology.*, v. 56, (1-4), pp. 207-260.

—————. 1986. Jurassic Paleotectonics in the West-Central Part of the Colorado Plateau, Utah and Arizona. *In*: Peterson, J.A., ed., Paleotectonics and Sedimentation in the Rocky Mountain Region, United States, *American Association of Petroleum Geologists Memoir* 41, pp. 563-596.

Péwé, T. L. 1968. Colorado River Guidebook: A Geologic and Geographic Guide from Lees Ferry to Phantom Ranch. Tempe, Arizona.

Podolny, W., Jr. and J.D. Cooper. 1974. Toward an Understanding of Earthquakes. *Highway Focus,* v. 6(1), 108 pp.

Potochnik, A.R. and S. J. Reynolds. 1986. Geology of Side Canyons of the Colorado, Grand Canyon National Park. Arizona Bureau of Geology and Mineral Technology *Fieldnotes,* v. 16 (1), pp. 1-8.

Powell, J. W. 1895. Canyons of the Colorado: Flood and Vincent. Reprinted as, *The Exploration of the Colorado River and its Canyons.* New York: Dover, 400 pp.

Powell, J.W. 1876a. Exploration of the Colorado River of the West. Smithsonian Institution, 291 pp.

Pratt, B.R. and N.P. James. 1986. The St. George Group (Lower Ordovician) of Western Newfoundland: Tidal Flat Island Model for Carbonate Sedimentation in Shallow Epeiric Seas. *Sedimentology,* v. 33, pp. 313-343.

Racey, J.S. 1974. Conodont Biostratigraphy of the Redwall Limestone at East-Central Arizona. M.S. thesis, Arizona State University, Tempe, 199 pp.

Ragan, D. M. and M.F. Sheridan. 1970. The Archean Rocks of the Grand Canyon, Arizona. Geological Society of America. *Abstracts with Programs,* Part 2, pp. 132-133.

Ransome, F. L. 1904. The Geology and Ore Deposits of the Bisbee Quadrangle, Arizona. *U. S. Geological Survey Professional Paper* 21, 168 pp.

Rawson, R.R. and C.E. Turner-Peterson. 1980. Paleogeography of Northern Arizona during the Deposition of the Permian Toroweap Formation. *In:* Fouch, T.D. and E.R. Magathan, eds., *Paleozoic Paleogeography of West Central United States.* Rocky Mountain Section, Society of Economic Paleontologists and Mineralogists, pp. 341-352.

————. 1979. Marine-Carbonate, Sabkha, and Eolian Facies Transitions Within the Permian Toroweap Formation, Northern Arizona. *In:* Baars, D.L., ed., *Permianland, Four Corners Geological Society Guidebook,* 9th Field Conference, pp. 87-99.

————. 1974. The Toroweap Formation—A New Look. *In:* Swann, G., T. Karlstrom, and R. Eastwood, eds. *,Guidebook to the Geology of Northern Arizona, Part I.* Flagstaff Publishing Company, Flagstaff, Arizona, pp. 155-190.

Read, C.B. and A.A. Wanek. 1961. Stratigraphy of Outcropping Permian Rocks in Parts of Northeastern Arizona and Adjacent Areas. *U.S. Geological Survey Professional Paper* 374-H: H1-H10.

Reches, Z. 1978a. Analysis of Faulting in Three Dimensional Strain Field. *Tectonophysics,* v. 47, pp. 109-129.

————. 1978b. Development of Monoclines, Part I, Structure of the Palisades Creek Branch of the East Kaibab Monocline, Grand Canyon, Arizona. *Geological Society of America Memoir* 151, pp. 235-271.

Reches, Z. and A. M. Johnson. 1978. Development of Monoclines, Part II, Theoretical Analysis of Monoclines. *Geological Society of American Memoir* 151, pp. 273-311.

Reed, V. S. 1974. Stratigraphy of the Hakatai Shale, Grand Canyon, Arizona. Abstract. Geological Society of America. *Abstracts with Programs,* v. 6 (5), p. 469.

Reeside, J.B. and Bassler, H. 1922. Stratigraphic Sections in Southwestern Utah and Northwestern Arizona, *United States Geological Survey Professional Paper* 129, pp. 53-77.

Reiche, Parry. 1938. An Analysis of Cross-lamination—The Coconino Sandstone. *Journal of Geology,* v. 46, pp. 905-932.

————. 1937. The Toreva Block, a Distinctive Landslide Type. *Journal of Geology,* v. 45, pp. 538-548.

Reineck, H.E. and F. Wunderlich. 1968. Classification and Origin of Flaser and Lenticular Bedding. *Sedimentology,* v. 11, pp. 189-228.

Reynolds, M. W. and D. P. Elston. 1986. Stratigraphy and Sedimentation of Part of the Proterozoic Chuar Group, Grand Canyon, Arizona. Geological Society of America. *Abstracts with Programs* (Rocky Mountain Section), v. 18, p. 405.

Rigby, J.K. 1976. Some Observations on Occurrences of Cambrian *Porifera* in Western North America and Their Evolution. *In:* Robinson, R.A. and A.J. Rowell, eds., *Paleontology and Depositional Environments: Cambrian of Western North America.* Brigham Young University Geological Studies, v. 23, pp. 51-60.

Ritter, S.M. 1983. Conodont Biostratigraphy of Devonian-Pennsylvanian Rocks of Iceberg Ridge, Mojave County, Northwest Arizona. M.S. thesis, Brigham Young University, Provo, 54 pp.

497

Rogers, J. D. and M. R. Pyles. 1979. Evidence of Catastrophic Erosional Events in the Grand Canyon of the Colorado River, Arizona. 2nd. Conference on Scientific Research in the National Parks. 26-30 November 1979, San Francisco, California.

Rouse, H., B.V. Bhoota, and E. Hsu. 1951. Design of Channel Expansions. *Transactions. American Society of Civil Engineers*, v. 116, pp. 347-363.

Rubin, D.M. and R. E. Hunter. 1985. Why Deposits of Longitudinal Dunes are Rarely Recognized in the Geologic Record. *Sedimentology*, v. 32, pp. 147-157.

Rubin, D.M. and R.E. Hunter. 1982. Bedform Climbing in Theory and Nature. *Sedimentology*, v. 29, pp. 121-138.

Rubin, D. M. And R. E. Hunter. 1983. Reconstructing Bedform Assemblages from Compound Crossbedding. *In:* Brookfield, M. E. and T. S. Ahlbrandts, eds., *Eolian Sediments and Processes.* Amsterdam: Elsevier Science Publishers, pp. 407-427.

Scarborough, R.B. 1989. Cenozoic Erosion and Sedimentation in Arizona. *In:* Jenney, J.P. and S. J. Reynolds, eds., Geologic Evolution of Arizona. *Arizona Geological Society Digest* v. 17.

Scarborough, R.B., C.M. Menges, and P.A. Pearthree. 1986. *Map of Late Pliocene-Quaternary (post-4-my) Faults, Folds, and Volcanic Outcrops in Arizona.* Map 22. Arizona Bureau of Geology and Mineral Technology.

Schmidt, J.C. and J.B. Graf. 1987. Aggradation and Degradation of Alluvial-sand Deposits, 1965-1986, Colorado River, Grand Canyon National Park, Arizona. *U.S. Geological Survey Open-File Report* 87-555, approximately 100 pp.

Schopf, W. J., T.D. Ford, and W. J. Breed. 1973. Microorganisms from the Late Precambrian of the Grand Canyon, Arizona. *Science*, v. 179, pp. 1319-1321.

Schreiber, B.C. 1986. Arid Shorelines and Evaporites. *In:* Reading, H.G., ed., *Sedimentary Environments and Facies.* London: Blackwell Scientific Publications, pp. 189-228.

Schumm, S. A. and Chorley, R. J. 1966. Talus Weathering and Scarp Recession in the Colorado Plateau. *Zeitschrift fur Geomorphologie*, V. 10, pp. 11-36

Schuster, R.L. and J.E. Costa. 1986. A Perspective on Landslide Dams. *In:* Schuster, R.L., ed., *Landslide Dams: Processes, Risk, and Mitigation, Proceedings,* Geotechnical Engineering Division of the American Society of Civil Engineers, Geotechnical Special Publication No. 3. New York: American Society of Civil Engineers, pp. 1-20.

Sears, J.W. 1973. Structural Geology of the Precambrian Grand Canyon Series. M. S. thesis, University of Wyoming, Laramie, 100 pp.

Sears, J. R. and R. A. Price. 1978. The Siberian Connection: A Case for the Precambrian Separation of the Siberian and North American Cratons. *Geology*, v. 6, pp. 267-270.

Seilacher, A. 1970. *Cruziana* Stratigraphy of Non-fossiliferous Paleozoic Sandstones. *In:* Crimes, T.P. and J.C. Harper, eds., *Trace Fossils.* Geological Journal Special Issue, pp. 447-476.

Sepkoski, J.J. 1982. Flat-pebble Conglomerates, Storm Deposits, and the Cambrian Bottom Fauna. *In:* Einsele, G. and A. Seilacher, eds., *Cyclic and Event Stratification.* New York: Springer-Verlag, pp. 371-385.

Sharp, R.P. 1940. Ep-Archean and Ep-Algonkian Erosion Surfaces, Grand Canyon, Arizona. *Geological Society of America Bulletin*, v. 51, pp. 1235-1270.

Shinn, E.A. 1983. Tidal Flat Environment. *In:* Scholle, P.A., D.G. Debout, and C.H. Moore, eds., Carbonate Depositional Environments. *American Association of Petroleum Geologists Memoir* 33, pp. 171-210.

Shinn, W.A., R.N. Ginsburg, and R.M. Lloyd. 1965. Recent Supratidal Dolomite from Andros Island, Bahamas. *In:* Dolomitization and Limestone Diagenesis—a Symposium. *Society of Economic Paleontologists and Mineralogists Special Publication* 13, pp. 89-111.

Shirley, D.H. 1987. Geochemical Facies Analysis of the Surprise Canyon Formation in Fern Glen Channelway, Central Grand Canyon, Arizona. M.S. thesis, Northern Arizona University, Flagstaff, 208 pp.

Shoemaker, E.M., R.L. Squires, and M.J. Abrams. 1975. The Bright Angel, Mesa Butte, and Related Fault Systems of Northern Arizona. *California Institute of Technology Jet Propulsion Laboratory Technical Report* 32-1597: 23-41. Also in: 1978. Bright Angel and Mesa Butte Fault Systems of Northern Arizona. *In:* Smith, R.B. and G.P. Eaton, eds., Cenozoic Tectonics and Regional Geophysics of the Western Cordillera. *Geological Society of America Memoir* 152, pp. 341-367.

498

Shoemaker, E.M., R.L. Squires and M.J. Abrams. 1974. The Bright Angel and Mesa Butte Fault Systems of Northern Arizona. *In*: Karlstrom, T.N.V., Gordon A. Swann and Raymond L. Eastwood, eds., Geology of Northern Arizona, Part 1, Regional Studies. Rocky Mountain Section, Geological Society of America Annual Meeting, Flagstaff, pp. 255-291.

Shride, A. F. 1967. Younger Precambrian Geology in Southern Arizona. *U. S. Geological Survey, Professional Paper* 566, 89 pp.

Skipp, B. 1969. Foraminifera. *In*: McKee, E.D. and R.C. Gutschick, eds., History of the Redwall Limestone of Northern Arizona. *Geological Society of America Memoir* 114, pp. 173-195.

Smith, P.B. 1970. New Evidence for Pliocene Marine Embayment along the Lower Colorado River Area, California and Arizona. *Geological Society of American Bulletin*, v. 81, pp. 1411-1420.

Smith, R.B. 1978. Seismicity, Crustal Structure, and Intraplate Tectonics of the Interior of the Western Cordillera. *In*: Smith, R.B. and G.P. Eaton, eds., Cenozoic Tectonics and Regional Geophysics of the Western Cordillera. *Geological Society of America Memoir* 152, pp. 111-144.

Smith, R.B., and M.L. Sbar. 1974. Contemporary Tectonics and Seismicity of the Western United States with Emphasis on the Intermountain Seismic Belt. *Geological Society of America Bulletin*, v. 85, pp.1205-1218.

Snyder, W.S., W.R. Dickinson, and M.L. Silberman. 1976. Tectonic Implications of Space-Time Patterns of Cenozoic Magmatism in the Western United States. *Earth and Planetary Science Letters*, v. 32, pp. 91-106.

Sorauf, J.E. 1962. Structural Geology and Stratigraphy of the Whitmore Area, Mohave County, Arizona. Ph.D. dissertation, University of Kansas, 361 pp.

Spamer, Earle E. 1984. Paleontology in the Grand Canyon of Arizona: 125 years of Lessons and Enigmas from the Past to the Present. *The Mosasaur*, v.2, pp. 45-118.

Stevens, Larry. 1985. The 67 Elephant Theory or Learning to Boat Big Water Hydraulics. *River Runner*, v. 5 (1), pp. 24-25.

Stevenson, G. M. 1973. Stratigraphy of the Dox Formation, Grand Canyon, Arizona. M. S. thesis, Northern Arizona University, Flagstaff. 225 pp.

Stevenson, G. M. and Beus, S. S. 1982. Stratigraphy and Depositional Setting of the Upper Precambrian Dox Formation in Grand Canyon. *Geological Society of American Bulletin*, v. 93, pp. 163-173.

Stewart, J.H., F.G. Poole, and R.F. Wilson. 1972. Stratigraphy and Origin of the Chinle Formation and Related Upper Triassic Strata in the Colorado Plateau Region. *U.S. Geological Survey Professional Paper* 690, 336 pp.

Stewart, J.H. and C.A. Suczek. 1977. Cambrian and Latest Precambrian Paleogeography and Tectonics in the Western United States. *In*: Stewart, J.H., C.H. Stevens, and A.E. Fritsche, eds., *Paleozoic- paleogeography of the Western United States*. Society of Economic Paleontologists and Mineralogists, Pacific Section, Paleogeography Symposium 1, pp. 1-18.

Stewart, J.H. 1972. Initial Deposits in the Cordilleran Geosyncline: Evidence of a Late Precambrian Continental Separation. *Geological Society of America Bulletin*, v. 83, pp. 1345-1360.

Stokes, William Lee. 1986. *Geology of Utah*. Occasional Paper No. 6. Salt Lake City: Museum of Natural History, 310 pp.

Stokes, W.L. 1968. Multiple Parallel-truncation Bedding Planes—a Feature of Wind-deposited Sandstone Formations. *Journal of Sedimentary Petrology*, v. 38, pp. 510-515.

Strahler, A.N. 1948. Geomorphology and Structure of the West Kaibab Fault Zone and Kaibab Plateau. *Geological Society of America Bulletin*, v. 59, pp. 513-540.

Strahler, A.N. 1940. Landslides of the Vermilion and Echo Cliffs. *Journal of Geomorphology*, v. 3, pp. 285-300.

Strand, R.I. 1986. Water Related Sediment Problems. Water Systems Management Workshop-1986, Session 4-1, U.S. Bureau of Reclamation, Denver, Colorado, pp. 1-30.

Sturgul, J. R. and Z. Grinshpan. 1975. Finite Element Model for Possible Isostatic Rebound in the Grand Canyon. *Geology*, v. 3, pp. 169-171.

Sturgul, J.R. and T.D. Irwin. 1971. Earthquake History of Arizona. *Arizona Geological Society Digest*, v. 9, pp. 1-23.

Teichert, C. 1965. Devonian Rocks and Paleogeography of Central Arizona. *U. S. Geological Survey Professional Paper* 464, 181 pp.

Thompson, P.A. 1972. *Compressible-fluid Dynamics*. New York: McGraw-Hill, 665 pp.

Turner, C.E. 1974. Facies of the Toroweap Formation, Marble Canyon, Arizona. M.S. thesis, Northern Arizona University, Flagstaff, 120 pp.

Van Eysinga, F.S.B. 1975. *Geological Time Table*. Amsterdam: Elsevier Science Publishers.

Van Gundy, C.E. 1951. Nankoweap Group of the Grand Canyon Algonkian of Arizona. *Geological Society of America Bulletin*, v. 62, pp. 953-959.

————. 1937. Jellyfish from the Grand Canyon Algonkian. *Science*, v. 85, p. 314.

Vidal, G. 1981. Aspects of Problematic Acid-resistant, Organic-walled Microfossils (Acritarchs) in the Proterozoic of the North Atlantic Region. *Precambrian Research*, v. 15, pp. 9-23.

Vidal, G. and T.D. Ford. 1985. Microbiotas from the Late Proterozoic Chuar Group (Northern Arizona) and Uinta Mountain Group (Utah) and Their Chronostratigraphic Implications. *Precambrian Research*, v. 28, pp. 349-389.

Walcott, C.D. 1920. Cambrian Geology and Paleontology, IV; Middle Cambrian Spongiae. *Smithsonian Miscellaneous Collection*, v. 67, pp. 261-364.

————. 1910. Cambrian Geology and Paleontology II; Abrupt Appearance of the Cambrian Fauna on the North American Continent. *Smithsonian Miscellaneous Collection*, v. 57, p. 14.

————. 1899. Precambrian Fossiliferous Formations. *Geological Society of America Bulletin*, v. 10, pp. 199-244.

————. 1894. Precambrian Igneous Rocks of the Unkar Terrane, Grand Canyon of the Colorado. *U. S. Geological Survey 14th Annual Report for 1892/3*, part 2, pp. 497-519.

————. 1890. The Fauna of the Lower Cambrian or *Olenellus* Zone. *U.S. Geological Survey, 10th Annual Report*, pp. 509-760.

————. 1889. A Study of a Line of Displacement in the Grand Canyon of the Colorado in Northern Arizona. *Geological Society of American Bulletin*, v. 1, pp. 49-64.

————. 1883. Pre-Carboniferous Strata in the Grand Canyon of the Colorado, Arizona. *American Journal of Science*, series 3, v. 16, pp. 437-442; 484.

————. 1880. The Permian and Other Paleozoic Groups of the Kanab Valley, Arizona. *American Journal of Science*, 3rd Series, v. 20, pp. 221-225.

Walen, M. B. 1973. *Petrogenesis of the Granitic Rocks of Part of the Upper Granite Gorge, Grand Canyon, Arizona*. M.S. thesis, Western Washington University, Bellingham.

Wanless, H.R. 1975. Carbonate tidal flats of the Grand Canyon Cambrian. *In:* Ginsburg, R.N., ed., *Tidal Deposits: A Casebook of Recent Examples and Fossil Counterparts*. New York: Springer-Verlag, pp. 269-277.

————. 1973. *Cambrian of the Grand Canyon: A Reevaluation of the Depositional Environment*. Ph. D. dissertation, Johns Hopkins University, Baltimore, 128 pp.

Wardlaw, B.R. 1986. Paleontology and Deposition of the Phosphoria Formation. *Contributions to Geology*, University of Wyoming, v. 24, pp. 107-142.

Wardlaw, B.R., and J.W. Collinson. 1978. Stratigraphic Relations of the Park City Group (Permian) in Eastern Nevada and Western Utah, *American Association of Petroleum Geologists Bulletin*, v. 62, pp. 1171-1184.

Webb, R.H., P.T. Pringle, and G.R. Rink. 1987. Debris Flows from Tributaries of the Colorado River in Grand Canyon National Park, Arizona. *U.S. Geological Survey Open-File Report* 87-118, 64 pp.

Webster, G.D. 1969. *Chester Through Derry Conodonts and Stratigraphy of Northern Clark on Southern Lincoln Counties, Nevada*. University of California Publications in Geological Services, v. 79, 119 pp.

Webster, G.D. 1984. Conodont Zonations Near the Mississippian-Pennsylvanian Boundary in Eastern Great Basin. *In:* Lintz, J., Jr., ed., *Western Geology Excursions (Geological Society of American Annual Meeting Guidebook)*. Department of Geological Sciences, Mackay School of Mines, Reno, Nevada, v. 1, pp. 78-82.

Wenrich, K.J., G.H. Billingsley, and P.W. Huntoon. 1986. Breccia Pipe and Geologic Map of the Northeastern Hualapai Indian Reservation and Vicinity, Arizona. U.S. *Geological Survey Open-File Report* 86-458C, 29 pp.

White, D. 1929. *Flora of the Hermit Shale, Grand Canyon, Arizona*. Washington D.C.: Carnegie Institute, Publication 405, pp. 1-221.

500

Wilson, E. D. 1962. Resumé of the Geology of Arizona. *Arizona Bureau of Mines Bulletin* 171, 140 pp.

Wilson, E.D., R.T. Moore, and J.R. Cooper. 1969. Geologic Map of Arizona. Arizona Bureau of Mines. Scale 1:500,000.

Wilson, I.G. 1972a. Aeolian Bedforms: Their Development and Origins. *Sedimentology*, v. 19, pp. 173-210.

————. 1972b. Ergs. *Sedimentary Geology*, v. 10, pp. 77-106.

Wilson, Richard F. 1974. Mesozoic Stratigraphy of Northeastern Arizona. *In*: T. Karlstrom, G. Swann, and R. Eastwood eds., *Geology of Northern Arizona*. Flagstaff: Geological Society of America, Rocky Mountain Section Meeting.

Wood, W.H. 1956. The Cambrian and Devonian Carbonate Rocks at Yampai Cliffs, Mohave County, Arizona. Ph.D. dissertation, University of Arizona, Tucson, 228 pp.

Woodward-Clyde Consultants. 1982. Geologic Characterization Report for the Paradox Basin Study Region, Utah Study Areas, Volume 1, Regional Overview. Battelle Memorial Institute Office of Nuclear Waste Isolation, ONWI-290.

Young, Richard A. 1989. Paleogene-Neogene Deposits of Western Grand Canyon, Arizona. *In*: Beus S., ed., *Centennial Field Guide Volume 2*. Boulder: Rocky Mountain Section of The Geological Society of America.

Young, R.A. 1982. Paleogeomorphic Evidence for the Structural History of the Colorado Plateau Margin in Arizona. *In*: Forst, E.D. and D.L. Martin, eds., *Mesozoic-Cenozoic Tectonic Evolution of the Colorado River Region, California, Arizona, and Nevada*. pp. 29-39. Cordilleran Publishers.

————. 1979. Laramide Deformation, Erosion and Plutonism Along the Southwestern Margin of the Colorado Plateau. *Tectonophysics*, v. 61, pp. 25-47.

————. 1970. Geomorphological Implications of Pre-Colorado and Colorado Tributary Drainage in the Western Grand Canyon Region. *Plateau*, v. 42, pp. 107-117.

————. 1966. Cenozoic Geology Along the Edge of the Colorado Plateau in Northwestern Arizona. Ph. D. dissertation, Washington University, Washington, 167 pp.

Young, R.A. and W.J. Brennan. 1974. Peach Springs Tuff, Its Bearing on Structural Evolution of the Colorado Plateau and Development of Cenozoic Drainage in Mohave County, Arizona. *Geological Society of America Bulletin*, v. 85, pp. 83-90.

Young, R.A. and E.D. McKee. 1978. Early and Middle Cenozoic Drainage and Erosion in West-Central Arizona. *Geological Society of America Bulletin*, v. 89, pp. 1745-1750.

501

INDEX

513